SPACEFLIGHT
A HISTORICAL ENCYCLOPEDIA

Spaceflight

A HISTORICAL ENCYCLOPEDIA

Patrick J. Walsh

GREENWOOD
An Imprint of ABC-CLIO, LLC

A B C ☰ C L I O

Santa Barbara, California • Denver, Colorado • Oxford, England

Library of Congress Cataloging-in-Publication Data

Walsh, Patrick J., 1963–
    Spaceflight : a historical encyclopedia / Patrick J. Walsh.
        p. cm.
    Includes bibliographical references and index.
    ISBN 978-0-313-37869-0 (set : print : alk. paper) — ISBN 978-0-313-37870-6
(set : ebook) — ISBN 978-0-313-37871-3 (volume 1 : print : alk. paper) —
ISBN 978-0-313-37872-0 (volume 1 : ebook) — ISBN 978-0-313-37873-7
(volume 2 : print : alk. paper) — ISBN 978-0-313-37874-4 (volume 2 : ebook) —
ISBN 978-0-313-37875-1 (volume 3 : print : alk. paper) — ISBN 978-0-313-37876-8
(volume 3 : ebook)
    1. Space flight—Encyclopedias.   2. Outer space—Exploration—Encyclopedias.   I. Title.
    TL788.W35   2010
    629.403—dc22   2009037030

14   13   12   11   10   1   2   3   4   5

This book is also available on the World Wide Web as an eBook.
Visit www.abc-clio.com for details.

ABC-CLIO, LLC
130 Cremona Drive, P.O. Box 1911
Santa Barbara, California 93116-1911

This book is printed on acid-free paper ∞
Manufactured in the United States of America

This work is dedicated with love and gratitude to
my parents, Helen E. and John E. Walsh.

CONTENTS

LIST OF ENTRIES

HADFIELD, CHRIS A.

(1959–)

Canadian Astronaut

A veteran of two spaceflights, Chris Hadfield was born in Sarnia, Canada on August 29, 1959, and grew up in Milton, Ontario, Canada. He pursued an interest in becoming a pilot at an early age, and won scholarships for flying both glider and powered aircraft during his teenage years.

In 1977, he graduated from Milton District High School, where he was honored at graduation for his academic skills as an Ontario Scholar.

Hadfield joined the Canadian Armed Forces in 1978. He attended Royal Roads Military College in Victoria, British Columbia, and received flight training at Portage La Prairie, Manitoba, where he was recognized as the leading pilot in his class in 1980. He then attended Royal Military College in Kingston, Ontario, where he earned a Bachelor of Science degree in mechanical engineering, graduating with honors in 1982.

He received basic training in jet aircraft in Moose Jaw, Saskatchewan, and was honored as the top overall graduate of his class in 1983. He was then trained as a fighter pilot in CF-5 and CF-18 aircraft at Cold Lake, Alberta, Canada.

While assigned to squadron 425 of the North American Aerospace Defense Command (NORAD), Hadfield served for three years as a pilot of CF-18 aircraft, and made the first CF-18 intercept of a Soviet "Bear" aircraft.

Hadfield went on to attend the United States Air Force Test Pilot School at Edwards Air Force Base in California, where he was honored with the Liethen-Tittle award as the outstanding graduate of his class in 1988, and subsequently became a U.S. Navy exchange officer, serving at the Strike Test Directorate at Patuxent River, Maryland. He was honored as the U.S. Navy Test Pilot of the Year in 1991.

Continuing his education concurrently with his military career, Hadfield attended the University of Tennessee, where he earned a Master of Science degree in aviation systems in 1992.

Hadfield flew more than 70 types of aircraft during his outstanding military career, and rose to the rank of colonel in the Canadian Air Force by the time of his retirement from the service in 2003.

The Canadian Space Agency (CSA) chose Hadfield for training as an astronaut as a member of its June 1992 selection of astronaut candidates. He was trained at NASA's Johnson Space Center (JSC) in Houston, Texas for assignment to flights aboard the U.S. space shuttle.

Hadfield first flew in space as a mission specialist aboard the space shuttle Atlantis during STS-74, the second docking of a space shuttle and the Russian space station Mir. Launched on November 12, 1995 from the Kennedy Space Center (KSC) in Florida, the STS-74 crew was commanded by Kenneth Cameron, and also included pilot James Halsell, and Hadfield's fellow mission specialists Jerry Ross and William McArthur, Jr.

The primary goal of the STS-74 flight was to deliver a Russian-built docking module to Mir. Hadfield played a key role in the delivery, when he operated Atlantis' Remote Manipulator System (RMS) robotic arm to move the docking module out of the shuttle's payload bay and position it for mating with the Orbiter Docking System—the shuttle's docking module.

After the docking was successfully achieved, the STS-74 and Mir 20 crews opened the hatches between the two vehicles in the early hours of November 15. The shuttle crew worked with Mir 20 crew members Sergei Avdeyev, Yuri Gidzenko, and Thomas Reiter to transfer supplies and equipment onto the station, and to move experiment samples and other materials from the station to the shuttle for the return trip to Earth. Atlantis returned to KSC on November 20, 1995.

In 1996, Hadfield became chief astronaut for the CSA, a capacity in which he served until 2000, when he began preparations for his second space mission.

On April 19, 2001, Hadfield began his second flight in space as a mission specialist aboard the shuttle Endeavour, during STS-100. Lifting off with commander Kent Rominger, pilot Jeffrey Ashby, and Hadfield's fellow mission specialists Scott Parazynski and John Phillips of NASA, European Space Agency (ESA) astronaut Umberto Guidoni of Italy, and Russian cosmonaut Yuri Lonchakov, Endeavour traveled to the International Space Station (ISS), where the STS-100 crew delivered and installed the Canadarm2 robotic arm.

The next generation of the Remote Manipulator System (RMS) robotic arms that were built by CSA engineers for each of the space shuttles in the U.S. shuttle fleet, the ISS Canadarm2 was the first of three parts of the station's Space Station Mobile Servicer System (SSMSS). The SSMSS includes the Candarm2, the Mobile

Base System (MBS) work platform, and the Special Purpose Dexterous Manipulator (SPDM) or "Canada Hand."

The Canadarm2 RMS was equipped with four color cameras and a system designed to automatically detect and avoid collisions with nearby objects. Constructed of 19-ply carbon fiber, the device weighed 3,968 pounds, and was designed with a larger range of motion than a human arm.

Hadfield made two spacewalks during the flight, while he and Scott Parazynski attached and activated the 57.7-foot Canadarm2 on the ISS Destiny Laboratory module. They made their first EVA on April 22, 2001, a day after Endeavour docked with the ISS, beginning their work with a 7 hour and 10 minute spacewalk in which they fastened the robotic arm in place.

They spent another 7 hours and 40 minutes in EVA on April 24, when they routed power cables between Canadarm2 and Destiny and installed a backup power system. Once the system was installed and operational, it was used in a series of novel experiments. It was first directed to maneuver itself off of the platform on which it had been delivered and hook itself onto a grapple fixture that Hadfield and Parazynski had installed. Then, later in the flight, Canadarm2 was directed to pick up the cradle on which it had previously sat in the shuttle's payload bay and pass the small platform to the shuttle's RMS robotic arm—thereby completing the first fully robotic transfer of an object in space.

Working with the resident ISS Expedition Two crew, STS-100 crew members also transferred 6,000 pounds of equipment and supplies onto the station, and used an IMAX camera to document the flight for later use in the 3D film "The International Space Station."

The STS-100 crew left the ISS on April 29, and returned to Earth on May 1, 2001.

Chris Hadfield has logged more than 20 days in space, including 14 hours and 50 minutes in EVA, during his two space shuttle flights.

After his second space mission, Hadfield moved to Star City, Russia, to serve as NASA's director of operations at the Gagarin Cosmonaut Training Center (GCTC). In addition to his administrative duties during that time, he also received training for flights aboard the Russian Soyuz TMA series of spacecraft. He served at GCTC until 2003, and subsequently returned to JSC to become chief of robotics for the NASA Astronaut Office.

HAIGNERÉ, CLAUDIE

(1957–)

French Astronaut; First French Woman to Fly in Space

The first French woman to fly in space and a veteran of visits to the space station Mir and the International Space Station (ISS), Claudie Haigneré was born in Le Creusot, France on May 13, 1957. She initially pursued a career as a medical doctor, receiving a Doctorate of Medicine with a certificate of specialized studies

With the launch of Soyuz TM-24 on August 17, 1996, Claudie Haigneré (seen here aboard the International Space Station with NASA astronaut Frank Culbertson) became the first French woman to fly in space. [NASA/ courtesy of nasaimages.org]

(C.E.S.) in biology and sports medicine from the University of Gijon, in Spain, in 1981, followed by a C.E.S. in aeronautical and space medicine in 1982, and a C.E.S. in rheumatology in 1984.

Known as Claudie Andre-Deshays prior to her marriage to fellow French astronaut Jean-Pierre Haigneré in 2001, she worked as a doctor in the rheumatology clinic and rehabilitation department of Cochin Hospital in Paris, specializing in rheumatology and sports trauma injuries, from 1984 to 1992. During the same period, she also worked at the Centre National de la Recherche Scientifique in Paris, where she helped to develop experiments that were conducted by French cosmonaut Jean-Loup Chrétien during his 1988 Soyuz TM-7 visit to the Mir space station, which was also known as the "Aragatz" mission.

The French space agency Centre National d'Études Spatiales (CNES) chose Haigneré for training as an astronaut as a member of its first group of astronaut candidates, on September 9, 1985. She was the only female member of the seven-member CNES Group One selection, whose members were chosen from a pool of 1,000 applicants.

Her initial assignments as a CNES cosmonaut included service as the lead coordinator of space physiology and medical programs for the space agency's Life Sciences division, and she helped to develop the program of scientific experiments that French cosmonaut Michel Tognini conducted during the joint French-Soviet Antares mission to Mir during the Soyuz TM-15 flight in 1992. She also served as the backup to Jean-Pierre Haigneré during his Soyuz TM-17 Altair mission to Mir in July 1993. In 1994, she contributed to the scientific work of European Space Agency (ESA) astronaut Ulf Merbold of Germany when he visited Mir aboard Soyuz TM-20 for the EuroMir '94 mission.

Continuing her education concurrently with her training as a cosmonaut, Haigneré earned a diploma in the biomechanics and physiology of movement, and in 1992, she completed a Doctorate degree in neuroscience.

After completing intensive training at the Gagarin Cosmonaut Training Center (GCTC) in Star City, Russia, Haigneré became the first French woman to fly in space when she lifted off on her first space mission on August 17, 1996. She launched aboard Soyuz TM-24 with commander Valery Korzun and flight engineer Alexander Kaleri, and traveled to the Mir space station for a brief stay, during which she carried out the Cassiopée program of scientific research for CNES.

At Mir, the Soyuz TM-24 crew was met by the station's resident cosmonauts Yuri Onufrienko and Yury Usachev and NASA astronaut Shannon Lucid.

During a visit of about a week, Haigneré carried out microgravity studies of the human cardiological and neurological systems, fluid dynamics investigations, analyses of structural dynamics in space, and studies of the role that gravity plays in the development of egg embryos while completing the Cassiopée project, which was the fifth cooperative French-Russian human spaceflight.

She returned to Earth with Onufrienko and Usachev in Soyuz TM-23 on September 2, 1996, while Korzun and Kaleri remained aboard the station with Lucid to begin their stay as the Mir 22 long-duration crew.

In her first trip into orbit, Haigneré accumulated 15 days, 18 hours, and 23 minutes in space.

After her milestone first space mission, she returned to Moscow to work for the French-Russian joint venture Starsem. The following year, she served as backup to Jean-Pierre Haigneré for the Soyuz TM-29 Perseus French-Russian mission, which was the final official flight to Mir (although subsequent attempts to commercialize the station led to one more flight to the station in 2000). During the flight, she served as crew interface coordinator at Russian Mission Control.

She continued to train at GCTC, and in July 1999 became the first woman to qualify for assignment as commander of a Soyuz spacecraft during its return from space.

She joined the ESA European Astronaut Corps in 1999 and contributed to the design and development of the life science research planned for the ESA Columbus module at the International Space Station (ISS).

On October 21, 2001, Haigneré began her second flight in space aboard Soyuz TM-33, traveling to the ISS with commander Viktor Afanasyev and flight engineer Konstantin Kozeyev. Her mission was also known as the ESA Andromeda project of scientific research. At the ISS, she and her crew mates were greeted by ISS Expedition 3 crew members Frank Culbertson of NASA and Rosaviakosmos cosmonauts Vladimir Dezhurov and Mikhail Tyurin.

Haigneré's Andromeda research was the first in a series of ISS science missions designed by the ESA and arranged in cooperation with CNES and Rosaviakosmos. The Andromeda scientific experiments included investigations in a variety of fields, including biology and life and materials sciences, and also incorporated educational and Earth observation activities.

At the end of her stay at the ISS, Haigneré returned to Earth with Afanasyev and Kozeyev in Soyuz TM-32 on October 31, 2001, after a flight of 9 days, 19 hours, and 59 minutes.

Claudie Haigneré spent more than 25 days in space during her two space missions.

She retired from the ESA astronaut corps in June 2002 to become France's minister for Research and New Technologies, a capacity in which she served until March 2004. She then served as France's minister for European Affairs from April 2004 to May 2005.

HAIGNERÉ, JEAN-PIERRE

(1948–)

French Astronaut

A veteran of two spaceflights, including a long-duration stay aboard the Russian space station Mir, Jean-Pierre Haigneré was born on May 19, 1948 in Paris, France.

He attended the French Air Force Academy at Salon-de-Provence, where he earned an engineering degree in 1971, and then received training as a fighter pilot. He was initially assigned as a fighter pilot in the Mirage 5 and Mirage IIIE aircraft in 1973, and subsequently became a squadron leader.

Haigneré was selected for training as a test pilot at the Empire Test Pilot School at Boscombe Down, England, where he earned "Hawker Hunter" and "Patuxent Shield" honors when he graduated in 1981.

That same year, he became project test pilot for the Mirage 2000N aircraft at the Bretigny-sur-Orge Flight Test Center. In 1983 he became chief test pilot at the center.

During his outstanding military career, Haigneré accumulated 5,500 hours of flying time in 105 different types of aircraft. He is a licensed test pilot and air transport pilot, and is qualified to pilot the Airbus A300 and A320 aircraft. He is a general in the French Air Force.

Haigneré was chosen by the French space agency Centre National d'Études Spatiales (CNES) in September 1985 for training as an astronaut. His initial assignments as a CNES astronaut included service as the leader of the agency's manned flight directorate, a capacity in which he participated in the preliminary development of the Hermes project, which sought to develop a manned European spacecraft. Haigneré also oversaw the Caravelle program of zero gravity aircraft flights.

Continuing his education concurrently with his career as an astronaut, Haigneré studied astrophysics at the university of Orsay.

In December 1990 Haigneré relocated to Star City, Russia to begin training at the Gagarin Cosmonaut Training Center (GCTC) for assignment to a flight aboard a Russian Soyuz spacecraft. He served as backup to fellow French cosmonaut Michel Tognini during Tognini's Soyuz TM-15 visit to Mir in July 1992.

Haigneré first flew in space with Russian cosmonauts Aleksandr Serebrov and Vasili Tsibliyev, lifting off on July 1, 1993 aboard Soyuz TM-17 to fly to Mir. He spent several weeks at the station, carrying out a program of scientific research as part of the joint French-Russian Altair project. He returned to Earth on July 22, 1993, with Mir 13 crew members Gennadi Manakov and Alexander Poleschuk, aboard Soyuz TM-16, after a flight of 20 days, 16 hours, and 4 minutes in space.

After his first spaceflight, Haigneré served as a member of the astronaut support crew for the EuroMir '95 and Cassiopée missions, and then returned to France to serve as a test pilot flying the zero gravity Airbus aircraft. He returned to the GCTC in Russia in 1997 to begin training for his second space mission, the joint French-Russian "Perseus" long-duration flight to Mir.

Haigneré joined the European Space Agency (ESA) astronaut corps in June 1998.

On February 20, 1999 he began a remarkable second space mission, lifting off with Russian cosmonaut Viktor Afanasyev and Ivan Bella—the first citizen of Slovakia to fly in space—aboard Soyuz TM-29.

Designed as the final mission to the Mir space station (although subsequent attempts to commercialize Mir would result in one additional flight), the Soyuz TM-29 flight brought the international crew to the station for a short visit with Mir 26 cosmonauts Gennady Padalka and Sergei Avdeyev, and Haigneré and Afanasyev then remained at the station with Avdeyev to become the Mir 27 crew, while Padalka and Bella returned to Earth aboard Soyuz TM-28.

Haigneré spent more than six months in space during his second visit to Mir, conducting scientific research and engineering and space station maintenance with his long-duration crew mates Afanasyev and Avdeyev. His scientific work covered biology, life sciences investigations, and physics experiments designed by scientists from France, Germany, and the ESA.

On April 16, 1999, he made a spacewalk of 6 hours and 19 minutes with Afanasyev to deploy experiments on the outside of the station.

Then, on August 28, 1999, Haigneré, Afanasyev, and Avdeyev closed out the Mir space station for what was then thought to be the last time, and returned to Earth aboard Soyuz TM-29. From launch to landing, Haigneré accumulated more than 188 days in space during his Mir 27 mission.

Jean-Pierre Haigneré has spent more than 209 days in space, including more than six hours in EVA, during his two remarkable space missions.

After his second flight, Haigneré served as leader of the ESA's astronaut division at the European Astronaut Center (EAC) in Cologne, France, and he subsequently became senior advisor to the ESA director of launchers.

HAISE, FRED W., JR.

(1933–)

U.S. Astronaut

Fred Haise was born in Biloxi, Mississippi on November 14, 1933. He graduated from Biloxi High School and attended Perkinston Junior College, where he received an Associate degree. He joined the U.S. Navy in 1952 and received his initial training as a pilot in Pensacola, Florida. He subsequently served at the U.S. Navy Advanced Training Command in Kingsville, Texas as a tactics and all weather flight instructor, and also served as a fighter pilot in the U.S. Marine Corps while attached to VMF-533 at Marine Corps Air Station Cherry Point, North Carolina.

He then served as a fighter interceptor pilot in the 185th Fighter Interceptor Squadron of the Oklahoma National Guard, from 1957 to 1959.

Haise later attended the University of Oklahoma, from which he received a Bachelor of Science degree, with honors, in aeronautical engineering in 1959.

Haise began his association with NASA as a research pilot, serving at the agency's Dryden Flight Research Center (DFRC) at Edwards Air Force Base in California and at the NASA Lewis Research Center (now known as the John H.

Glenn Research Center at Lewis Field) in Cleveland, Ohio. The results of his research were published in several technical papers, including one that he presented to the Tenth Symposium of the Society of Experimental Test Pilots in 1966.

In 1961, he began a year's duty with the U.S. Air Force, serving as a tactical fighter pilot with the 164th Tactical Fighter Squadron. He also attended the Air Force Aerospace Research Pilot School at Edwards Air Force Base, where he was recognized as the outstanding graduate of his class.

During his remarkable military career, Haise accumulated 9,300 hours of flying time, including 6,200 hours in jet aircraft.

NASA selected him for training as an astronaut in April 1966 as part of its Group 5 selection of pilot astronauts. His initial assignments as an astronaut included service as the backup lunar module pilot for the Apollo 8 lunar orbit mission and the Apollo 11 mission, which featured the first lunar landing. He subsequently served as the backup commander for the Apollo 16 lunar landing mission.

On April 11, 1970, Haise lifted off aboard Apollo 13 with commander James Lovell and command module pilot John "Jack" Swigert. Lovell and Haise were to have been the fifth and sixth individuals to walk on the surface of the Moon, but an explosion in an oxygen tank aboard the Apollo 13 service module put crew in immediate peril, and the planned lunar landing had to be abandoned.

Immediately activating all its resources to aid the crew and return them safely to Earth, NASA engineers, astronauts, and officials threw themselves into the tasks of evaluating the damage from the accidental explosion and working out a new plan for the suddenly perilous flight.

In the aftermath of the accident, the crew recognized the dire nature of their circumstances and rapidly began the process of shifting from their damaged Apollo 13 command module Odyssey, where life support systems were rapidly deteriorating, to the lunar module Aquarius. Originally intended to operate for about 40 hours while ferrying Lovell and Haise to the lunar surface, landing, and then bringing them back to rendezvous with Odyssey, the lunar module was pressed into service as a temporary home for all three astronauts for four days.

While mission managers simulated virtually every aspect of the crew's experience in space and devised methods for making the most of the spacecraft's remaining supplies of precious commodities such as oxygen, electricity, fuel—and perhaps most importantly, time—the astronauts valiantly maintained their poise, conserving food and water and describing their circumstances calmly and in great detail.

A major problem arose when the lunar module's environmental control system became overwhelmed and required immediate repair. Working with controllers on the ground, the crew was able to improvise a solution that allowed them to adapt filters originally intended for use in the service and command modules to work with the lunar module system.

The astronauts also had to endure a nerve-wracking series of course corrections that required brief firings of the lunar module's engine to keep them on the proper trajectory for their return to Earth. Again, acting with considerable skill and remarkable precision, and with constant guidance and support from the ground,

Lovell, Haise, and Swigert were able to maintain the prescribed course despite reduced visibility due to debris from the explosion.

After four difficult days on a "free return trajectory," the course chosen as most likely to save the crew, and which resulted in their first completing their journey to and around the Moon before they headed back toward Earth, the crew faced the most harrowing part of their ordeal as they transferred from Aquarius to Odyssey to prepare for their re-entry into the Earth's atmosphere. They first jettisoned the Apollo 13 service module, which gave them a close look at the stunning damage wrought by the accident (and which they photographed for later analysis by investigators, who would determine the cause as an explosion in one of the craft's oxygen tanks).

There followed the tedious, tense process of restoring the Odyssey command module, which required careful adherence to a sequence that had been determined in ground simulations in order to preserve enough electrical power to enable the spacecraft to withstand re-entry safely. In a flawless performance by both the astronauts on the ground and the crew in space, Odyssey was brought back to life in the proper sequence, and Lovell, Haise, and Swigert were able to leave the lunar module, which was not equipped with a heat shield and could not be used for the return to Earth, and transfer back into the command module.

Aquarius was jettisoned, and everyone associated with the flight, as well as millions of individuals following the astronauts' progress on television and radio, drew a deep breath as the crew plunged back into the Earth's atmosphere aboard the Apollo 13 command module. Fortunately, the outside of the Odyssey command module had not sustained substantial enough damage in the accident to impair the performance of its heat shield, and the astronauts splashed down safely in the Pacific Ocean on April 17, 1970.

The ordeal left the crew exhausted and ill, and Haise contracted an infection that stoked his temperature to a dangerous 103 degrees, but all three astronauts would recover fully in time. Dubbed a successful failure for the safe return of the crew, the conduct of the astronauts and all those involved in the Apollo 13 flight remains a testament to the dedication and professionalism that informed the best instincts of NASA during the first era of space exploration.

After the Apollo 13 flight, Haise attended Harvard Business School and continued his NASA career. In 1973, he was named technical assistant to the manager of the Space Shuttle Orbiter Project, and he then served as commander of one of the crews for the space shuttle approach and landing (ALT) tests in 1977. He left the space agency in 1979 to become vice president of space programs at Grumman Aerospace Corporation, and was subsequently named president of Grumman Technical Services, Inc. of Titusville, Florida, and Northrop Worldwide Aircraft Services of Lawton, Oklahoma.

He has also continued to pursue his interest in novel forms of flight as an honorary member of the National World War II Glider Pilots Association.

Among many honors he received for his courageous performance during the Apollo 13 mission, Haise was awarded the Presidential Medal for Freedom by President Richard Nixon in 1970.

Fred Haise spent 5 days, 22 hours, and 54 minutes in space during his remarkable Apollo 13 flight.

HALSELL, JAMES D., JR.

(1956–)

U.S. Astronaut

James Halsell was born in West Monroe, Louisiana on September 29, 1956. In 1974, he graduated from West Monroe High School and then attended the United States Air Force Academy, where he received a Bachelor of Science degree in engineering in 1978.

He received his undergraduate pilot training at Columbus Air Force Base in Mississippi, and was qualified to fly F-4 aircraft carrying both conventional and nuclear weapons. In 1980, he was assigned to Nellis Air Force Base in Las Vegas, Nevada, and then served at Moody Air Force Base in Valdosta, Georgia from 1982 to 1984.

Continuing his education concurrently with his military service, he attended Troy University, where he earned a Master of Science degree in management in 1983. He next attended the Air Force Institute of Technology at Wright-Patterson Air Force Base in Dayton, Ohio, from which he earned a Master of Science degree in space operations in 1985.

Halsell went on to attend the Air Force Test Pilot School at Edwards Air Force Base in California, where he was awarded the Liethen-Tittle Trophy as the top graduate in his class in 1986. He remained at Edwards until 1990, serving as a test pilot flying F-4, F-16, and SR-71 aircraft.

He had risen to the rank of colonel in the U.S. Air Force by the time he retired from the service.

Halsell first flew in space in July 1994 as pilot of the space shuttle Columbia during STS-65, the second flight of the International Microgravity Laboratory (IML-2)

The STS-65 crew worked around the clock in alternating shifts to complete more than 80 experiments in areas including life sciences and materials processing, completing a program of scientific research that had been designed by hundreds of scientists representing six space agencies. The flight also advance international cooperation in space research through the expansion of telescience, as researchers on Earth monitored the experiments aboard the shuttle in real time while Columbia was in orbit. The 15-day flight set a new mark for the longest shuttle mission, and the crew made 236 orbits aboard Columbia before returning to Earth on July 23, 1994.

Halsell also served as pilot during the STS-74 flight of the space shuttle Atlantis, which achieved the second docking of a shuttle with the Russian Space Station Mir. During that flight, the STS-74 crew delivered a docking module designed to ease future shuttle-Mir dockings. The module also served as a cargo carrier, with two solar arrays for the station making the trip inside the docking module. The solar arrays were scheduled to be moved onto the station by the Mir 20 crew during a later spacewalk.

The linked spacecraft opened their hatches at 4:02 A.M. EST on November 15. In addition to delivering the docking module, the STS-74 and Mir 20 crews transferred equipment and supplies and more than 1,000 pounds of water between the shuttle and the space station during the productive flight. Atlantis landed on November 20, 1995 at Kennedy Space Center (KSC), after 129 orbits completed in a little over eight days.

Halsell made his third flight in space in April 1997, as commander of STS-83 aboard the space shuttle Columbia. The ill-fated first flight of the Microgravity Science Laboratory (MSL-1), STS-83 ran into trouble early when one of Columbia's three fuel cells began to behave erratically, indicating a rise in differential voltage in one of the device's groupings of cells. During a shuttle flight, the astronauts rely on the fuel cells to generate electricity and water for drinking, which the cells provide via the chemical reaction of liquid hydrogen and liquid oxygen. Although enough electricity can be generated by even just one fuel cell to allow the crew to operate in space and to make a safe landing, but NASA's rules for shuttle flights mandate that all three fuel cells must work well enough to ensure that the crew will be safe during the mission and during re-entry and landing.

Despite a variety of measures to correct the problem with the malfunctioning fuel cell during STS-83, the malfunction persisted and the Mission Management Team decided on April 6 to shut down the balky cell and shorten the flight. STS-83 was only the third flight in the history of the shuttle program to end early; the two previous occasions were STS-2 in 1981 and STS-44 in 1991.

The crew was able to complete some of their planned MSL-1 experiments before landing safely, on April 8, 1997 at KSC after 3 days, 23 hours, 12 minutes, and 39 seconds.

Pleased with the safe return but also taking into account the long training of the crew prior to the flight and the voluminous amount of work by the scientists and engineers who contributed to the design of the involved experiments and equipment, NASA officials decided to make a second attempt to fly the mission, again aboard Columbia, with the same crew and payloads and a virtually identical mission profile, at a later date. Thus, Halsell again served as commander of the MSL-1 re-flight, STS-94, when he made his fourth flight in space, in July 1997. The unique mission was the first instance in which an entire mission was re-flown (as opposed to earlier instances in which individual payloads had been re-flown).

Free of the irritating fuel cell malfunction of the first flight, the STS-94 crew made the most of their time in space, working around the clock in alternating shifts to operate the MSL-1 while conducting 33 experiments in biotechnology, combustion, and materials processing. Using the Spacelab module and the MSL-1 equipment, they tested techniques and facilities targeted for later use on the International Space Station (ISS) and carried out experiments designed by scientists at NASA, the European Space Agency (ESA), the German space agency Deutsches Zentrum für Luft- und Raumfahrt e.V. (DARA), and the National Space Development Agency (NASDA) of Japan.

Columbia landed at KSC on July 17, 1997 after a flight of 15 days, 16 hours, 44 minutes, and 34 seconds.

In 1998, Halsell spent six months in Star City, Russia while serving as NASA's director of operations at the Gagarin Cosmonaut Training Center (GCTC).

On May 19, 2000, he began a fifth flight in space as commander of STS-101, the third shuttle flight devoted to the construction of the ISS.

During STS-101, the crew of Atlantis repaired and updated the ISS Zarya and Unity modules, replacing faulty equipment, installing fans, fire extinguishers, smoke detectors, and electronic components.

Atlantis undocked from the station on May 26 and landed on May 29, 2000.

James Halsell accumulated more than 1,250 hours in space during five shuttle flights.

HARBAUGH, GREGORY J.

(1956–)

U.S. Astronaut

Gregory Harbaugh was born in Cleveland, Ohio on April 15, 1956 and grew up in Willoughby, Ohio, where he graduated from Willoughby South High School in 1974. He attended Purdue University, from which he received a Bachelor of Science degree in aeronautical and astronautical engineering in 1978.

He began his NASA career that same year, working at the Johnson Space Center (JSC) in Houston in engineering and technical management roles. As a member of the staff at Mission Control, he supported shuttle flight operations for most of the first 25 missions of the space shuttle program. His assignments have included responsibility for overseeing the huge amounts of data received from the shuttle during the various stages of a mission, and in that capacity he served as Lead Data Processing Systems Officer for STS-9 and STS-41D. He also handled orbit data processing during STS-41B and STS-41C, and was responsible for the data acquired during ascent and re-entry during STS-41G. He also served as a senior flight controller for several shuttle missions.

Continuing his education concurrently with his JSC duties, Harbaugh attended the University of Houston at Clear Lake and received a Master of Science degree in physical science in 1986. He also holds a commercial pilot's license with an instrument rating, and has accumulated more than 1,600 hours of flying time.

Harbaugh became an astronaut in 1988. His technical duties at NASA expanded as a result, and his new assignments included development of the Hubble Space Telescope servicing mission and telerobotics systems development.

He first flew in space as a mission specialist during the STS-39 flight of the space shuttle Discovery, which launched on April 28, 1991. The 40th flight of the space shuttle program, STS-39 was the first shuttle mission to carry both classified payloads for the U.S. Department of Defense and unclassified payloads. Harbaugh operated the shuttle's Remote Manipulator System (RMS) during the flight and was responsible for the Infrared Background Signature Survey spacecraft. High winds at the originally scheduled landing site, Edwards Air Force Base in California, caused Discovery's landing to be diverted to the Kennedy Space Center (KSC) in Florida, where the shuttle touched down on May 6, 1991 after a flight of a little more than eight days.

In January 1993 he again flew in space, this time as a mission specialist aboard the shuttle Endeavour during STS-54. The crew deployed the fifth Tracking and Data Relay Satellite (TDRS-F) on the first day of the flight, and then, with fellow mission specialist Mario Runco, Jr., Harbaugh made a spacewalk of 4 hours and 28 minutes in the shuttle's open payload bay to try out a variety of EVA procedures. Endeavour landed at KSC on January 19, 1993 after nearly six days in space.

Harbaugh's third spaceflight was as a member of the STS-71 crew, during the historic first docking of a space shuttle with the Russian space station Mir. The flight was also the milestone 100th human spaceflight launched from Cape Canaveral in Florida. Launched on June 27, 1995 aboard the shuttle Atlantis, the STS-71 crew included Russian cosmonauts Anatoly Solovyov and Nikolai Budarin, who would remain on the station as the Mir 19 crew. On the return trip to Earth, the Atlantis crew included Mir 18 cosmonauts Vladimir Dezhurov and Gennady Strekalov and NASA astronaut Norman Thagard, who had been living on Mir since March.

The docking went well, and during their five days of linked flight, Atlantis and Mir together constituted the largest spacecraft yet flown in orbit, with a combined weight of about 500,000 pounds. The flight also marked the beginning of a period of cooperation between American and Russian crews that would lead to the commonplace international cooperation of the International Space Station (ISS) program. Atlantis and Mir undocked on July 4, and the landmark mission ended on July 7, 1995 with the shuttle's landing at KSC.

Harbaugh began his fourth space mission, the second devoted to servicing the Hubble Space Telescope (HST), with the nighttime launch of STS-82 aboard the shuttle Discovery on February 11, 1997. Among the most complex missions undertaken during the entire shuttle program, the HST service and repair flights required enormous stamina on the part of alternating spacewalking teams and minute attention to detail over long periods of working in space.

The crew made five spacewalks, including an unscheduled EVA at the end of the repair program, and used more than 150 tools during their work on the HST. Working in two alternating teams, mission specialists Mark Lee and Steven Smith took turns with Harbaugh and Joseph Tanner while accumulating a total of 33 hours and 11 minutes in EVA during the remarkable flight.

Lee and Smith made the first EVA, on February 14, logging 6 hours and 42 minutes while replacing several of the telescope's scientific instruments. Harbaugh and Tanner made their first spacewalk the following day, replacing the HST's engineering and science tape recorder and installing the optical control electronics enhancement kit and fine guidance sensor. During their 7 hour and 27 minute EVA they also discovered cracks in the telescope's thermal insulation. Lee and Smith then added another 7 hours and 15 minutes to the EVA total on February 15 while replacing more instruments and components.

Harbaugh and Tanner completed the original mission profile with a 6 hour and 34 minute spacewalk on February 16, and Lee and Smith made an unscheduled fifth EVA of 5 hours and 17 minutes to attach thermal insulation blankets to the telescope to protect key data processing, electronics, and telemetry instrumentation. Discovery returned to Earth in a night landing on February 21, 1997.

After his fourth flight, Harbaugh served as manager of NASA's Extravehicular Activity Project Office, which is tasked with the development of spacesuits, training, procedures, equipment, and tools for future EVAs on the shuttle, the ISS, and future planetary exploration. He left NASA in March 2001.

Gregory Harbaugh accumulated 34 days and 2 hours in space, including more than 18 hours in EVA, during his four space shuttle missions.

HARTSFIELD, HENRY W., JR.

(1933–)
U.S. Astronaut

One of the astronauts originally chosen for the pioneering Manned Orbiting Laboratory (MOL) program of the U.S. Air Force, Henry Hartsfield is a veteran of three space shuttle flights.

He was born on November 21, 1933 in Birmingham, Alabama. He graduated from West End High School in Birmingham and then attended Auburn University, where he received a Bachelor of Science degree in physics in 1954. While at Auburn, he participated in the U.S. Air Force ROTC program, and he was commissioned in the Air Force in 1955. He served in Bitburg, Germany as a member of the 53rd Tactical Fighter Squadron, and attended the Air Force Test Pilot School at Edwards Air Force Base in California.

After serving as an instructor at the school following graduation, he was selected to become an astronaut for the Air Force's MOL program in 1966. Designed to place astronauts in space for long-duration missions, the MOL program was an early precursor to the subsequent U.S. Skylab program, the Soviet Salyut and Mir space stations, and the International Space Station (ISS).

Hartsfield began his NASA career when the MOL astronauts were transferred to the civilian space agency in August 1969.

Continuing his education concurrently with his astronaut career, Hartsfield undertook graduate work in physics at Duke University and graduate work in astronautics at the Air Force Institute of Technology. In 1971, he received a Master of Science degree in engineering from the University of Tennessee.

During his distinguished career as a pilot, he has accumulated more than 7,400 hours of flying time, including 6,150 hours in jet aircraft including the F-86, F-100, F-104, F-105, F-106, T-33, and T-38.

His initial NASA assignments included service as a member of the astronaut support crew for the Apollo 16 Moon landing mission and for the Skylab 2, Skylab 3, and Skylab 4 missions. He also served a backup pilot for the STS-2 and STS-3 space shuttle flights.

In 1977, he retired from the U.S. Air Force after 22 years of active service, continuing his NASA career as a civilian.

He first flew in space as pilot for STS-4 aboard the shuttle Columbia, launched on June 27, 1982 with mission commander (and Apollo 16 veteran) Thomas Mattingly. The flight was the fourth orbital test flight of the shuttle program and provided an opportunity for an innovative test exercise when mission controllers evaluating the shuttle's performance found that rainwater had seeped into the protective coating of several of Columbia's heat shield tiles. Fearing that the water could freeze and damage the tiles, the scientists on Earth had Hartsfield and Mattingly maneuver the shuttle so that its waterlogged tiles would face the sun, which evaporated the water and solved the potentially dangerous problem.

In addition to proving the shuttle safe for future flights, which would carry larger crews and stay in space for longer and more complex missions, the STS-4 flight also deployed a classified payload for the U.S. Department of Defense and several scientific experiments.

On August 30, 1984 Hartsfield lifted off on his second space mission, as commander of STS-41D, the first flight of the space shuttle Discovery. The crew deployed three commercial satellites during the mission and operated the Office of Application and Space Technology (OAST-1) solar wing, a test designed to prove that large solar arrays could be used to provide power for future space facilities, including the ISS. In a unique exercise, the crew also used the shuttle's robotic arm, the Remote Manipulator System (RMS), to remove ice particles from the outside of the shuttle, leading mission controllers to dub the astronauts "icebusters." Discovery landed at Edwards Air Force Base in California on September 5, 1984.

Hartsfield was also commander of the STS-61A Spacelab mission, which launched aboard the shuttle Challenger on October 30, 1985. The first spaceflight to include eight astronauts on a single spacecraft, STS-61A featured an international crew, with payload specialists Reinhard Furrer and Ernst Messerschmid of West Germany (at the time, Germany was still divided into East and West, prior to the country's reunification after the fall of the Soviet Union) and European Space Agency (ESA) astronaut Wubbo Ockels, who became the first citizen of The Netherlands to fly in space. Designed by German scientists, the STS-61A Spacelab operations included 75 experiments in microgravity research, life sciences, materials science, communications, and navigation.

Hartsfield continued to serve NASA in a variety of management positions after his third spaceflight. In 1986, he was appointed deputy chief of the agency's Astronaut Office, and the following year he became deputy director for Flight Crew Operations. He served as director of the Technical Integration and Analysis Division at NASA headquarters in Washington, D.C. in 1989, and then moved to the Marshall Space Flight Center in Alabama in 1990 to map out future space station operations as deputy manager for operations (and later, as deputy manager of the office) in the Space Station Projects Office.

His central role in the development of NASA's future space station operations continued with his 1991 appointment as the Man-Tended Capability Phase Manager for Space Station Freedom Program and Operations at the Johnson Space Center. (Space Station Freedom later became the International Space Station.) In 1993, he was named the Manager of International Space Station (ISS) Independent Assessment, responsible for overseeing the initial development phase of the ISS. Those responsibilities were expanded in 1996 to include independent assessment of all programs of NASA's Human Exploration and Development of Space Enterprise.

Henry Hartsfield spent 20 days and 3 hours in space during his three shuttle missions.

HAUCK, FREDERICK H.

(1941–)

U.S. Astronaut

Frederick "Rick" Hauck was born in Long Beach, California on April 11, 1941. In 1958, he graduated from St. Albans School in Washington, D.C., and he then attended Tufts University, where he earned a Bachelor of Science degree in physics in 1962. He also participated in the U.S. Navy ROTC program while working on his degree and he received his commission upon graduation.

He served as a communications officer aboard the USS *Warrington* (DD-843) from 1962 to 1964, and then attended the U.S. Naval Postgraduate School in Monterey, California. He also studied Russian at the Defense Language Institute in Monterey, prior to attending the Massachusetts Institute of Technology (MIT) as a student in the Navy's Advanced Science Program. In 1966, he graduated from MIT with a Master of Science degree in nuclear engineering.

After receiving flight training at the Naval Air Station in Pensacola, Florida, Hauck received his pilot wings in 1968 and was assigned to Attack Squadron 35, Air Wing 15 aboard the USS *Coral Sea* (CVA-43).

He flew 114 combat and combat support missions during the Vietnam War.

Assigned to Attack Squadron 42 in August 1970, Hauck served as a visual weapons delivery instructor in A-6 aircraft and was selected to receive test pilot training at the U.S. Naval Test Pilot School in Patuxent River, Maryland, from which he graduated in 1971.

Hauck went on to serve as a project test pilot for systems designed to facilitate automatic landings on aircraft carriers while assigned to the Carrier Suitability Branch of the Flight Test division at the Naval Air Test Center. In 1974, he was deployed aboard the USS *Enterprise* (CVN-65) as operations officer to Commander Carrier Air Wing 14, and he then served as executive officer for Attack Squadron 145, in 1977. NASA selected him for astronaut training in January 1978.

Hauck accumulated more than 5,000 hours of flying time during his military career, and rose to the rank of captain in the U.S. Navy by the time of his retirement from the service in 1990.

He first flew in space as pilot of the shuttle Challenger, as part of the STS-7 crew in June 1983. The mission was particularly notable for the presence of Sally Ride among the crew members, as she became the first American woman to fly in space. The crew successfully deployed two satellites, one for Canada and one for Indonesia, and conducted a variety of scientific experiments.

He flew his second space mission as commander of the shuttle Discovery during STS-51A, in November 1984. The flight began with the successful deployment of the Canadian TELESAT-H (ANIK) communications satellite and the SYNCOM IV-1 (LEASAT-1) defense communications satellite. Then, using Manned Maneuvering Units (MMUs) to retrieve their targets, mission specialists Joseph Allen and Dale Gardner captured the PALAPA-B2 and WESTAR-VI satellites, which had failed to reach their proper orbits when they were originally deployed. With satellite launches and rescues complete, Discovery landed at the Kennedy Space Center (KSC) in Florida on November 16, 1984.

After his second spaceflight, Hauck helped to develop NASA's plans to integrate a liquid-fueled Centaur rocket into the shuttle, and in May 1985 he was named to command the future shuttle flight that would deploy the European Space Agency's (ESA's) Ulysses solar probe, which was scheduled to be boosted into orbit by a Centaur. The loss of the space shuttle Challenger in January 1986 changed the course of both assignments, however, as the Shuttle Centaur program was subsequently canceled, and the Ulysses mission was postponed indefinitely during the long period of recovery following the Challenger loss.

Later in 1986, Hauck was named associate administrator for external relations at NASA.

On September 29, 1988, Hauck led America's return to spaceflight as commander of Discovery during STS-26, the first shuttle flight after the Challenger accident. The crew successfully deployed a tracking and data relay satellite during the mission, and performed a number of scientific experiments during the four-day flight, returning to Earth at Edwards Air Force Base in California on October 3, 1988.

Following his third space shuttle mission, he served in the Office of the Chief of Naval Operations as director of the U.S. Navy's Space Systems Division, where he was responsible for developing and overseeing the budget for the Navy's space projects. He retired from the Navy in 1990 and subsequently became president and chief executive officer of AXA Space, Inc., an underwriting manager for space telecommunications insurance, headquartered in Bethesda, Maryland.

Hauck is also a member of the Board of Trustees of his alma mater, Tufts University, and served on the U.S. Department of Transportation's Commercial Space Transportation Advisory Committee for nearly a decade. He has continued to put his spaceflight and management expertise to use as a member of the External Requirements Assessment Team for NASA's Second Generation Reusable Launch Vehicle Program and as chairperson of the NRC Committee on Precursor Measurements Necessary to Support Human Operations on the Surface of Mars.

Rick Hauck spent more than 18 days in space during his career as an astronaut.

HAWLEY, STEVEN A.

(1951–)

U.S. Astronaut

Steven Hawley was born in Ottawa, Kansas, on December 12, 1951. In 1969, he graduated from Salina (Central) High School in Salina, Kansas, and he then attended the University of Kansas, where he graduated with the highest distinction with a Bachelor of Arts degree in physics and a Bachelor of Arts in astronomy in 1973.

Widely recognized for his scientific research as an undergraduate, Hawley received a number of academic scholarships while pursuing his education at the University of Kansas, and was honored as the university's outstanding physics major in his senior year. His primary area of study was aimed at determining the chemical composition of gaseous nebulae and emission-line galaxies through the use of spectrophotometry.

During the summer recess of his sophomore year he pursued his research interests as an assistant at the U.S. Naval Observatory in Washington, D.C., and he spent the following two summers in Green Bank, Virginia as a research assistant at the National Radio Astronomy Observatory.

In 1977, he graduated from the University of California with a Ph.D. in astronomy and astrophysics. At the time of his selection by NASA for astronaut training, he was working at the Cerro Tololo Inter-American Observatory in La Serena, Chile as a post-doctoral research associate.

Hawley's NASA assignments have been exceptional in both their variety and accomplishment, as he has served on six separate teams that have received NASA's Group Achievement Award, for work including software testing at the Shuttle Avionics Integration Laboratory (SAIL), the second test and checkout of the space shuttle at the Kennedy Space Center (KSC), strategic planning at the Johnson Space Center (JSC), the agency's ESIG 300 Integration Project, the 1995 space shuttle program workforce review, and a contracting project in 1997.

He also served as a member of the astronaut support crew for three of the first four space shuttle missions, and in 1984 he became technical assistant to the director of flight crew operations.

Hawley first flew in space during STS-41D in 1984, the first flight of the space shuttle Discovery. The crew deployed three commercial satellites during the mission and operated a solar wing designed by the Office of Application and Space Technology (OAST-1) in a test designed to prove that large solar arrays could be used to provide power for future space facilities, including the International Space Station (ISS). The crew also used an IMAX movie camera to record their activities, and, using the shuttle's robotic arm, the Remote Manipulator System (RMS), removed ice particles from the outside of the shuttle. The unique de-icing exercise led mission controllers to give the astronauts the nickname "icebusters." Discovery landed at Edwards Air Force Base in California on September 5, 1984 after 96 orbits in 6 days and 57 minutes.

Hawley's second flight was as a mission specialist aboard the shuttle Columbia, during STS-61C in January 1986. The mission lifted off on January 12 after six launch attempts were foiled by poor weather and technical problems. The crew included Florida Congressman Bill Nelson, the second public official to fly in space (Senator Jake Garn of Utah had been the first, in 1985). The crew deployed the RCA Americom satellite SATCOM KU-1 during the flight and conducted experiments in astrophysics and materials processing before returning to Earth at Edwards Air Force Base in California on January 18, 1986.

As things turned out, STS-61C was the last successful shuttle flight for almost three years. Ten days after Hawley and his fellow Columbia crew mates touched down in California, the space shuttle Challenger was lost during launch at the KSC, on January 28, 1986.

Hawley was honored for his work during NASA's long, painful recovery from the Challenger accident. He helped to plan the 1988 return to flight mission as deputy chief of the Astronaut Office.

His third space mission began on April 24, 1990, with the launch of the shuttle Discovery on STS-31, the deployment of the Hubble Space Telescope (HST). The 35th flight of the space shuttle program, STS-31 brought Discovery to a much higher than usual orbit of 380 miles to facilitate the launch of the HST. The crew also used a handheld IMAX camera to record their activities and conducted experiments in protein crystal growth, polymer membrane processing, and the impact of microgravity and magnetic fields on an ion arc.

After STS-31, Hawley transferred from the JSC in Houston to NASA's Ames Research Center in California, where he served as associate director for two years, until his return to JSC in August 1992 to become deputy director of flight crew operations.

Hawley's fourth spaceflight was on STS-82, in February 1997. Launched aboard the shuttle Discovery in a nighttime lift-off on February 11, STS-82 was the second HST maintenance mission. The HST service and repair flights were among the most complex missions undertaken during the entire shuttle program, requiring enormous stamina and careful attention to detail. Hawley played a key role in the retrieval of the large HST telescope at the start of the mission, using the shuttle's robotic arm, the RMS, to capture the device in the early morning hours of February 13 and then place it in Discovery's payload bay, where Lee and Smith took turns with Harbaugh and Tanner during a series of five space walks totaling 1 day, 9 hours, and 11 minutes to complete their service and repair tasks. With the maintenance complete, Hawley again operated the RMS to re-deploy the HST, which was then boosted into a higher orbit. Discovery returned to Earth with a night landing at KSC on February 21, 1997.

He resumed his duties as deputy director of flight crew operations at JSC after completing the STS-82 mission.

On July 23, 1999, Hawley launched on a fifth space mission, aboard shuttle Columbia during STS-93. He participated in the successful deployment of the Chandra X-ray Observatory, the third in NASA's series of Great Observatories (the program began with the deployment of the Hubble Space Telescope and also included the Compton Gamma Ray Observatory). STS-93 also gave Hawley an opportunity to study the solar system with a broadband ultraviolet telescope as the crew captured ultraviolet images of the Earth, Moon, Mercury, Venus, and Jupiter using the Southwest Ultraviolet Imaging System.

Steven Hawley has spent more than 32 days in space during his five space shuttle missions.

In 2002, he was named director of flight crew operations, and subsequently assumed the role of director of the astromaterials research and exploration science directorate at JSC, responsible for the acquisition and study of astromaterials and research into the human exploration of space.

HELMS, SUSAN J.

(1958–)

U.S. Astronaut

A veteran of five space missions, including a long-duration stay aboard the International Space Station (ISS), Susan Helms was born in Charlotte, North Carolina on February 26, 1958. In 1976, she graduated from Parkrose Senior High School in Portland, Oregon and then attended the United States Air Force Academy, where she received a Bachelor of Science degree in aeronautical engineering in 1980.

Commissioned in the U.S. Air Force, she served as an F-16 weapons separation engineer with the Air Force Armament Laboratory at Eglin Air Force Base in Florida. In 1982, she was named lead engineer for F-15 weapons separation, and the following year she was honored as the Air Force Armament Laboratory Junior Engineer of the Year.

Helms went on to attend Stanford University, where she earned a Master of Science degree in aeronautics/astronautics in 1985, and was subsequently assigned to teach aeronautics at the Air Force Academy as an assistant professor.

In 1987, she returned to California to attend the Air Force Test Pilot School at Edwards Air Force Base. Upon graduation from the school the following year, she was recognized as a distinguished graduate.

Helms trained as a flight test engineer for a year and was honored with the R. L. Jones Award as the outstanding flight test engineer of her class. Putting her training and expertise to good use, she next worked at the Aerospace Engineering Test Establishment at the Canadian Forces Base at Cold Lake, Alberta, Canada, as a U.S. Air Force Exchange Officer. During her service in Canada she oversaw the development of a Flight Control System Simulation for the CF-18 aircraft and also served as a CF-18 flight test engineer and project officer. The Canadian armed forces honored her with the Aerospace Engineering Test Establishment Commanding Officer's Commendation in 1990.

During her exceptional military career, Helms has risen to the rank of brigadier general in the U.S. Air Force.

She became an astronaut in 1991 and first flew in space as a mission specialist during the STS-54 flight of the space shuttle Endeavour, in January 1993. Launched on January 13, the STS-54 crew deployed the fifth Tracking and Data Relay Satellite (TDRS-F).

In a test tied to the future development of the ISS, STS-54 marked the first time that a fuel cell was shut down and restarted while a shuttle was in orbit. At the close of the successful mission, Endeavour landed at the Kennedy Space Center (KSC) in Florida on January 19, 1993, after nearly six days in space.

During her second spaceflight, STS-64 in September 1994, Helms flew aboard the shuttle Discovery and was responsible for operating the Remote Manipulator

System (RMS), the shuttle's robotic arm. The crew successfully tested an innovative optical radar system, the Lidar in Space Technology Experiment (LITE), and deployed and retrieved a SPARTAN free-flyer satellite. The mission also featured the first untethered spacewalk by American astronauts in 10 years.

The radar exercise was part of NASA's "Mission to Planet Earth" initiative. An experimental system that utilized laser pulses instead of radio waves, the LITE instrument was trained on a variety of targets, including cloud structures, dust clouds, and storm systems, among others. Groups in 20 countries around the globe collected data using ground- and aircraft-based radar instruments to help verify the data collected during the LITE experiment.

As operator of the RMS, Helms played a key role in the deployment and retrieval of the Shuttle Pointed Autonomous Research Tool for Astronomy (SPARTAN-201), which collected data about the solar wind and the Sun's corona during two days of free flight, and the Shuttle Plume Impingement Flight Experiment (SPIFEX), which was carried on the end of the shuttle's robotic arm while collecting data about the potential impact of the shuttle's Reaction Control System thrusters on space structures like the Mir space station or the future ISS. Discovery landed at Edwards Air Force Base on September 20, 1994.

Helms made her third flight in space as payload commander for STS-78. A milestone mission in international cooperation, the flight launched aboard the shuttle Columbia on June 20, 1996. The primary payload for the mission, the Life and Microgravity Spacelab (LMS), was the work of scientists from 10 countries and the personnel of five space agencies, including NASA, the European Space Agency (ESA), the French space agency Centre National d'Études Spatiales (CNES), the Canadian Space Agency (CSA), and the Italian space agency Agenzia Spaziale Italiana (ASI). The STS-78 crew conducted more than 40 LMS experiments during the flight, which at nearly 17 days in length, was the longest shuttle flight up to that time.

In addition to the scientific work done by the crew on the shuttle, the mission also achieved a thorough testing of the kind of communications systems necessary to link researchers around the world to astronauts in space, a necessity for the ISS. Columbia returned to Earth on July 7, 1996.

In May 2000 Helms made her fourth flight in space, traveling on the shuttle Atlantis during STS-101, the third shuttle flight devoted to the construction of the ISS. Her fellow crew members for the mission included James Voss and Russian cosmonaut Yury Usachev, who would, in 2001, live and work with Helms aboard the ISS for nearly six months as the ISS Expedition 2 crew.

During STS-101, the crew of Atlantis repaired and updated the ISS Zarya and Unity modules, replacing faulty equipment, installing fans, fire extinguishers, smoke detectors, and electronic components.

Helms played a key role in the ISS refurbishing by making repairs to the station's Functional Cargo Block. She was also responsible for the onboard computer network and served as mission specialist responsible for the shuttle's rendezvous with the ISS.

Atlantis undocked from the station on May 26 and landed on May 29, 2000 after more than nine days in space.

On March 8, 2001, Helms began a remarkable fifth space mission as a member of the ISS Expedition 2 crew. Launched aboard the space shuttle Discovery during STS-102, she returned to the space station with Expedition 2 crew mates James Voss and Yury Usachev to begin a six-month stay.

The flight began with an amazing highlight for Helms and Voss as the two astronauts made an epic, record-shattering spacewalk of 8 hours and 56 minutes, the longest EVA in space shuttle history, while they prepared the ISS Unity module to be linked to the Leonardo Multi-Purpose Logistics Module so that the crew could transfer supplies and equipment to the station. During the long spacewalk, Helms also set a new record for the longest EVA by a female astronaut.

Discovery remained docked with the ISS for about nine days, and then, when it undocked and returned to Earth, Helms, Voss, and Usachev remained on the ISS to settle into their long stay as the Expedition 2 crew. During their ISS flight, they installed the Space Station Remote Manipulator System (SSRMS). Similar to the robotic arm on the space shuttle, the SSRMS was nicknamed "Canadarm2" in honor of its country of origin, Canada. Helms used the SSRM to install the space station's airlock, which was delivered to the ISS during STS-104. The crew also conducted medical tests and scientific experiments during their 163 days onboard the station.

At the end of their long, productive mission, Helms, Voss, and Usachev returned to Earth aboard the space shuttle Discovery during its STS-105 flight, which landed on August 21, 2001. From launch to landing, the Expedition 2 crew spent 167 days, 6 hours, and 41 minutes in space.

During her remarkable career as an astronaut, Susan Helms spent 211 days in space, including her record-setting 8 hour and 56 minute EVA.

She left NASA in July 2002 to return to the U.S. Air Force, initially serving at the Air Force Space Command in Colorado Springs, Colorado and subsequently serving as commander of the 45th Space Wing and director of the Eastern Range at Patrick Air Force Base in Florida.

HENRICKS, TERENCE T.

(1952–)

U.S. Astronaut

Terence "Tom" Henricks was born in Bryan, Ohio on July 5, 1952. In 1970, he graduated from Woodmore High School and then attended the United States Air Force Academy, where he received a Bachelor of Science degree in civil engineering in 1974.

Henricks received his initial pilot training at Craig Air Force Base in Selma, Alabama, where he was recognized as the distinguished graduate of his class, and then attended F-4 Fighter Weapons School, where he received the school's outstanding flying award. He was subsequently assigned to fly the F-4 in fighter squadrons in England and Iceland. Receiving additional training in F-16 aircraft, he was recognized as Top Gun of his F-16 Conversion Course. He was assigned to Nellis Air Force Base in Las Vegas, Nevada in 1980.

Continuing his education concurrently with his military service, Henricks earned a Master of Science degree in public administration from Golden Gate University in 1982.

In 1983, he was selected to attend the Air Force Test Pilot School at Edwards Air Force Base in California. Following graduation, he was assigned as an F-16C test pilot and Chief of the 57th Fighter Weapons Wing Operating Location at Edwards.

During his distinguished career as a pilot, Henricks has accumulated more than 6,000 hours of flying time in 30 different types of aircraft, and he holds an FAA commercial pilot rating. He is also recognized as a Master Parachutist, and has made 749 parachute jumps. Prior to his retirement in 1997, he rose to the rank of colonel in the U.S. Air Force.

Henricks became an astronaut in 1986. His initial assignments at NASA included the reevaluation of landing sites around the world for the space shuttle program, serving as lead astronaut of the Shuttle Avionics Integration Laboratory at the Johnson Space Center (JSC) in Houston, and serving as chief of the Operations Development Branch of the Astronaut Office.

He first flew in space as pilot of STS-44, the 10th flight of the space shuttle Atlantis. Launched after dark on November 24, 1991, the flight featured both classified payloads for the U.S. Department of Defense and unclassified payloads. The crew suffered some tense moments when, near the end of the mission, one of the shuttle's three orbiter inertial measurement units failed. Crucial to the spacecraft's navigation, the failed device forced mission controllers to shorten the planned 10-day mission by two days and to divert the shuttle's planned landing from the Kennedy Space Center (KSC) in Florida to Edwards Air Force Base in California. Despite the difficulties, Atlantis landed safely at Edwards on December 1, 1991.

Henricks's next spaceflight was as pilot of the shuttle Columbia, during the STS-55 Spacelab mission. Launched April 26, 1993, STS-55 was the second Spacelab flight dedicated to German scientific research. The launch was preceded by a pad abort on March 22. Just three seconds before the shuttle was scheduled to lift off, Columbia's computers detected an incomplete ignition in the number three Space Shuttle Main Engine (SSME). The fault was later traced to a liquid oxygen leak; at the time, however, the T-3 abort was a nerve-wracking ordeal for crew and mission controllers alike.

Once in orbit, the crew worked around the clock in alternating shifts to complete 88 experiments. The flight was extended by one day to give the astronauts more time to finish their work, and the landing was diverted from KSC to Edwards Air Force Base because of extensive cloud cover at the Florida site. When the shuttle touched down, adding its flight of just under 10 days to the cumulative total of the entire shuttle program, the total accumulated time in space for the shuttle fleet (including the lost Challenger) passed the one year mark. The total spaceflight hours for all shuttle missions up to and including STS-55 was 365 days, 23 hours, and 48 minutes.

On July 13, 1995 Henricks began his third spaceflight, this time as commander of STS-70 aboard the shuttle Discovery. The crew deployed the sixth Tracking and

Data Relay Satellite on the first day of the flight, completing the network of TDRS advanced tracking and communications satellites which began with the deployment of TDRS-1 during STS-6 in April 1983. The crew also conducted a number of scientific experiments during the nine day mission, returning to Earth on July 22, 1995 after 142 orbits.

Henricks led an international crew on his fourth spaceflight, as commander of STS-78. A milestone mission in international cooperation, the flight launched aboard the shuttle Columbia on June 20, 1996. The crew included pilot Kevin Kregel, flight engineer Susan Helms, mission specialists Richard Linnehan and Charles Brady, Jr., and payload specialists Jean-Jacques Favier of France and Robert Thirsk of Canada.

The primary payload for the flight, the Life and Microgravity Spacelab (LMS), was the work of scientists from 10 countries and the personnel of five space agencies, including NASA, the European Space Agency (ESA), the French space agency Centre National d'Études Spatiales (CNES), the Canadian Space Agency (CSA), and the Italian space agency Agenzia Spaziale Italiana (ASI). The crew conducted more than 40 LMS experiments during the flight, which at nearly 17 days, was the longest shuttle flight up to that time.

In addition to the scientific work done by the crew on the shuttle, the mission also achieved a thorough testing of the kind of communications systems necessary to link researchers around the world to astronauts in space, a necessity for the International Space Station (ISS).

In a fitting end to the landmark flight's celebration of international cooperation, and a fitting symbolic finish to its commander's remarkable fourth space mission, after Columbia landed at KSC, Henricks and Kregel participated in the passing of the Olympic Torch during a ceremony at the space center.

Tom Henricks spent more than 42 days in space during his four shuttle missions and was the first person to log more than 1,000 hours in space as a space shuttle commander and pilot.

He left NASA in 1997 to pursue a private sector career.

HIEB, RICHARD J.

(1955–)

U.S. Astronaut

Richard Hieb was born in Jamestown, North Dakota on September 21, 1955. In 1973, he graduated from Jamestown High School, and he then attended Northwest Nazarene College, where he received a Bachelor of Arts degree in math and physics in 1977. In 1979, he earned a Master of Science degree in aerospace engineering from the University of Colorado.

After graduation, he began his career at NASA working on the development of crew procedures and activities. His work at the space agency coincided with the beginning of the space shuttle program, and he worked at Mission Control in Houston during STS-1 and subsequent flights, helping to develop procedures for orbital rendezvous and related operations.

Hieb became an astronaut in 1986. He first flew in space in 1991, as a mission specialist aboard the shuttle Discovery during STS-39, the 40th mission of the space shuttle program. Launched on April 28, STS-39 was the first shuttle flight to carry both classified and unclassified payloads. Hieb's duties during the mission included responsibility for the Infrared Background Signature Satellite (IBSS), which he deployed and retrieved from Discovery's payload bay using the shuttle's robotic arm, the Remote Manipulator System (RMS). The shuttle made 134 orbits during the eight-day flight.

In 1992, he flew again in space, during STS-49, the first flight of the space shuttle Endeavour. Launched on May 7, Hieb and his STS-49 crew mates achieved several milestones: they made a satellite retrieval that featured the first-ever three-person spacewalk, they were the first shuttle crew to make four EVAs, and they made two spacewalks that were longer than any EVAs previously attempted. Hieb and Pierre Thuot participated in three of the EVAs, including the historic three-person spacewalk, which they shared with Thomas Akers.

The crew located and captured the INTELSAT-VI satellite, which had been stranded in a useless orbit since its launch in 1990, and attached a motor to propel in into the proper orbit. Mission specialists Hieb, Thuot, and Akers made a record-setting spacewalk of 8 hours and 29 minutes, and Kathryn Thornton set a new record for the longest spacewalk by a female astronaut up to that time. At the end of its superb first flight, Endeavour returned to Earth at Edwards Air Force Base in California on May 16, 1992.

On July 8, 1994, Hieb lifted off on his third space mission, this time as payload commander aboard the space shuttle Columbia during STS-65, the second flight of the International Microgravity Laboratory (IML-2).

As payload commander, he was responsible for overseeing a program of scientific research that had been designed by hundreds of scientists representing six space agencies. The crew worked around the clock in alternating shifts to complete more than 80 experiments in areas including life sciences and materials processing, and advanced international participation and cooperation in space research through the expansion of telescience, as researchers on Earth monitored the experiments aboard the shuttle while it was in orbit. Mission controllers at the Spacelab Mission Operations Control in Huntsville, Alabama, sent over 25,000 commands to the Spacelab instruments during the flight, more than during any previous mission. The 15-day flight set a new mark for the longest shuttle mission, and the crew made 236 orbits aboard Columbia before returning to Earth on July 23, 1994.

Richard Hieb has spent more than 31 days in space, including more than 17 hours in EVA, during his three shuttle missions.

HILMERS, DAVID C.

(1950–)

U.S. Astronaut

David Hilmers was born in Clinton, Iowa on January 28, 1950. In 1968, he graduated from Central Community High School in Dewitt, Iowa, and then attended

Cornell College in Mount Vernon, Iowa, where he graduated summa cum laude with a Bachelor of Arts degree in mathematics in 1972.

Awarded a post-graduate fellowship by the National Collegiate Athletic Association (NCAA), Hilmers was also recognized as a member of Phi Beta Kappa and as the outstanding scholar athlete of the Midwest Conference in 1971. He was selected as a member of the first all-conference football team in 1971, and was honored for his track and field and overall athletic skills in 1972.

He began his career in the U.S. Marine Corps in July 1972, attending the Marine Corps Basic School and Naval Flight Officer School, and was subsequently assigned to the Marine Corps Air Station at Cherry Point, North Carolina, where he flew A-6 Intruder aircraft as a member of VMA(AW)-121.

In 1975, he began a new tour of day with the 6th Fleet, deployed in the Mediterranean Sea, serving with the 1st Battalion, 2nd Marines as an air liaison officer.

Continuing his education concurrently with his military service, Hilmers attended the U.S. Naval Postgraduate School, where he earned a Master of Science degree in electrical engineering as a distinguished graduate in 1977, and an electrical engineering degree the following year. He then served in Iwakuni, Japan as a member of the 1st Marine Aircraft Wing, and was assigned to the 3rd Marine Aircraft Wing in El Toro, California. NASA selected him for astronaut training in 1980. He had risen to the rank of colonel in the U.S. Marine Corps by the time of his retirement from the military.

Hilmers first flew in space in October 1985, during STS-51J, the first flight of the space shuttle Atlantis. The second shuttle flight devoted to the classified activities of the U.S. Department of Defense, the four-day mission came to a close with the shuttle landing at Edwards Air Force Base in California on October 7, 1985.

His next spaceflight assignment was as a member of the STS-61F crew, which was scheduled to deploy the Ulysses solar probe for the European Space Agency (ESA). The Ulysses launch was indefinitely delayed after the January 1986 loss of the space shuttle Challenger (the probe was eventually launched during STS-41 in 1990), and NASA subsequently abandoned the effort to develop a Centaur upper stage rocket for deploying the Ulysses probe. The combination of events led to a change in assignments for Hilmers, who had been part of the Shuttle Centaur team in addition to being scheduled to fly in space aboard the canceled STS-61F. His revised duties included the development of plans for crew abort during the ascent phase of shuttle launches before he began training for his next space mission.

Hilmers's next mission, STS-26, was the first shuttle flight after the Challenger disaster. As a member of the crew that brought America back into space after the long, wrenching recovery following the accident, he launched aboard the shuttle Discovery on September 29, 1988. The crew deployed the Tracking and Data Relay Satellite TDRS-C and performed a variety of experiments during the flight, which landed on October 3, 1988.

His third space mission, STS-36, was the sixth shuttle flight dedicated to the classified activities of the U.S. Department of Defense. After a week's delay due to the illness of mission commander John Creighton, STS-36 launched February 28,

1990, orbiting the Earth 72 times before landing at Edwards Air Force Base on March 4.

On January 22, 1992, Hilmers flew in space for a fourth time, as a mission specialist during the STS-42 International Microgravity Laboratory (IML) mission. A milestone in international scientific cooperation in space, the flight featured 55 major IML experiments designed by scientists from six international organizations representing 11 countries.

The STS-42 crew worked around the clock in alternating shifts to conduct the experiments, observing the impact of microgravity on a variety of life forms and materials. The mission was extended by one day to allow time for the crew to complete their work, and Discovery made a total of 129 orbits in a little more than eight days, landing at Edwards Air Force Base on January 30, 1992.

David Hilmers spent more than 20 days in space during his four space shuttle flights.

He retired from NASA in 1992 and subsequently enrolled at the Baylor College of Medicine in Houston, Texas as a medical student.

HITEN

(1990–1993)

Unmanned Japanese Spacecraft

The Hiten satellite was designed to test flight systems and technologies for future lunar and planetary exploration. Built by the Institute of Space and Aeronautical Science (ISAS) at the University of Tokyo, Hiten was launched from the Kagoshima Space Center in Japan on January 24, 1990.

Originally referred to as Muses-A, Hiten was renamed in honor of the Buddhist angel Hiten, who loved music and was able to fly, and whose veil was named Hagoromo, the name given to the small orbiter carried by the Hiten spacecraft. The Hagoromo orbiter was released near the Moon, but its transmitter failed on February 21, 1990 before it could be maneuvered into lunar orbit.

Hiten's planned navigational maneuvers and aerobraking exercises were carried out successfully, and its mission was extended to include further exercises, a lunar orbit, and a hard landing on the lunar surface. Those additional objectives were all accomplished, culminating in Hiten's crash landing on the Moon on April 10, 1993.

HOFFMAN, JEFFREY A.

(1944–)

U.S. Astronaut

Jeffrey Hoffman was born in Brooklyn, New York on November 2, 1944. In 1962, he graduated from Scarsdale High School, and he then attended Amherst College, where he was recognized for his scientific research. He graduated summa cum laude from Amherst in 1966 with a Bachelor of Arts degree in astronomy.

After graduating from Amherst, he was awarded pre-doctoral fellowships by the Woodrow Wilson Foundation and by the National Science Foundation, and went on to attend Harvard University, where he earned a Ph.D. in astrophysics in 1971. While at Harvard he designed and built a balloon-borne, low-energy gamma ray telescope.

He received a postdoctoral visiting fellowship from the National Academy of Sciences for 1971–72, a Harvard University Sheldon International Fellowship for 1972–73, and a postdoctoral fellowship from NATO in 1973–74. He pursued his postdoctoral research at Leicester University in the United Kingdom, where he designed equipment for testing X-ray beams, and he served as project scientist for the medium-energy X-ray experiment on the EXOSAT satellite of the European Space Agency (ESA).

In 1975, Hoffman became project scientist for the orbiting HEAO-1 A4 hard x-ray and gamma ray experiment at the Center for Space Research at the Massachusetts Institute of Technology (MIT), playing a central role in the design and development of the experiment prior to its launch in 1977 and in the analysis of the data it returned. His study of X-ray bursts is reflected in more than 20 scientific papers he has authored or co-authored.

Hoffman became an astronaut in 1979. He served on the astronaut support crew for STS-5 and as a spacecraft communicator (Capcom) on STS-8 and STS-82, and he worked on the development of a high-pressure spacesuit. He was also a co-founder of the Astronaut Office Science Support Group.

He made his first flight in space as a mission specialist aboard the space shuttle Discovery, during STS-51D. Launched on April 12, 1985, the busy mission featured the launch of two satellites, the Canadian TELESAT-I (ANIK C-1) and the SYNCOM IV-3 (LEASAT-3), the first spaceflight of an American public official, Senator Jake Garn of Utah, and a number of scientific experiments.

When a lever that was designed to start a sequence that would fire a motor designed to propel the satellite into orbit failed to engage on the SYNCOM IV-3, mission controllers extended the mission for two days while they improvised a way to fix the problem. Hoffman and fellow mission specialist David Griggs made the first unscheduled spacewalk of the space shuttle program when they put the proposed solution to the test. They attached two flap-like devices (which subsequently became nicknamed "the flyswatters" in NASA lore) to the end of the shuttle's robotic arm, the Remote Manipulator System (RMS), and then mission specialist Rhea Seddon used the RMS to catch the stuck lever with the improvised flaps to try to force the lever into the proper position. Despite the great skill and exceptional coordination of the crew and mission controllers, the lever simply would not work, and the repair of SYNCOM IV-3 had to wait until STS-51I, later in the year. Discovery returned to Earth at the Kennedy Space Center in Florida on April 19, 1985 in a harsh landing that led to a blown tire and extensive damage to the shuttle's braking system. Fortunately, the crew was not injured.

Continuing his education concurrently with his career as an astronaut, Hoffman attended Rice University, where he earned a Master's degree in materials science in 1988.

On December 2, 1990, Hoffman lifted off aboard the shuttle Columbia for STS-35 in a mission that represented the culmination of eight years of work. The first shuttle flight devoted to astronomical observations, STS-35 drew upon Hoffman's scientific expertise and equipment and flight planning abilities. Using the instruments of the ASTRO-1 astronomical observatory, the crew worked in alternating shifts to make around the clock ultraviolet and X-ray observations. Unexpected drama arose when the data display units on the observatory's ultraviolet telescopes failed, but again Hoffman's expertise came to the rescue as he and the crew manually fine-tuned the instruments according to the aiming instructions of mission support teams on the ground at the Marshall Space Flight Center. The shuttle had to make over 200 maneuvers to accommodate the proper pointing of the telescopes to achieve the mission's objectives.

Hoffman served as payload commander for STS-46, his third space mission, which launched on July 31, 1992 aboard the space shuttle Atlantis. Dedicated to international flight experiments, STS-46 featured the first flight in space of an Italian citizen, payload specialist Franco Malerba, and the first spaceflight of a citizen from Switzerland, Claude Nicollier. Nicollier deployed one of the flight's two primary payloads, the European Retrievable Carrier (EURECA) of the European Space Agency (ESA), which would remain in orbit until its retrieval during STS-57 in 1993.

The flight also featured the first major test of tethered spaceflight, in a joint project of NASA and the Italian space agency Agenzia Spaziale Italiana (ASI), with which Hoffman had been involved since 1987.

The concept of joining two spacecraft with a tether for linked flight had been tested in space as early as the Gemini program in 1966. Prior to STS-46, teams of engineers from NASA and ASI had collaborated for more than a decade to design the equipment, the plan, and the procedures that were to have been used during the Tethered Satellite System (TSS-1) test aboard Atlantis. The exercise was designed to establish a link between the shuttle and a satellite that would be deployed at a different orbital altitude from Atlantis, for tethered flight at a distance of as much as 12.5 miles.

Despite all the careful preparations, however, a mechanical hitch foiled the test when the tether jammed after it had been unwound to a distance of just 840 feet. The crew attempted for several days to release the stuck line, but the experiment finally had to be written off as a failure. The TSS satellite was retrieved and returned to Earth.

Hoffman's fourth spaceflight was as a mission specialist during STS-61, one of the most complex and ambitious missions yet attempted, the first servicing of the Hubble Space Telescope (HST). His crew mates on the flight included mission commander Richard Covey, pilot Kenneth Bowersox, payload commander Story Musgrave, and fellow mission specialists Kathryn Thornton, Thomas Akers, and Claude Nicollier of the European Space Agency (ESA).

Endeavour caught up with the Hubble Space Telescope on the third day of the flight, and Nicollier used the shuttle's robotic arm, the RMS, to position the HST in Endeavour's payload bay. Then, working in teams during an amazing series of spacewalks over the course of five days, Hoffman and Musgrave alternated with Akers and Thornton to service and repair the telescope.

Hoffman and Musgrave made the first EVA, working on the HST for 7 hours and 54 minutes while they replaced the instrument's Rate Sensing Units (RSUs)—the devices containing the HST's gyroscopes—and their underlying electronics and electrical hardware. After a day off while Akers and Thornton made their first trip outside the shuttle, Hoffman and Musgrave replaced the telescope's Wide Field/Planetary Camera, one of the HST's five scientific instruments, in about a quarter of the time mission planners had originally set aside for the task. They then used the remainder of their 6 hour and 48 minute spacewalk to install two new magnetometers on the telescope.

Akers and Thornton took their turn again the next day, and on the eighth day of the flight, Hoffman and Musgrave made the fifth spacewalk of the mission. They installed the electronics for the new solar arrays, placed protective covers over the telescope's original magnetometers, and installed the Goddard High Resolution Spectrograph Redundancy kit. Their 7 hours and 21 minutes EVA raised the total spacewalking time for the flight to a remarkable 35 hours and 28 minutes. The HST was re-deployed the following day, and Endeavour landed at the Kennedy Space Center in Florida in the early hours of December 13, 1993.

On February 22, 1996 Hoffman began a fifth flight in space, as a mission specialist aboard the shuttle Columbia on STS-75. The mission profile called for the second test of the Tethered Satellite System, TSS-1R, which had originally been tried out during STS-46 in 1992. Jointly designed by NASA and the Italian space agency Agenzia Spaziale Italiana (ASI), the TSS experiment called for tethered flight between the shuttle and a satellite in the Earth's atmosphere. After a day's delay because of difficulties with the experiment's on-board computer, the deployment of the TSS satellite began well, returning useful data until the satellite was almost fully extended to its planned distance of 12.8 miles. Unfortunately, the tether broke just prior to the completion of the deployment, ending the experiment, but the effort still yielded important data, including confirmation of the fact that tethers can produce electricity while in linked flight in the Earth's ionosphere.

The mission also featured the third flight of the United States Microgravity Payload (USMP-3), and the crew worked around the clock to complete their scientific research. During STS-75, Hoffman added another 15 days, 17 hours, 40 minutes, and 21 seconds to his career total time in space, and became the first astronaut to pass 1,000 hours of service aboard the space shuttle. The flight came to a close with Columbia's landing on March 9, 1996.

Jeffrey Hoffman spent more than 50 days and 11 hours in space, including more than 25 hours in EVA, during his five shuttle flights.

Following his fifth flight in space, Hoffman became the head of the payload and habitability branch of the Astronaut Office. In 1997, he was named NASA's European representative in Paris, acting as a liaison between the space agency and its European partners. In 2001, he joined the faculty of the Massachusetts Institute of Technology (MIT) as a professor in the department of aeronautics and astronautics.

HOROWITZ, SCOTT

(1957–)

U.S. Astronaut

Scott "Doc" Horowitz was born in Philadelphia, Pennsylvania on March 24, 1957. In 1974, he graduated from Newbury Park High School in Newbury Park, California, and then attended California State University at Northridge, where he received a Bachelor of Science degree in engineering in 1978. He then attended the Georgia Institute of Technology, earning a Master of Science degree in aerospace engineering in 1979, and a Doctorate degree in aerospace engineering in 1982.

An outstanding student and engineer, he won first place in the design competition of the American Society of Mechanical Engineers (ASME), and was honored with the outstanding doctoral research award for 1981–82 when he graduated from the Georgia Institute of Technology.

He began his career at the Lockheed-Georgia Company in Marietta, Georgia, working on aerospace technology projects as an associate scientist. He also joined the U.S. Air Force, receiving his undergraduate pilot training at Williams Air Force Base in Arizona and then serving at Williams as a flight instructor in T-38 aircraft and working in the base's Human Resources Laboratory. In 1985, he was recognized as an Outstanding T-38 Instructor Pilot, and he was named a Master T-38 Instructor Pilot in 1986.

Horowitz was then assigned to the 22nd Tactical Fighter Squadron as an operational F-15 Eagle fighter pilot, serving at Bitburg Air Base in Germany. In 1990, he attended the United States Air Force Test Pilot School at Edwards Air Force Base in California, where he was recognized as a distinguished graduate of class 90A. He remained at Edwards after graduation, serving as a test pilot attached to the 6512th Test Squadron while flying A-7 and T-38 aircraft.

He also taught graduate courses in aircraft design, aircraft propulsion, and rocket propulsion at Embry Riddle University as an adjunct professor from 1985 to 1989, and in 1991 he taught graduate courses in mechanical engineering as a professor at California State University at Fresno.

Horowitz accumulated over 5,000 hours of flying time during

Scott Horowitz during training for STS-105, July, 2001. [NASA/courtesy of nasa images.org]

his distinguished military career, and rose to the rank of colonel in the U.S. Air Force by the time he retired from the service.

NASA selected Horowitz for astronaut training in 1992, and he first flew in space as pilot of the space shuttle Columbia during STS-75. Launched on February 22, 1996, STS-75 featured the third flight of the United States Microgravity Payload (USMP-3) and the second flight of the Tethered Satellite System (TSS), which had originally been tried out during STS-46 in 1992. Jointly designed by NASA and the Italian space agency Agenzia Spaziale Italiana (ASI), the TSS experiment called for tethered flight between the shuttle and a satellite in the Earth's atmosphere. After a day's delay because of difficulties with the experiment's on-board computer, the deployment of the TSS satellite began well, returning useful data until the satellite was almost fully extended to its planned distance of 12.8 miles. Unfortunately, the tether broke just prior to the completion of the deployment, ending the experiment, but the effort still yielded important data, including confirmation of the fact that tethers can produce electricity while in linked flight in the Earth's ionosphere.

The crew worked around the clock to complete their USMP-3 experiments, returning to Earth at KSC on March 9, 1996.

Horowitz again served as pilot during his second flight in space, aboard the shuttle Discovery during STS-82, the second servicing of the Hubble Space Telescope (HST). STS-82 began with a nighttime launch on February 11, 1997.

Among the most complex missions undertaken during the entire shuttle program, the HST service and repair flights required enormous stamina on the part of alternating spacewalking teams, and minute attention to detail over long periods of working in space.

The crew made five spacewalks, including an unscheduled EVA at the end of the repair program, and used more than 150 tools during their work on the HST. Working in two alternating teams, mission specialists Mark Lee and Steven Smith took turns with Gregory Harbaugh and Joseph Tanner while accumulating a total of 33 hours and 11 minutes in EVA during the remarkable flight. Discovery returned to Earth in a night landing on February 21, 1997.

During STS-101 in May 2000, Horowitz piloted the shuttle Atlantis during the third shuttle mission dedicated to the construction of the International Space Station (ISS). His fellow crew members for the mission included Susan Helms, Jim Voss, and Russian cosmonaut Yury Usachev, who would, in 2001, live and work aboard the ISS for nearly six months, as the ISS Expedition 2 crew.

During STS-101, the crew of Atlantis repaired and updated the ISS Zarya and Unity modules, replacing faulty equipment, installing fans, fire extinguishers, smoke detectors, and electronic components.

Atlantis undocked from the station on May 26 and landed on May 29, 2000 after 155 orbits and a flight of 9 days, 20 hours, and 9 minutes in space.

On August 10, 2001, Horowitz lifted off on a fourth space mission, as commander of STS-105 aboard Discovery. Traveling to the ISS to deliver the ISS Expedition 3 crew, the STS-105 astronauts also used the Leonardo Multi-Purpose Logistics Module (MPLM) for the second time, facilitating the transfer of nearly three tons of equipment and supplies to the ISS and the removal of months' worth of refuse.

Launched with the STS-105 Discovery crew, Frank Culbertson, Vladimir Dezhurov, and Mikhail Tyurin transferred to the ISS to begin their four-month stay as the ISS Expedition 3 crew. The ISS Expedition 2 crew—James Voss, Susan Helms, and Russian cosmonaut Yury Usachev—returned to Earth with Horowitz and the shuttle crew, landing on August 22, 2001.

Scott Horowitz spent more than 47 days in space during his career as an astronaut.

In October 2004 he retired from NASA to join A.T.K.-Thiokol as director of space transportation and exploration, but returned to NASA in September 2005 to serve at NASA's headquarters in Washington, D.C. as associate administrator of the agency's Exploration Systems Mission Directorate.

HUBBLE SPACE TELESCOPE (HST).

See Great Observatories Program

HUNGARY: FIRST CITIZEN IN SPACE.

See Intercosmos Program

HUSBAND, RICK D.

(1957–2003)

U.S. Astronaut

A veteran of two space shuttle flights and the commander of the STS-107 final flight of the space shuttle Columbia, Rick Husband was born in Amarillo, Texas on July 12, 1957. In 1975, he graduated from Amarillo High School and then attended Texas Tech University, where he received a Bachelor of Science degree in mechanical engineering in 1980. An outstanding student, he was honored with the outstanding engineering student award at Texas Tech. He also participated in the U.S. Air Force ROTC program at the university and was honored as a distinguished graduate of the program.

After receiving his commission as a second lieutenant in the Air Force, Husband received his initial pilot training at Vance Air Force Base in Oklahoma, where he was also honored as a distinguished graduate. He trained in F-4 aircraft at Homestead Air Force Base in Florida in 1981, and flew the F-4E at Moody Air Force Base in Georgia when he was assigned there the following year. He returned to Homestead Air Force Base in 1985 to attend F-4 Instructor School, where he was again honored as a distinguished graduate, and then served as an instructor pilot and academic instructor at George Air Force Base in California. In 1987, he was honored as F-4 Tactical Air Command Instructor Pilot of the Year.

Husband was later recognized as a distinguished graduate of the Air Force Test Pilot School at Edwards Air Force Base in California. Following his graduation from the school, he was assigned to test the F-4 and each of the five models of the F-15 aircraft, and he served as program manager for the Pratt & Whitney F-100-PW-229 increased performance engine.

Continuing his education concurrently with his career as a military test pilot, Husband attended California State University at Fresno, where he received a Master of Science degree in mechanical engineering in 1990.

His next military assignment, in 1992, brought him to Boscombe Down, England, where he served as an exchange test pilot serving with the Royal Air Force. His duties during that deployment included project pilot for the Tornado GR1 and GR4 aircraft, and he also flew as a test pilot in a variety of other British aircraft, including the Buccaneer, Hawk, Harvard, Hunter, Jet Provost, and Tucano.

During his exceptional military career, Husband accumulated more than 3,800 hours of flight time in more than 40 different types of aircraft. He was a colonel in the U.S. Air Force at the time of his death, aboard the space shuttle Columbia, in 2003.

In December 1994, NASA selected Husband for astronaut training. His initial assignments at the agency included service as the Astronaut Office representative for advanced projects. In that capacity he helped to develop upgrades for the space shuttle, participated in planning missions to return to the Moon and to travel to Mars, and worked on the development of a Crew Return Vehicle. His other duties within the Astronaut Office also included service as Chief of Safety. Over the course of his NASA career he was a member of two teams that were recognized with the agency's Group Achievement Award, the X-38 Development Team and the Orbiter Upgrade Definition Team.

Husband first flew in space as pilot of STS-96, which launched aboard the space shuttle Discovery on May 27, 1999. The historic first docking of a shuttle and the International Space Station (ISS), STS-96 brought replacement parts and equipment to the ISS in preparation for its eventual habitation by its first long-duration crew. The STS-96 crew transferred 3,000 pounds of equipment and supplies to the station during six days of linked flight, and mission specialists Tamara Jernigan and Daniel Barry installed two cranes and equipment for use in future EVAs during a 7 hour and 55 minute spacewalk. Husband and mission commander Kent Rominger brought Discovery back to Earth at the Kennedy Space Center (KSC) in Florida in a night landing during the early hours of June 6, 1999.

His second space mission was as commander of STS-107, aboard the space shuttle Columbia. Launched on January 16, 2003, the STS-107 crew worked around the clock in two shifts to complete 80 scientific experiments.

Tragically, Columbia had been seriously damaged at launch (more seriously than NASA officials realized during the mission) and was destroyed by the stresses of re-entry into the Earth's atmosphere at the end of its 16-day mission. Commander Husband and his fellow Columbia crew mates were killed when the shuttle broke apart over the southern United States, just 16 minutes before the spacecraft was scheduled to land.

Rick Husband accumulated 25 days, 11 hours, and 39 minutes in space during his two shuttle flights. He is survived by his wife and their two children.

In honor of his service, Husband was posthumously awarded the Congressional Space Medal of Honor, the NASA Space Flight Medal, the NASA Distinguished Service Medal, and the Defense Distinguished Service Medal.

HUYGENS PROBE

(1997–2005)

Unmanned European Space Agency Craft Designed to Study Saturn's Moon Titan

An atmospheric probe designed by the European Space Agency to explore Saturn's moon Titan, Huygens was launched aboard NASA's Cassini spacecraft on October 15, 1997.

The mission profile for Huygens included the return of data such as the density, pressure, and temperature of Titan's atmosphere and information about its chemistry and meteorology, as well as photographic coverage of its surface.

Huygens entered orbit around Titan on December 25, 2004 and successfully completed its mission when it set down on the surface on January 14, 2005, the first man-made object to be landed in the outer solar system, and the farthest landing from Earth.

Although it was expected to operate on the surface for a maximum of 30 minutes, the Huygens probe was able to transmit surface data to Cassini for more than an hour. Among its findings, the spacecraft indicated that the Titan terrain features a thin frozen crust about 10 centimeters deep, with a less dense layer of clay-like material beneath.

INDIAN SPACE RESEARCH ORGANIZATION (ISRO)

(1962–)
National Space Agency of India

Headquartered in Bangalore, the Indian Space Research Organization (ISRO) is the national space agency of India.

The roots of India's space program began with the Indian National Committee for Space Research (INCOSPAR), which was founded in 1962 at the suggestion of Homi Bhabha, the director of the country's department of Atomic Energy. The first director of INCOSPAR was Vikram Sarabhai, a visionary scientist who is recognized as the leading figure in the early history of the nation's space program.

Initially, the engineers and scientists of the Indian space program gained experience in spaceflight through the launch of unmanned sounding rockets manufactured in the United States, France, Great Britain, and the Soviet Union. These launches, which began on November 21, 1963 and took place throughout the 1960s, lifted off from the Thumba Equatorial Rocket Launching Station. To accompany the launch facility and provide world-class support infrastructure for the nation's space program, the Space Science and Technology Center was opened at Thumba in 1965.

India's first launch vehicles were the Rohini series of sounding rockets, which led to the development of the nation's first Satellite Launch Vehicle (SLV). The Indian program of launcher development was a particularly remarkable achievement in that the rockets used for launching satellites into orbit were pure launch vehicles that had not been derived from missiles initially designed for military use, as was the case with a majority of other nations that have developed an independent launch capability.

As the engineers of the Indian space program worked to develop the nation's first "home-grown" launch vehicle, they also set out to build satellites for communications and remote sensing applications that could be launched by other nations until the first Indian-built launch vehicle was ready.

Indian Space Research Organization (ISRO) chairman Vikram Sarabhai (left) and NASA Administrator Thomas Paine sign an agreement for a cooperative satellite project, September 18, 1969. [NASA/courtesy of nasa images.org]

Aryabhata, the first Indian satellite, was launched on April 19, 1975 by the Soviet Union at its Kapustin Yar launch facility. Featuring an innovative design that shaped the vehicle roughly like a sphere but which actually included 26 panels, 24 of which were outfitted with solar panels that provided a total 46 watts of electricity, the Aryabhata spacecraft weighed 360 kilograms. It was equipped with instruments to perform X-ray astronomy studies, to make measurements of solar neutrons and Gamma rays, and to study electrons in the ionosphere. Mission managers tracked its progress from the Shar Center in Sriharikota.

The success of the Aryabhata project provided invaluable expertise and experience for the nation's space program and established several leading personalities among Indian scientists, including Udipi Ramachandra Rao, who served as project director and as principal investigator for the X-ray astronomy experiment, Satya Prakash, principal investigator of the Ionospheric Electron Trap, and Ranjan Daniel, principal investigator of the experiment involving the measurement of solar neutrons and Gamma rays.

On July 18, 1980, India became the seventh nation in the world to establish an independent launch capability, with the launch of the Rohini 1 satellite. Rohini 1 lifted off from the Sriharikota Space Center atop an SLV-3 launch vehicle, the result of the long development effort that began as the logical next step forward from the earlier sounding rockets.

A small, lightweight sphere measuring 0.6 meters in diameter and weighing 35 kilograms, Rohini 1 achieved an initial orbit of 302 kilometers perigee by 919 kilometers apogee, with a period of 96.9 minutes. It orbited the Earth for nearly a year, until re-entering the atmosphere on May 20, 1981.

Two other Rohini satellites followed the first launch; Rohini 2 launched on May 31, 1981, and Rohini 3 lifted off on April 17, 1983.

In April 1984 the nation saw its first citizen launched into space when, as a part of the Soviet Union's Intercosmos program, Rakesh Sharma visited the Soviet Salyut 7 space station. Sharma launched aboard Soyuz T-11 on April 3, 1984 with Soviet cosmonauts Yuri Malyshev and Gennady Strekalov.

The manned Intercosmos flights had arisen from the Soviet need to replace Soyuz spacecraft docked for long periods in space with a replacement spacecraft before the docked vehicle exceeded its safe operations limit. The resulting Soyuz switching missions were also used to provide opportunities for citizens of nations friendly to or aligned with the Soviet Union to fly in space.

Sharma's visit to Salyut 7 was a source of national pride for India, and he proved an interesting guest for his crew mates and the resident cosmonauts aboard the space station as he demonstrated the use of yoga as a means of adapting to the weightlessness of space. He returned to Earth with Malyshev and Strekalov on April 11, 1984, after a flight of 8 days and 42 minutes.

In 1988, the nation's first remote sensing satellite, IRS 1A, was launched by the Soviet Union. In following years, ISRO would provide both satellites and launch vehicles for flights originating from Indian launch facilities, as the agency's e ngineers developed the program's next generation of launchers.

The Polar Satellite Launch Vehicle (PSLV) was designed in cooperation with the French space agency Centre National d'Études Spatiales (CNES) with the goal of being able to launch larger, more functional satellites

A second launcher, the Augmented Satellite Launch Vehicle (ASLV) was designed as an experimental vehicle that could be used to test the technologies necessary for the PSLV and similar development projects.

After several failures, the first successful launch using the ASLV was achieved in 1992. The PSLV also failed in its first attempt, but was successfully launched in 1994 and ultimately became the foundation of the ISRO launch capability, successfully launching communications and remote sensing satellites.

To further broaden its launch capabilities, ISRO began development on the Geostationary Satellite Launch Vehicle (GSLV) in the early 1990s. The GSLV was designed to propel satellites into geostationary transfer orbit. The new launcher required technology that the agency had not yet developed in-house, which led ISRO to seek the necessary technology from Russia. The U.S. government effectively blocked the transfer, however, on the grounds that it would provide India with technology that could also be used in the production of nuclear weapons. An agreement was eventually reached in which Russia provided completed components rather than the technology necessary to produce them.

ISRO has continued to develop the GSLV, which was first flight tested in 2001, and has also pursued the development of additional next-generation launch vehicles. The agency has successfully launched satellites for other nations, too, notably launching a reconnaissance satellite for Israel, for example, in 2005.

Indian-designed satellites launched by ISRO cover a variety of applications, such as communications (the Indian National Satellite or INSAT series), remote sensing (the Indian Remote Sensing or IRS series), reconnaissance (the Technology Experiment Satellite series), and meteorology (the METSAT series, which was renamed the Kalpana series in 2003 to honor NASA astronaut Kalpana Chawla, who was born in Karnal, India and who died in the loss of the U.S. space shuttle Columbia during STS-107 on February 1, 2003).

On October 21, 2008, ISRO launched India's first lunar mission, Chandrayaan-1 ("Moon craft" in ancient Sanskrit), from the Satish Dhawan Space Centre in

Sriharikota. A primary goal of the two-year mission is to gather high quality mapping data about the Moon and the minerals beneath the lunar surface.

INTERCOSMOS PROGRAM

(1967–1990s)

International Space Initiative of the Soviet Union

The leaders of the Soviet Union launched the Intercosmos program to give the nation's political allies a role in the successful Soviet space program. The program initially allowed Soviet bloc nations to participate in the development of scientific satellites, and ultimately provided opportunities for non-Soviet cosmonauts to fly in space for the first time during visits to the Salyut 6, Salyut 7, and Mir space stations.

In the program's initial framework, which was agreed upon in April 1967, eight countries were chosen to participate with the Soviet Union in the development of a series of unmanned scientific satellites. The original signatory nations were Bulgaria, Cuba, Czechoslovakia, East Germany, Hungary, Mongolia, Poland, and Romania. Over the course of the ensuing decades, each of these nations would also see the first of their citizens fly in space under the auspices of the manned portion of the Intercosmos program.

The unmanned Intercosmos program began with the launch of Cosmos 261 on December 20, 1968, and continued with several launches per year throughout the 1970s and 1980s and into the 1990s. Studies carried out by the vehicles launched as part of the series included investigations of the Sun and of the Earth's ionosphere and magnetosphere, among many others, and representatives of the Intercosmos partner nations were provided opportunities to participate in a range of scientific activities, from the engineering and development of space vehicles to the analysis and study of the scientific data gathered during each mission.

The first manned Intercosmos flight, Soyuz 28, took place in March 1978. It featured a brief visit to the Salyut 6 space station by guest cosmonaut Vladimír Remek of Czechoslovakia.

For the non-Soviet cosmonaut visiting the Soviet space station, typical activities during the brief stay including the conducting of scientific experiments designed in his home country and the making of much-publicized statements praising the Soviet space program, and, by extension, the nation's political system.

The first nine Intercosmos flights involved visits to the Salyut 6 space station, which was in orbit from September 29, 1977 to July 29, 1982. Two Intercosmos missions visited Salyut 7, which was in orbit from April 19, 1982 to February 7, 1991. Three flights nominally recognized as part of Intercosmos, although they occurred later and in a different political climate from the early missions (and just one featured the politically-inspired "first flight" of a non-Soviet cosmonaut), were made to the space station Mir.

Among the formidable challenges encountered by Soviet engineers at the advent of the space station era was the continuous need to replace the Soyuz spacecraft that would transport cosmonauts to the orbiting stations. Past experience dictated that the reliability of the systems and equipment of the Soyuz vehicle could only be counted on for a period of about 90 days, so a steady stream of replacement spacecraft had to visit the space station, so that the arriving vehicle could be left at the facility and the previously flown Soyuz could be used to bring the second crew back to Earth.

Converting this need for frequent visits to Salyut 6 and Salyut 7 into an opportunity to send politically acceptable foreign cosmonauts on brief trips into space, the Soviets launched the first international mission with Soyuz 28, on March 2, 1978. Vladimir Remek of Czechoslovakia (which became the Czech Republic after the fall of the Soviet Union) and Soviet cosmonaut Aleksei Gubarev visited Yuri Romanenko and Georgi Grechko, who were at the time in the midst of a then-record stay of 96 days on Salyut 6. Remek and Gubarev returned to Earth on March 10.

Miroslav Hermaszewski became the first Polish citizen in space during the flight of Soyuz 30, from June 27, 1978 to July 5, 1978. Hermaszewski and the mission's Soviet commander, Pyotr Klimuk, visited Salyut 6 cosmonauts Vladimir Kovalyonok and Alexander Ivanchenkov, who would go on to set a new endurance record of more than 139 days in space.

Cosmonaut Valeri Bykovsky and Sigmund Jähn, the first East German citizen to fly in space, also visited Kovalyonok and Ivanchenkov in Salyut 6 during their Soyuz 31 mission, from August 26 to September 3, 1978.

The next Intercosmos flight was a frightening misadventure that nearly resulted in tragedy. After their launch on April 10, 1979, Bulgarian Georgi Ivanov and Soviet cosmonaut Nikolai Rukavishnikov had almost reached the Salyut 6 station when a routine firing of the main engine on their Soyuz 33 spacecraft ended prematurely. The failure meant their mission would have to be cut short; more importantly, it meant that they would have to rely entirely on their backup engine. Fortunately, the backup engine did work, and they returned to Earth after a harrowing re-entry on April 12, having spent just under two days in space.

Coupled with a concurrent but unrelated failure in one of the fuel tanks aboard the Salyut 6 space station, the Soyuz 33 failure resulted in a long delay before the next mission featuring a non-Soviet cosmonaut.

That flight, the fifth Intercosmos mission, began on Soyuz 36 on May 26, 1980. Flying into space with Soviet mission commander Valeri Kubasov, Bertalan Farkas became the first Hungarian in space. Unlike the near-miss Soyuz 33 mission, the flight of Soyuz 36 successfully reached Salyut 6. Kubasov and Farkas visited Leonid Popov and Valeri Ryumin, who were in the midst of a new record 184-day stay at the station and who would also play host to two more Intercosmos missions before they left Salyut 6 in October 1979.

Tuân Pham of Vietnam visited Salyut 6 with Soviet cosmonaut Victor Gorbatko during the flight of Soyuz 37, from July 23 to July 30, 1980, in what was easily the most controversial flight of the Intercosmos program. An Air Force veteran of the Vietnam War, Pham's military experiences were promoted at some length during his involvement with the Intercosmos program, and his flight coincided with an American boycott of the 1980 Summer Olympics, which were hosted that year by the Soviet Union in Moscow. The already difficult political atmosphere between the two superpowers was exacerbated by accounts of Pham's prowess in fighting against the United States during the war.

In September 1980, Arnaldo Tamayo-Mendez became the first Cuban citizen to fly in space, visiting Salyut 6 with Yuri Romanenko aboard Soyuz 38 from September 18 to September 26.

Soviets Vladimir Kovalyonok and Viktor Savinykh began the last long-endurance mission aboard Salyut 6 in March 1981.

Their first Intercosmos visitor was Jugderdemidiyn Gurragcha of Mongolia, who arrived with Soviet cosmonaut Vladimir Dzhanibekov aboard Soyuz 39 shortly after Kovalyonok and Savinykh in March 1981.

Soyuz 40, commanded by Leonid Popov, launched with Dumitru Prunariu, the first Romanian citizen in space, on May 14, 1981. As with the other Intercosmos flights, Popov and Prunariu briefly visited Salyut 6 and returned to Earth in a little less than eight days' time from their launch. Theirs was the final manned mission to Salyut 6 and the final use of the first generation Soyuz spacecraft. Kovalyonok and Savinykh returned to Earth on May 26, therein marking the end of the space station's occupation.

Although Intercosmos flights continued after the end of Salyut 6, a great deal changed in the year between the Soyuz 40 flight and the launch of Salyut 7, on April 19, 1982. The American space shuttle program became fully operational during that period (the first shuttle lifted off in April 1981), and the international cooperation that would mark the shuttle era was already being made obvious by NASA.

Perhaps in response, or maybe just as a precursor of the political climate that would eventually lead to the disintegration of the totalitarian Soviet state by the decade's end, the Soviet government de-emphasized the rhetoric and reckless propaganda of the earlier flights in favor of a more stately approach that encompassed a wider array of participating countries.

Thus, an opportunity was born for Jean-Loup Chrétien to become the first French citizen in space, during the 10th Intercosmos flight, on Soyuz T-6, from June 24 to July 2, 1982. With Soviet cosmonauts Vladimir Dzhanibekov and Aleksandr Ivanchenkov, Chrétien visited Salyut 7 residents Anatoly Berezovoy and Valentin Lebedev, who would achieve a new endurance record for the Soviet Union by staying in orbit aboard Salyut 7 for 211 days.

Rakesh Sharma became the first citizen of India to fly in space on April 3, 1984, when he lifted off with Soviet cosmonauts Yuri Malyshev and Gennady Strekalov aboard Soyuz T-11. They visited Leonid Kizim, Vladimir Solovyov, and Oleg Atkov, who went on to set a new endurance record of 236 days on Salyut 7. Sharma, Malyshev, and Strekalov returned to Earth on April 11.

The last crew to live on Salyut 7 left the station in November 1985. The crew of Soyuz T-15 visited the station in 1986 and made repairs that made the facility at least theoretically inhabitable, but that same Soyuz T-15 mission was also the first to visit the then-new space station Mir, whose remarkable adaptability clearly made the Salyut generation of stations obsolete. Salyut 7 was never again occupied; it remained in orbit for another five years, until February 7, 1991.

Following their earlier, abortive adventure during the flight of Soyuz 33 in 1979, the citizens of Bulgaria were given a second chance to cheer their nation's space efforts when Alexander Aleksandrov became the second citizen from Bulgaria to fly in space. Aboard Soyuz TM-5, Aleksandrov visited Mir with Soviet cosmonauts Viktor Savinykh and Anatoli Solovyov and spent 10 days in space, returning to Earth on June 17.

The awkward political implications so often inherent in earlier Intercosmos missions reasserted themselves in the Soyuz TM-6 flight, which took place from August 29 to September 7, 1988. Although it successfully made Abdul Ahad Mohmand the first citizen of Afghanistan to fly in space, the mission was marred by a lack of proper preparation and a series of frightening difficulties that nearly prevented the crew's safe return to Earth.

The normal pre-launch preparation and training period was drastically reduced so the flight could take place prior to the end of the Soviet occupation of Afghanistan. The link between the spaceflight and the war was made explicit in the selection of Mohmand, a pilot in the Afghani Air Force. In any case, the rush may have played a role in the difficulties that cropped up at the mission's end.

Along with Soviets Vladimir Lyakhov and Valeri Polyakov, Mohmand launched into space on August 29. The crew was able to successfully dock with Mir, and at the end of their stay, Lyakhov and Mohmand bid farewell to Polyakov, who was to remain on Mir for an extended mission.

On the return trip, a computer fault caused Lyakhov and Mohmand to miss their first opportunity for re-entry. A second attempt was marred by further complications, and Lyakhov was unable to correct the situation. He and Mohmand then spent a difficult 24 hours awaiting a third try. Ground controllers were finally able to overcome the software glitch that had caused the initial difficulty, and Lyakhov and Mohmand landed safely on September 7, 1988.

Amid all the political and technological unpleasantness of the TM-6 mission, Mohmand gave the experience a genuinely meaningful larger perspective when he read from the Koran.

In November 1988, Jean-Loup Chrétien made his second Intercosmos flight, this time aboard Soyuz TM-7. He had been the first French citizen to fly in space when he participated in the tenth Intercosmos mission in 1982. Following their launch on November 26, Chrétien and Soviet cosmonauts Alexander Volkov and Sergei Krikalyov visited Vladimir Titov and Musa Manarov aboard the Mir space station. Titov and Manarov were nearing the end of their new endurance record of more than 365 days in space; they would return with Chrétien on December 21, trading places with Volkov and Krikalev, who became the new Mir crew.

The mission differed significantly from previous Intercosmos flights, which was probably an indication of both the changing political climate on Earth, and the fast-approaching era of less self-conscious international cooperation in space, typified by the U.S. space shuttle program and the later International Space Station (ISS).

Political considerations notwithstanding, Chrétien achieved a significant milestone during his second trip into space. On December 9, he became the first citizen of a country other than the Soviet Union or the United States to make an EVA. His accomplishment was magnified by the circumstances of his spacewalk: he and Volkov labored long and hard to deploy a flexible experimental extension to Mir, spending a total of more than six hours outside the space station.

INTERNATIONAL SPACE STATION (ISS)

(1984–)

Multinational Space Station in Earth Orbit

The most complex international scientific undertaking of the 20th century, the orbital outpost known as the International Space Station (ISS) began with a deceptively simple sounding proposition put forth by U.S. President Ronald Reagan during his January 25, 1984 State of the Union address.

Giving NASA a mandate to build a "permanently manned space station" by the end of the 1980s and extending an opportunity to nations friendly to the United States to participate in the project, President Reagan announced the beginning of Space Station Freedom.

Implicit in the invitation to friendly nations to cooperate in the station's design and operation was an understanding that the project would demonstrate the technical prowess of the United States and its allies, in direct contrast to the capabilities of the Soviet Union, which was at that time developing the Mir space station.

Several U.S. allies expressed early interest in the Space Station Freedom project. In 1985, the Canadian Space Agency (CSA), the European Space Agency (ESA), and the National Space Development Agency (NASDA) of Japan were the first to formally agree to participate in the station's development, which was headed by the United States' National Aeronautics and Space Administration (NASA).

Championed initially by NASA Administrator James Beggs and Deputy Administrator Hans Mark, the concept of an internationally developed, continuously manned space station was given its greatest push forward by Reagan's decision to embrace the concept and to present it personally to key United States allies. The president made the project a major focus of his participation in the June 1984 London Economic Summit.

Concerns about financial commitments, the role and responsibilities of each partner nation, and the degree of potential benefit each participant might reasonably expect to gain led to a protracted period of discussion and negotiation that continued throughout most of the 1980s.

Progress on the proposed program of international cooperation moved slowly throughout the end of the Cold War and the dissolution of the Soviet Union in the early 1990s. The resulting alterations in the global political climate gave the idea new life, however, and in 1993, during the administration of U.S. President Bill Clinton, the original concept was modified and expanded, and the Russian Federation was invited to join the original partners.

The International Station, June 3, 1999. The Unity node sits atop the Zarya module. [NASA/courtesy of nasaimages.org]

Initially dubbed Space Station Alpha but then officially re-named the International Space Station, the 1993 incarnation of the project was expanded throughout the 1990s to eventually include five partner agencies—NASA, CSA, ESA, the Japan Aerospace Exploration Agency (JAXA, the successor of NASDA), and the

Russian Federal Space Agency (RKA)—and a sixth space agency participant, the Brazilian space agency Agencia Espacial Brasileira (AEB).

U.S.-Russian Cooperation in Human Space Flight Program

As the first part of their joint participation in the ISS, the United States and Russia began a program of shared space missions known as the Cooperation in Human Space Flight Program, which also encompassed the Shuttle-Mir Program of docking missions involving the U.S. space shuttle and the Russian space station Mir. The overall effort got underway with the February 1994 STS-60 flight of the space shuttle Discovery, whose crew included Sergei Krikalyov, the first Russian cosmonaut to fly on a U.S. space shuttle.

Later flights included the first shuttle-Mir rendezvous (STS-63), the first U.S. astronaut to serve as a long-duration Mir crew member (Norman Thagard), nine shuttle dockings, and six long-duration Mir stays by U.S. astronauts that established a continuous U.S. presence of 907 consecutive days on the Russian space station.

Highlights of the Shuttle-Mir Program (U.S.-Russian "First Phase" of ISS Program):

—STS-60; Discovery, February 3–11, 1994:

- First flight of a Russian cosmonaut on an American shuttle

—STS-63; Discovery, February 3–11, 1995:

- First rendezvous and fly-around of a space shuttle and the Russian space station Mir
- Second Russian cosmonaut to fly on an American space shuttle

—Soyuz TM-21; Launched March 14, 1995:

- Delivered U.S. astronaut Norman Thagard to Mir, as part of the Mir 18 crew with Russian cosmonauts Vladimir Dezhurov and Gennady Strekalov, for a stay of nearly four months until their return during STS-71 in July 1995

—STS-71; Atlantis, June 27–July 7, 1995:

- First docking of a space shuttle and the Russian space station Mir
- Delivered Mir 19 long-duration crew
- Returned Mir 18 long-duration crew, including NASA astronaut Norman Thagard

—STS-74; Atlantis, November 12–20, 1995:

- Second docking of a space shuttle and the Russian space station Mir
- Delivered Russian docking module to Mir

—STS-76; Atlantis, March 22–31, 1996:

- Third docking of a space shuttle and the Russian space station Mir
- Start of continuous U.S. presence aboard Mir (delivered Shannon Lucid)

—STS-79; Atlantis, September 16–26, 1996:

- Fourth docking of a space shuttle and the Russian space station Mir
- First exchange of U.S. Mir crew members (delivered John Blaha, returned Shannon Lucid)

—STS-81; Atlantis, January 12–22, 1997:

- Fifth docking of a space shuttle and the Russian space station Mir
- Second exchange of U.S. Mir crew members (delivered Jerry Linenger, returned John Blaha)

—STS-84; Atlantis, May 15–24, 1997:

- Sixth docking of a space shuttle and the Russian space station Mir
- Third exchange of U.S. Mir crew members (delivered Michael Foale, returned Jerry Linenger)
- During his Mir stay, Linenger and Vasili Tsibliyev made the first U.S.-Russian spacewalk, on April 29, 1997

—STS-86; Atlantis, September 25–October 6, 1997:

- Seventh docking of a space shuttle and the Russian space station Mir
- Fourth exchange of U.S. Mir crew members (delivered David Wolf, returned Michael Foale)
- STS-86 crew mates Scott Parazynski and Vladimir Titov made the first shuttle-based U.S.-Russian spacewalk, on October 1, 1997

—STS-89; Endeavour, January 22–31, 1998:

- Eighth docking of a space shuttle and the Russian space station Mir
- Fifth and final exchange of U.S. Mir crew members (delivered Andrew Thomas, returned David Wolf)

—STS-91; Discovery, June 2–12, 1998:

- Ninth and final docking of a space shuttle and the Russian space station Mir
- Returned U.S. Mir crew member Andrew Thomas at end of his 130 days aboard the station, completing a total of 907 consecutive days of continual U.S. presence aboard Mir.

With the successful conclusion of the first phase of their ISS-related cooperation in space, the United States and Russia focused on construction of the new space station. Mir remained in orbit for several more years while the ISS was being built; the legendary Russian station was de-orbited on March 23, 2001 after more than 15 years in space.

ISS Construction

Designed as separate modules to be launched independently and assembled in orbit with pieces delivered and installed by future shuttle crews, construction of the ISS began with the November 20, 1998 launch of the Zarya Functional Cargo

Block (FCB) module, which lifted off from the Baikonur Cosmodrome in Kazakh-stan atop a Russian Proton launch vehicle.

A series of construction missions, including the launch of the Zvezda Ser-vice Module in December 1998, the arrival of the U.S. Destiny Laboratory Mod-ule during the STS-98 flight of the space shuttle Atlantis in February 2001, and the April 2001 arrival of the CSA-developed and built Canadarm2, followed at regular intervals until the loss of the space shuttle Columbia in 2003.

The Russian Zarya module serves as the systems core of the ISS. It houses the station's propulsion system and electrical equipment and the instruments that keep the ISS properly oriented in orbit. The Zvezda module, which was also devel-oped by the engineers of the Russian space program, serves as the station's primary living area. The U.S. Destiny laboratory was the first ISS research facility.

During its first seven years in space, the ISS was equipped with two airlocks (Quest and Pirs), several truss segments, and support structures such as the sta-tion's solar arrays and the Mobile Base System, which enables crew members to position the Canadarm2 robotic arm at various points around the outside of the space station. To facilitate the station's expansion, Unity Nodes 1 and 2 serve as links for additional modules and as connection points for support infrastructure, including electric, communications, and life support systems.

The original schedule for completing the ISS called for the station to be fully assembled by about 2005, but the February 1, 2003 accident that resulted in the loss of the STS-107 Columbia forced ISS planners to alter their timetable drastically.

Although the U.S. space shuttle was designated the primary delivery vehicle for a majority of the major structural components of the ISS, unmanned Russian Progress resupply vehicles were also given a prominent role in the delivery of equip-ment and supplies necessary to maintain a constant presence on the station. The Russian Soyuz series of spacecraft, upgraded to a TMA series version, became the primary crew transport vehicle in the aftermath of the Columbia accident.

In addition to delivering large structural components of the station, space shuttle visits to the ISS were also designed to deliver large quantities of supplies, to facilitate the exchange of long-duration crews, and to remove materials no longer needed at the complex. Delivery and removal activities are made more efficient by the use of the Multi-Purpose Logistics Modules (MPLMs).

Designed and built by the Italian space agency ASI in cooperation with Italian aerospace company Alenia Aerospazio, three MPLMs—Leonardo, Raffaello, and Donatello—were supplied to NASA for use during shuttle flights to the ISS. Car-ried in the shuttle's payload bay, the MPLM was designed to be docked to the station, where its contents could be quickly unloaded by shuttle and ISS crew members, and then removed for return to Earth aboard the shuttle or optionally left at the station to provide additional work space. The first MPLM, Leonardo, was first used during the STS-102 shuttle flight in March 2001.

By July 2006, following the STS-121 second "return to flight" shuttle mis-sion of the space shuttle Discovery, plans for visits to the ISS by a variety of ad-ditional spacecraft included proposed visits by a number of vehicles still under development by several space agencies. These included, among others, the ESA Automated Transfer Vehicle (ATV), the JAXA H-II Transfer Vehicle (HTV), and the Russian Kliper space shuttle.

Orbiting the Earth at an average altitude of about 220 miles (360 kilometers) and traveling around the planet at an average speed of over 27,685 kilometers per hour, the ISS completed more than 37,000 orbits by the end of its seventh year in space.

The first crew to occupy the station, beginning on November 2, 2000, was commanded by NASA astronaut William Shepherd. His crewmates were Russian cosmonauts Yuri Gidzenko and Sergei Krikalyov of Rosaviakosmos. A succession of long-duration crews have alternated in residence at the station since, each living and working aboard the ISS for a period of about six months and all coming from the United States or Russia until the arrival of ESA astronaut Thomas Reiter, who became a member of the ISS Expedition 13 crew in July 2006.

In addition to resident crew members, by mid-2006 the station had also been visited by more than 120 astronauts from a dozen nations and by several space tourists (or spaceflight participants) whose visits were arranged by financial agreements with the Russian space agency.

The cost of NASA's involvement in the ISS over the projected 30-year lifespan of the project, excluding ISS-related space shuttle flights, has been estimated at more than $50 billion.

INTERNATIONAL SUN-EARTH EXPLORER (ISEE)/ INTERNATIONAL COMETARY EXPLORER (ICE)

(1978–1997)

Unmanned U.S. Spacecraft Designed to Study the Sun and Several Comets

The International Sun-Earth Explorer program was designed to investigate the interaction between the Sun and the Earth at the outer boundary of the Earth's magnetosphere, to examine the structure of the solar wind near the Earth, to investigate the plasma sheets, and to expand scientists' knowledge of cosmic rays and emissions from solar flares.

The program involved three spacecraft, ISEE 1, ISEE 2, and ISEE 3, the first two of which were "mother-daughter" craft designed to work in conjunction with each other and in coordination with the ISEE 3, which was designed as a heliocentric craft, the first to use the halo orbit. The three spacecraft carried a variety of instruments designed to make measurements of plasmas, energetic particles, waves, and fields.

Launched on August 12, 1978, ISEE 3 was maneuvered out of the halo orbit and into a heliocentric orbit on December 22, 1983, on its way to an encounter with the comet Giacobini-Zinner. To mark this new phase of its mission, ISEE 3 was renamed International Cometary Explorer (ICE).

The ICE spacecraft accomplished its primary objective on September 11, 1985 when it came into contact with the plasma tail of the Giacobini-Zinner comet and returned data that allowed scientists to study the interaction between the comet's atmosphere and the solar wind.

In March 1986 ICE became the first craft to make close examination of two comets when it collected data about the comet Halley.

NASA approved an extended mission for ICE in 1991, giving the spacecraft the tasks of investigating coronal mass ejections, continuing its study of cosmic rays, and making special observations as opportunities arose. The project continued until May 1997.

IRWIN, JAMES B.

(1930–1991)

U.S. Astronaut; Eighth Human Being to Walk on the Moon

The eighth human being to walk on the surface of the Moon, James Irwin was born in Pittsburgh, Pennsylvania on March 17, 1930. He graduated from East High School in Salt Lake City, Utah, and then attended the United States Naval Academy, where he received a Bachelor of Science degree in Naval Science in 1951.

He was commissioned in the U.S. Air Force after graduation and received his initial training as a pilot at Hondo Air Base and Reese Air Force Base in Texas. He served at Air Defense Command headquarters as chief of the Advanced Requirements Branch, and, continuing his education concurrently with his military career, he attended the University of Michigan, where he received both a Master of Arts degree in aeronautical engineering and a Master of Arts in instrumentation engineering in 1957.

In 1961, Irwin graduated from the U.S. Air Force Experimental Test Pilot School, and he also attended the Air Force Aerospace Research Pilot School, from which he graduated in 1963. He served at Edwards Air Force Base in California as a member of the F-12 Test Force and in the AIM 47 Project office at Wright-Patterson Air Force Base in Ohio.

During his outstanding military career, Irwin logged over 7,015 flight hours, including 5,300 hours in jet aircraft. He rose to the rank of colonel in the U.S. Air Force by the time of his retirement from the service in 1972.

NASA chose Irwin as one of its Group Five selection of astronauts in April 1966. His initial duties at the space agency included work on the development of the lunar module, and he also served as a member of the Apollo 10 support crew and as backup to Alan Bean as lunar module pilot for Apollo 12.

On July 26, 1971, Irwin lifted off with commander David Scott and command module pilot Alfred Worden at the start of the Apollo 15 lunar landing mission, the fourth flight to land crew members on the surface of the Moon.

Leaving Worden in lunar orbit in the command module Endeavor, Irwin and Scott landed on the Moon in the lunar module Falcon on July 30, 1971. They

James Irwin on the surface of the Moon during Apollo 15, August 2, 1971. [NASA/courtesy of nasaimages.org]

touched down in the Hadley Rille valley of the Moon's Apennine Mountains. Then, making the first use of the Lunar Roving Vehicle (LRV), they spent 6 hours, 32 minutes, and 42 seconds exploring the area around their landing site. During that first moonwalk (and moondrive) on July 31, they also set up a package of experiments.

On August 1, they ventured outside the Falcon again for another long trek across the lunar surface. It was during that second EVA that Irwin and Scott discovered the "Genesis rock," a sample they dug out of the lunar dust that had been undisturbed for millions of years which yielded important information that later helped scientists to form theories about how the Moon and the Earth had formed. The discovery of the pristine rock was subsequently deemed to be one of the single most important finds of the entire Apollo program.

They also gathered the deepest core sample of any of the lunar landing flights during their second EVA, and spent a total of 7 hours, 12 minutes, and 14 seconds on the lunar surface.

Irwin and Scott made their third and final excursion on the Moon on August 2, when they drove the LRV to Hadley Rille and walked into the long valley.

They conducted several science demonstrations, including a gravity experiment in which Scott dropped a feather and a hammer, which each hit the lunar surface at the same moment despite the difference in their weights. In their third moonwalk, the Apollo 15 commander and lunar module pilot added another 4 hours, 49 minutes, and 50 seconds to their EVA total.

Irwin and Scott established new records for the longest stay on the lunar surface up to that time, as their Falcon lunar module sat in Hadley Rille for 66 hours and 54 minutes before they blasted off and returned to lunar orbit to reunite with Worden in the command module Endeavor.

As the seventh and eighth astronauts to walk on the Moon, Scott and Irwin spent a total of 18 hours, 34 minutes, and 46 seconds in EVA during their three excursions on the lunar surface.

The Apollo 15 crew also deployed the first satellite ever released in lunar orbit.

On August 5, during the return trip to Earth, Irwin added 39 minutes and 7 seconds to his spacewalking total when he performed a stand up EVA, standing in the hatch of Endeavor while Worden ventured outside the spacecraft to collect several cartridges of exposed film from the scientific instrument module (SIM) of the Apollo 15 service module.

Irwin, Scott, and Worden splashed down in the Pacific Ocean on August 4, 1971 after a flight of 12 days, 17 hours, and 12 minutes. Only two of the three parachutes on Endeavor deployed during re-entry, but the two chutes proved sufficient for slowing their descent, and they splashed down safely.

James Irwin spent 12 days, 17 hours, and 12 minutes in space during the Apollo 15 mission, including 18 hours, 34 minutes and 46 seconds on the surface of the Moon, and a total of 19 hours, 13 minutes, and 53 seconds in EVA.

Among the many awards he received for his participation in the fourth lunar landing flight, Irwin was honored with the United Nations Peace Medal, the Air Force Association's David C. Schilling Trophy, the Robert J. Collier Trophy, the Haley Astronautics Award of the American Institute of Aeronautics and Astronautics (AIAA), the Kitty Hawk Memorial Award, and the Arnold Air Society's John F. Kennedy Trophy.

Following his remarkable experiences during Apollo 15, Irwin left NASA and the U.S. Air Force in July 1972 and founded the High Flight Foundation of Colorado Springs, Colorado. As chairman of the board of High Flight, he made frequent appearances in which he described how his involvement in the space program helped to deepen his Christian faith.

In the decade following his walk on the Moon, Irwin led several explorations of Mount Ararat in Turkey in an effort to find evidence of Noah's Ark, the vessel described in the Bible as the mechanism of salvation for Noah and his family as they escaped the flood that consumed the known world in ancient times. Irwin was injured during one of the expeditions, in 1982.

James Irwin died on August 9, 1991 after suffering a heart attack. He was survived by his wife Mary Ellen and their five children.

ISRAEL: FIRST CITIZEN IN SPACE.

See Ramon, Ilan

ITALY: NATIONAL SPACE AGENCY.

See Agenzia Spaziale Italiana

IVANCHENKOV, ALEKSANDR S.

(1940–)

Soviet Cosmonaut

A veteran of two space missions, including one long-duration flight in which he spent more than 139 days in space, Aleksandr Ivanchenkov was born in Ivanteyevka, Russia on September 28, 1940. He attended the Moscow Aviation Institute, where he graduated in 1964 with an engineering degree.

He was chosen for training as a cosmonaut as a member of the civilian specialist group five selection of cosmonaut candidates in March 1973.

Ivanchenkov's first assignments as a cosmonaut included service as a member of the second backup crew for the Soviet Soyuz 19 portion of the Apollo-Soyuz Test Project (ASTP) flight, which flew in July 1975.

He began his first flight in space as flight engineer for the second Soyuz 29/ Salyut 6 long duration mission, with commander Vladimir Kovalyonok, on June 15, 1978. They spent four and a half months in space during the remarkable flight, conducting scientific experiments that included biological investigations and materials processing work, and they made extensive observations of the Earth's atmosphere, the Moon, Venus, Mars, and Jupiter with a telescope capable of operating in infrared and ultraviolet wavelengths.

They also received visits from two Intercosmos crews during their long stay at Salyut 6. The Soviet Intercosmos program made use of short Soyuz switching flights, in which a spacecraft docked at a space station would be replaced by a fresh Soyuz when the docked craft neared the end of its safe operations limit, to provide spaceflight opportunities for the citizens of nations aligned with or friendly to the Soviet Union.

Soviet commander Pyotr Klimuk and guest cosmonaut Miroslav Hermaszewski, the first citizen of Poland to fly in space, were the first Intercosmos crew to visit Ivanchenkov and Kovalyonok, on June 27, 1978. The second visiting crew, commanded by veteran Soviet cosmonaut Valeri Bykovsky and including Sigmund Jähn of East Germany (Germany was still divided into East and West at the time of Soyuz 31, only being reunited in the 1990s after the fall of the Soviet Union), arrived at the station on August 26 aboard Soyuz 31. In each case, the visiting crews visited for about a week and then returned to Earth.

Ivanchenkov and Kovalyonok also made a spacewalk during their long stay, venturing outside of Salyut 6 on July 29 for 2 hours and 5 minutes. Ivanchenkov retrieved scientific experiments from the outside of the station while his crew mate followed his progress from the station's airlock. At one point a passing meteor momentarily blinded both cosmonauts, but neither was injured in the incident and Ivanchenkov successfully completed his planned activities.

After setting a new record for the longest stay in space up to that time, at 139 days, 14 hours, and 48 minutes in space, Ivanchenkov and Kovalyonok returned to Earth in Soyuz 31 on November 2, 1978.

Ivanchenkov made his second flight in space as flight engineer for the Soyuz T-6 Intercosmos flight, which lifted off on June 24, 1982 with commander Vladimir Dzhanibekov and guest cosmonaut Jean-Loup J. M. Chrétien, the first citizen of France to fly in space.

The flight was a departure from the usual routine for Intercosmos flights, as it featured three cosmonauts rather than two, and the guest cosmonaut was the first from a Western European nation. Soyuz T-6 also differed from previous Soyuz switching flights in that the crew both arrived at and departed from the Salyut 7 space station in Soyuz T-6, since another flight was scheduled to arrive shortly afterward to be switched with the Soyuz already docked at the station.

Ivanchenkov and his crew mates landed on July 2, 1982.

He subsequently trained for a planned future flight aboard the Soviet space shuttle Buran, but the nation's space shuttle program was ultimately abandoned in the wake of the economic difficulties that followed the dissolution of the Soviet Union.

Aleksandr Ivanchenkov spent more than 147 days in space, including more than 2 hours in EVA, during his career as a cosmonaut. He left the cosmonaut corps in 1993.

IVINS, MARSHA S.

(1951–)

U.S. Astronaut

Marsha Ivins was born in Baltimore, Maryland on April 15, 1951. In 1969, she graduated from Nether Providence High School in Wallingford, Pennsylvania and then attended the University of Colorado, where she earned a Bachelor of Science degree in aerospace engineering in 1973.

She began working for NASA at the Johnson Space Center (JSC) in Houston in July 1974. Initially assigned as an engineer involved in the development of displays and controls for the space shuttle and in projects involving man-machine interface engineering, she played a key role in the design of the Orbiter Head-Up Display.

An experienced pilot, Ivins holds a multi-engine Airline Transport Pilot License with a Gulfstream-1 type rating. She holds single engine airplane, land, sea, and glider commercial licenses, and airplane, instrument, and glider flight instructor ratings.

In 1980, NASA assigned her to the Shuttle Training Aircraft as a flight engineer, and to fly the NASA administrative aircraft (Gulfstream-1) aircraft as a co-pilot.

During her career as a pilot, Ivins has accumulated more than 6,400 hours of flying time in a variety of civilian and NASA aircraft.

Ivins was selected for astronaut training in 1984, and her technical assignments since that time have included review of shuttle safety and reliability issues, upgrades to the shuttle cockpit, software verification in the Shuttle Avionics Integration Laboratory (SAIL), and service as a spacecraft communicator (Capcom) at Mission Control during shuttle missions. She has also served at the Kennedy Space Center (KSC) in Florida as head of the Astronaut Support Personnel team and as crew representative for a variety of issues, including space station and space shuttle stowage, habitability, logistics, and transfer issues, shuttle flight crew equipment issues, the shuttle's photographic system, and as crew representative to NASA's Future Exploration team.

She first flew in space as a mission specialist during STS-32 aboard the space shuttle Columbia in January 1990. The 11-day flight set a new record for the longest shuttle mission up to that time (previously the longest flight had been that of the STS-9 crew, from November 28 to December 8, 1983). The crew deployed the SYNCOM IV-F5 defense communications satellite (also known as LEASAT 5), and successfully retrieved the Long Duration Exposure Facility (LDEF), an orbiting cylinder containing 57 experiments designed to test the effects of long exposure to the space environment on a variety of materials. The LDEF was deployed during STS-41C in April 1984 and was originally intended to be retrieved on an earlier flight, but the January 1986 Challenger accident had delayed its retrieval until STS-32.

Columbia landed after dark at Edwards Air Force Base on January 20, 1990.

Ivins made her second flight in space during STS-46, which launched on July 31, 1992 aboard the space shuttle Atlantis. Dedicated to international flight experiments, STS-46 featured the first flight in space of an Italian citizen, payload specialist Franco Malerba, and the first spaceflight of a citizen of Switzerland, Claude Nicollier. Nicollier deployed one of the flight's two primary payloads, the European Retrievable Carrier (EURECA) of the European Space Agency (ESA), which would remain in orbit until its retrieval during STS-57 in 1993.

The flight also featured the Tethered Satellite System (TSS-1), a joint project of NASA and the Italian space agency Agenzia Spaziale Italiana (ASI), which was designed to be the first major test of tethered spaceflight.

The concept of joining two spacecraft with a tether for linked flight had been tested in space as early as the Gemini program in 1966. Prior to STS-46, teams of engineers from NASA and ASI had collaborated for more than a decade to design the equipment, the plan, and the procedures that were to have been used during the TSS-1 test aboard Atlantis. The exercise was designed to establish a link between the shuttle and a satellite that would be deployed at a different orbital altitude from Atlantis, for tethered flight at a distance of as much as 12.5 miles.

Despite all the careful preparations, however, a mechanical hitch foiled the test when the tether jammed after it had been unwound to a distance of just 840 feet.

The crew attempted for several days to release the stuck line, but the experiment finally had to be written off as a failure. The TSS satellite was retrieved and returned to Earth. Atlantis returned to Earth on August 8, 1992, landing at the KSC in Florida after nearly eight days in space.

On her third space mission, Ivins served aboard the shuttle Columbia as a mission specialist during STS-62. Launched on March 4, 1994, the crew of five—commander John Casper, pilot Andrew Allen, and mission specialists Ivins, Sam Gemar, and Pierre Thuot—conducted a remarkable 60 experiments in a wide variety of disciplines during the productive 14-day flight. They operated the second U.S. Microgravity Payload (USMP-2) and the second Office of Aeronautics and Space Technology payload (OAST-2), working with scientists on Earth while performing experiments in materials processing, protein crystal growth, human physiology, biotechnology, atmospheric ozone monitoring, and space technology and flight, among others.

In one exercise designed to study the glow that surrounds the shuttle during spaceflight, Columbia was lowered to an altitude of 105 nautical miles, the lowest orbital altitude yet flown during the space shuttle program. The flight was also the second-longest shuttle mission up to that time; Columbia landed at KSC on March 18, 1994.

Ivins's fourth flight in space was the 10-day STS-81, which launched aboard the shuttle Atlantis in January 1997. The fifth docking of a shuttle and the Russian space station Mir, STS-81 brought Jerry Linenger to the space station, where he began his long stay with the Mir 22 crew, and brought John Blaha back to Earth as part of the ongoing program of cooperation between the United States and Russia that maintained a continuous American presence aboard Mir.

In addition to the crew exchange, the shuttle also delivered more than three tons of food, water, equipment, and supplies to the space station during five days of linked flight. Atlantis also carried the SPACEHAB pressurized module for crew experiments during STS-81, the second time a SPACEHAB module was carried during a shuttle flight. The mission ended on January 22, 1997, with Atlantis' safe landing at KSC after 160 orbits.

On February 7, 1991, Ivins began a fifth space mission, lifting off from KSC on the space shuttle Atlantis at the start of the STS-98 International Space Station (ISS) assembly mission. The STS-98 crew delivered the U.S. Destiny Laboratory module to the ISS and installed the state-of-the-art 28-foot long cylindrical workspace at the station. The addition of the Destiny module made the ISS the largest space station in history, increasing its total mass to about 112 tons.

Ivins played a key role in the Destiny installation, operating Atlantis' Remote Manipulator System (RMS) robotic arm while mission specialists Robert Curbeam and Thomas Jones made three space walks totaling almost 20 hours to attach the lab to the station.

Atlantis was docked to the ISS for seven days, working with the ISS Expedition 1 crew to transfer supplies and equipment to the station. The shuttle returned to Earth on February 20, 2001 at Edwards Air Force Base, after a flight of more than 12 days.

Marsha Ivins has accumulated more than 1,318 hours in space during her career as an astronaut.

After her fifth mission, she served in the Astronaut Office space station and shuttle branches for crew equipment, habitability, and stowage, and headed the Astronaut Office Exploration Branch and NASA's Operations Advisory Group.

JÄHN, SIGMUND W.

(1937–)

German Cosmonaut; First Citizen of Germany to Fly in Space

The first citizen of Germany to fly in space, Sigmund Jähn was born in Morgenröthe-Rautenkranz, Germany on February 13, 1937. After completing his secondary education, he received vocational training for a career in the printing industry prior to enlisting in the East German Air Force in 1955. At the time, Germany was divided into East and West, prior to the nation's reunification in 1990.

In 1970, he graduated from the Monino Air Force Academy in the Soviet Union, and he subsequently returned to Germany to serve as a pilot instructor. During his military career, he rose to the rank of major general in the East German Air Force.

Jähn was chosen for training as a cosmonaut in 1976, as a member of the first Intercosmos group of cosmonaut candidates.

The Intercosmos program was conceived as a means of involving citizens of nations aligned with or friendly to the Soviet Union in the country's space program. The manned portion of the Intercosmos program arose from the need to switch Soyuz spacecraft docked at space stations during long-duration missions. Because the early versions of the Soyuz spacecraft were limited to 90 days in space before their on-board systems began to deteriorate, a fresh Soyuz craft had to replace a Soyuz docked in space before the end of the 90-day safe operation window.

Prior to his Intercosmos flight, Jähn relocated to Star City, Russia, where he received intensive training in preparation for his space mission.

On August 26, 1978, Jähn became the first citizen of East Germany to fly in space, when he launched with Soviet cosmonaut Valeri Bykovsky aboard

Soyuz 31 to travel to the Salyut 6 space station. At Salyut 6, they visited with Vladimir Kovalyonok and Alexander Ivanchenkov, who were in the midst of a then-record four and a half month stay aboard the station.

Jähn and his fellow cosmonauts participated in ceremonies and television broadcasts celebrating the cooperative nature of the mission, and then, after a stay of a little less than a week, Jähn and Bykovsky returned to Earth aboard Soyuz 29, leaving their Soyuz 31 spacecraft at the station to accomplish the switching of the Soyuz spacecraft.

Jähn spent 7 days, 20 hours, and 49 minutes in space during his Intercosmos visit to Salyut 6.

Following his landmark space mission, Jähn attended the Zentralinstitut fur Physikder Erde in Potsdam, where he earned a doctorate degree in physics in 1983. When East and West Germany were reunited after the dissolution of the Soviet Union, he became a consultant to the German space agency, a capacity in which he served until his retirement in 2002.

In 2001, the asteroid formerly known as 17737 was re-named in his honor.

JAPAN AEROSPACE EXPLORATION AGENCY (JAXA)

(2003–)

National Space Agency of Japan

The Japan Aerospace Exploration Agency (JAXA) was formed on October 1, 2003 by the consolidation of the three existing national space organizations of Japan.

Drawing upon the expertise and technologies previously developed by The Institute of Space and Astronautical Science (ISAS), the National Aerospace Laboratory of Japan (NAL), and the National Space Development Agency of Japan (NASDA), JAXA oversees the development of satellites, spacecraft, launch vehicles, and space transportation systems, conducts space science research and space and aeronautic engineering research, and coordinates Japan's participation in the International Space Station (ISS).

The long and colorful history of Japan's space program began with a 1955 rocket launch experiment by the Institute of Industrial Science at the University of Tokyo. Highlights of that early era include the June 1958 launch of the two-stage KAPPA rocket, which reached an altitude of 60 kilometers, and the flight of rocket K-8-1 in July 1960, which reached an altitude of 190 kilometers and made the first measurement of ion density.

The Kagoshima Space Center was established at the University of Tokyo in 1963, and the Institute of Space and Aeronautical Science was founded at the university the following year. The aeronautical institute became the Institute of Space and Astronautical Science, with an expanded mandate as a joint research organization encompassing other Japanese universities, in 1981.

The ISAS achieved many milestones in the history of Japan's space endeavors, including the launch of the nation's first weather observation satellite, MT-135-1,

The Japanese Kibo laboratory module for the International Space Station is seen in NASA's Space Station Processing Facility at the Kennedy Space Center in Florida, August 27, 2003. [NASA/courtesy of nasaimages.org]

in July 1964 and the launch of the first Japanese scientific satellite, SHINSEI (MS-F2), in September 1971.

In February 1970, the organization successfully propelled the OHSUMI satellite into orbit, giving Japan the honor of being only the fourth nation in history (following the Soviet Union, the United States, and France) to place a satellite in orbit.

Japan launched a number of successful satellites in the following years. The ISAS launched the EXOS series of satellites to study the upper atmosphere and the ASTRO series to study the radiation emissions of a variety of celestial objects. In 1985, the ISAS launched the Halley's Comet fly by satellites Sakigake and Suisei, followed in 1990 by the Hiten lunar orbiter mission.

The NAL was created in 1955 as the National Aeronautical Laboratory. The NAL focused on aviation and aerospace research and developed test facilities for the nation's burgeoning space activities.

Charged with the task of developing the personnel, facilities, vehicles, systems, equipment, and procedures necessary to the nation's space program, NASDA was established on October 1, 1969.

In February 1977 NASDA launched Japan's first geostationary satellite, the engineering test satellite Kiku-2. NASDA was also responsible for the selection of Japanese mission specialists and payloads for the U.S. space shuttle program.

Under an agreement with the McDonnell Douglas Corporation of the United States, NASDA developed a launch vehicle based on the McDonnell Douglas Delta launcher. The resulting N-I rocket, which could lift a maximum payload of 286 pounds (130 kilograms), was subsequently replaced by the N-II, which featured a lift capacity of 770 pounds (350 kilograms). The agency ultimately developed the entirely Japanese-built H-I launcher, which included two upper-stage rockets developed by NASDA. First launched in August 1986, the H-I could lift payloads as heavy as 1,200 pounds (550 kilograms).

In addition to developing the first multi-stage launch vehicle to incorporate significant Japanese engineering expertise and technology (and its subsequent descendants), NASDA also launched satellites for broadcast systems, communications, Earth observations, engineering tests, marine observations, and meteorological studies.

NASDA first used the Tanegashima launch complex in 1975, and the agency established its primary base of operations at the Tsukuba Space Center.

Originally developed by the ISAS and NASDA, the $480 million Japanese Kaguya spacecraft (officially known as the SELenological and ENgineering Explorer, or SELENE) was launched on September 14, 2007 and began orbiting the Moon on October 3, 2007. In its study of the lunar environment, Kaguya released two small satellites and collected data with an array of 13 instruments. Pleased with the mission's progress, JAXA officials extended the Kaguya project when it reached its scheduled completion date in October 2008.

Seven Japanese citizens had flown in space by the end of 2008, beginning with broadcast journalist Toyohiro Akiyama, who visited the space station Mir in December 1990 aboard Soyuz TM-11.

In 1997, Takao Doi became the first Japanese astronaut to perform a spacewalk when he served aboard the space shuttle Columbia during STS-87. Doi actually made two EVAs during the flight, the second of which was an unscheduled EVA to retrieve a SPARTAN free-flying satellite that had defied the crew's previous attempts to retrieve it with the space shuttle's robotic arm.

Doi marked another milestone in the Japanese space program in March 2008 when he traveled to the ISS during STS-123. The STS-123 crew delivered the first component of JAXA's ISS science facility, the Kibo Laboratory.

Mamoru Mohri flew aboard the space shuttle Endeavour during STS-47 in 1992 and STS-99 in February 2000.

Chiaki Mukai served aboard the shuttle Columbia during STS-65 in 1994, and aboard Discovery in 1998, during STS-95.

Koichi Wakata was a member of the STS-72 crew of the shuttle Endeavour in January 1996, and also served aboard STS-92, the 100th mission of the space shuttle program, in October 2000.

During NASA's return-to-flight STS-114 mission in 2005, Soichi Noguchi made three spacewalks as a mission specialist aboard the shuttle Discovery. He and Stephen Robinson tested in-flight repairs for the outside of the shuttle and performed maintenance and repairs at the ISS during their EVAs.

In June 2008, JAXA astronaut Akihiko Hoshide visited the ISS as a mission specialist during STS-124. The STS-124 crew delivered the pressurized component of the Japanese ISS Kibo Laboratory.

Japan was one of the earliest supporters of the development of the ISS. The Kibo module at the space station features a pressurized laboratory and a facility for experiments and observations requiring exposure to the space environment. From its early planning stages, a key objective for Kibo has been extensive study of the Earth's environment, in a search for solutions to problems such as the depletion of the ozone layer, global warming, and desertification. Reflecting the ideals inherent in its creation, the module was appropriately named Kibo, the Japanese word for hope.

JARVIS, GREGORY B.

(1944–1986)

U.S. Astronaut

Gregory Jarvis was born in Detroit, Michigan on August 24, 1944. In 1962, he graduated from Mohawk Central High School in Mohawk, New York and then attended the State University of New York at Buffalo, where he earned a Bachelor of Science degree in electrical engineering in 1967. He earned a Master's degree in electrical engineering at Northeastern University in Boston, Massachusetts in 1969, and also completed all of the necessary course work for a Master's degree in management science at West Coast University, in Los Angeles, California.

During his time at Northeastern, he put his electrical engineering expertise to practical use while working on circuit designs for the SAM-D missile at Raytheon Corp.

In the summer of 1969, he joined the U.S. Air Force and was assigned to El Segundo, California as a communications payload engineer in the office of the Air Force Satellite Communications Program. His duties included design and sourcing for advanced tactical communications satellites, including FLTSATCOM. He had risen to rank of captain in the U.S. Air Force by the time he left the service in 1973.

Jarvis went on to join the space and communications group of Hughes Aircraft Company, where he worked on the company's MARISAT satellite program as a communications subsystem engineer. He was responsible for the integration and test of the MARISAT F-3 satellite, which was placed in geosynchronous orbit in 1976. That same year, Jarvis joined the company's Systems Applications Laboratory, where he helped to define advanced UHF and SHF communications systems, and in 1978 he began work on the initial planning for the SYNCON IV/LEASAT satellite program as a member of the Hughes Advanced Program Laboratory.

Jarvis played a key role in the development of the LEASAT satellite program for more than half a decade. His engineering responsibilities included management of the spacecraft's power, thermal, and harness subsystems, and he was also instrumental in the development of the satellite's electrical bus system and cradle and the test and integration of the F-1, F-2, and F-3 versions of the spacecraft. The F-1 and F-2 LEASAT satellites were successfully placed in geosynchronous orbits.

Chosen by NASA for training as a payload specialist astronaut in July 1984, Jarvis was to have made his first flight in space aboard the shuttle Challenger during STS-51L, which launched on January 28, 1986 at the Kennedy Space Center (KSC) in Florida.

Tragically, payload specialist Jarvis and his fellow Challenger crew mates were killed when a fault in one of the shuttle's huge solid rocket boosters caused a fuel leak that ignited and caused a massive explosion just 73 seconds after liftoff. He was 41 at the time of his death.

Jarvis was survived by his wife, Marcia. He was posthumously awarded the Congressional Space Medal of Honor.

JEMISON, MAE C.

(1956–)

U.S. Astronaut; First African American Woman to Fly in Space

The first African American woman to fly in space, Mae Jemison was born in Decatur, Alabama on October 17, 1956. In 1973, she graduated from Morgan Park High School in Chicago and then attended Stanford University, where she earned a Bachelor of Science degree in chemical engineering in 1977 while simultaneously fulfilling the requirements necessary for a Bachelor of Arts degree in African and Afro-American Studies.

She was a recipient of National Achievement scholarships from 1973 to 1977, and also traveled extensively, making trips to Jamaica as Stanford's representative to Carifesta '76, to Cuba, as a member of an American Medical Student Association (AMSA) study group, to Kenya, where she conducted health studies on a grant from the International Travelers Institute, and to Thailand, where she worked in a refugee camp.

In 1981, Jemison received a Doctorate degree in medicine from Cornell University.

Dr. Mae Jemison, the first African American woman to fly in space. [NASA/courtesy of nasaimages.org]

Following her graduation from Cornell, she completed a one-year internship at Los Angeles County/USC Medical Center and then worked at the INA/Ross Loos Medical Group in Los Angeles.

She became an Area Peace Corps Medical Officer for Sierra Leone and Liberia in West Africa in January 1983. In that capacity, she managed health care delivery for U.S. Peace Corps and U.S. Embassy personnel, supervised medical staff, and oversaw pharmacy and laboratory operations. She also developed curricula and self-care manuals and taught volunteer health training, and created and implemented guidelines for public health and safety issues.

Her research work included studies involving the development of a Hepatitis B vaccine, schistosomaisis, and rabies, in projects carried out in conjunction with the National Institutes of Health and the Centers for Disease Control and Prevention.

Jemison returned to the United States in June 1985, and later that year began work as a general practitioner and employee of CIGNA Health Plans of California. She was working in that capacity and simultaneously taking graduate engineering classes when NASA selected her for astronaut training in June 1987.

On September 12, 1992, Jemison lifted off aboard the space shuttle Endeavour as a mission specialist during STS-47, the 50th flight of the space shuttle program. The crew for the milestone flight was commanded by Robert Gibson and also included pilot Curtis Brown, payload commander Mark Lee, mission specialists Jan Davis and Jay Apt, and payload specialist Mamoru Mohri of Japan. With the STS-47 mission, Lee and Davis became the first married couple to fly in space, and Mamorou Mohri became the first Japanese citizen to fly on the space shuttle (the journalist Toyohiro Akiyama had been the first Japanese citizen to fly in space, during a visit to the Soviet space station Mir in 1990).

The primary payload for STS-47 was Spacelab-J, a joint scientific program designed by NASA and the National Space Development Agency (NASDA) of Japan. Using the pressurized Spacelab module, the crew conducted 43 experiments in materials science and life sciences during the mission. Endeavour returned to Earth on September 20, 1992, landing at the Kennedy Space Center (KSC) in Florida after a flight of 7 days, 22 hours, 30 minutes, and 23 seconds (and 126 orbits).

Jemison left NASA in March 1993.

In honor of her remarkable achievements, an alternate public school, the Mae C. Jemison Academy, was established in Detroit, Michigan in 1992.

Jemison's STS-47 flight has rightly made her a role model, and she is also widely recognized for her involvement in a wide array of organizations devoted to the betterment of human life across the globe. Among her many activities, she is a board member of the World Sickle Cell Foundation, an honorary board member of the Center for the Prevention of Childhood Malnutrition, and a clinical teaching associate at the University of Texas Medical Center.

JERNIGAN, TAMARA E.

(1959–)

U.S. Astronaut

A veteran of five space shuttle missions, Tamara Jernigan was born in Chattanooga, Tennessee on May 7, 1959. In 1977, she graduated from Santa Fe High School in Santa Fe Springs, California and then attended Stanford University, where she graduated with honors in 1981, earning a Bachelor of Science degree in physics. She pursued an interest in athletics during her undergraduate years and was a member of Stanford's varsity volleyball team, an interest she maintains as a member of the United States Volleyball Association.

She began her career at the NASA Ames Research Center at Moffett Field in California, working as a research scientist in the agency's Theoretical Studies Branch. She has conducted research on a variety of astronomical phenomena, including areas of star formation, gamma ray bursts, and shockwaves in the interstellar medium.

Continuing her education concurrently with her NASA career, she earned a Master of Science degree in engineering science at Stanford in 1983, and then attended the University of California at Berkeley, where she received a Master of Science degree in astronomy in 1985. In 1988, she earned a Doctorate degree in space physics and astronomy from Rice University in Houston, Texas.

Jernigan was selected for astronaut training in 1985, and her technical assignments at NASA since that time have included service as spacecraft communicator (Capcom) at Mission Control during five space shuttle flights. She has also served as lead astronaut for the development of flight software, as chief of the Mission Development Branch of the Astronaut Office, and as deputy chief of the Astronaut Office.

She first flew in space as a mission specialist during STS-40 in June 1991, a milestone flight in the study of aerospace medicine and biomedical research

Although it was the fifth Spacelab mission, STS-40 was the first to concentrate entirely on the life sciences. The crew conducted a wide range of tests designed to study the effects of the space environment on six body systems, including the cardiovascular and cardiopulmonary systems (the heart, lungs and blood vessels), the renal and endocrine systems (the kidneys and hormone-secreting organs and glands), the blood system (blood plasma), the immune system (the white blood cells), the musculoskeletal system (muscles and bones), and the neurovestibular system (the brain and nervous system, the eyes, and inner ear).

All together, the crew, which included three medical doctors (James Bagian, Rhea Seddon, and Drew Gaffney) specially selected for their areas of expertise, conducted 18 medical experiments using human, rodent, and jellyfish test subjects and returned more medical data than any previous NASA mission.

Columbia returned to Earth at Edwards Air Force Base in California on June 14, 1991.

Jernigan's second spaceflight was STS-52, which launched on October 22, 1992 aboard the shuttle Columbia. The landmark flight featured a wide array of international scientific payloads for sponsors that included the Italian space agency Agenzia Spaziale Italiana (ASI), the Canadian Space Agency (CSA), and the European Space Agency (ESA).

Jointly sponsored by NASA and ASI, the Laser Geodynamic Satellite (LAGEOS) II satellite was propelled into orbit by the Italian Research Interim Stage (IRIS), which was used for the first time during STS-52.

The crew also conducted a series of experiments that were bundled together as the U.S. Microgravity Payload-1 (USMP-1), carried in Columbia's cargo bay. These activities included the Lambda Point Experiment (LPE), the French-sponsored

Material Pour L'Etude Des Phenomenes Interessant La Solidification Sur Terre Et En Orbite (MEPHISTO), and the Space Acceleration Measurement System (SAMS).

The international flavor of the mission was also reflected in the Canadian Experiment-2 (CANEX-2), whose experiments included the Space Vision System (SVS), which entailed the deployment, on the ninth day of the mission, of the Canadian Target Assembly (CTA) satellite, a study of the impact of low-Earth orbit exposure on various materials, the Queen's University Experiment in Liquid-Metal Diffusion (QUELD), a study of phase partitioning in liquids, the Sun Orbiter Glow-2 (OGLOW-2) experiment, and a series of Space Adaptation Tests and Observations (SATO).

The ESA was represented by the Attitude Sensor Package in the shuttle's cargo bay. Mounted on a "Hitchhiker" plate, the sensor package included a modular star sensor, a yaw Earth sensor, and a low altitude conical Earth sensor.

STS-52 came to an end with Columbia's return to Earth, with a landing at the Kennedy Space Center (KSC) in Florida on November 1, 1992.

Jernigan made her third flight in space as payload commander of STS-67, in March 1995. Launched aboard Endeavour in the early morning of March 2, STS-67 featured the second flight of the Astro observatory. The crew used the three Astro-2 telescopes to make around the clock ultraviolet observations during the 16-day mission, the longest shuttle mission up to that time. Endeavour returned to Earth at Edwards Air Force Base on March 18, 1995.

Jernigan's fourth flight in space, STS-80, began on November 19, 1996 and landed 17 days, 15 hours, 53 minutes, and 18 seconds later, on December 7, setting a new record for the longest shuttle flight.

Commanded by Kenneth Cockrell, the STS-80 crew also included Kent Rominger, who served as pilot of the shuttle Columbia, and Jernigan's fellow mission specialists Thomas Jones and Story Musgrave.

The STS-80 crew launched and retrieved two free-flying spacecraft: the Orbiting and Retrievable Far and Extreme Ultraviolet Spectrometer-Shuttle Pallet Satellite II (ORFEUS-SPAS II), which made its second shuttle flight during the mission, and the Wake Shield Facility-3 (WSF-3), which was deployed for the third time. In two weeks of observations, the ORFEUS-SPAS made 422 studies of about 150 astronomical bodies; and the Wake Shield Facility successfully cultivated seven thin film semiconductor materials for use in advanced electronics during its three days of free flight.

Despite the remarkable achievements of the flight, Jernigan and Jones suffered the personal disappointment of having to forego the chance to make two spacewalks for which they had trained as part of the original STS-80 mission profile when the outer hatch of Columbia's airlock jammed. The astronauts had been scheduled to test out equipment and procedures for future use on the International Space Station (ISS), but the malfunctioning hatch caused wary mission managers to call off the exercises to avoid the chance of harm to the crew or the shuttle.

Jernigan and Jones were able to conduct one of their scheduled tests while remaining on the shuttle, trying out the Pistol Grip Tool, which is similar to a handheld drill. Post-flight investigations revealed the cause of the malfunction that prevented their EVAs: a tiny screw had lodged in the actuator mechanism that secures the hatch.

The record-setting STS-80 flight ended with Columbia's landing on December 7, 1996, at KSC.

After her fourth flight in space, Jernigan served as assistant to the chief of the Astronaut Office for space station activities. In 1997, she shared in a NASA Group Achievement Award for her work on the EVA Development Test Team.

On May 27, 1999, she began a remarkable fifth spaceflight, aboard the space shuttle Discovery during STS-96, the historic first docking of a shuttle and the ISS. The STS-96 crew delivered replacement parts and equipment to the space station and prepared the facility for its eventual habitation by the first long-duration crew.

Jernigan and fellow mission specialist Daniel Barry made a 7 hour and 55 minute spacewalk during the flight while installing two cranes and equipment for use by future spacewalkers at the station.

Crew members also transferred 3,000 pounds of equipment and supplies during six days of docked operations at the ISS, before Discovery undocked and returned to Earth, landing at KSC during the early hours of June 6, 1999.

Tamara Jernigan spent more than 63 days in space, including 7 hours and 55 minutes in EVA, during five space missions.

Since her fifth flight, she has served as lead astronaut for external maintenance of the ISS. She is also active in a variety of civic and athletic pursuits. In 2000, she received the Explorer's Club Lowell Thomas Award, and she is also a lifetime member of the Girl Scouts.

JETT, BRENT W., JR.

(1958–)

U.S. Astronaut

Brent Jett was born in Pontiac, Michigan on October 5, 1958. He graduated from Northeast High School in Oakland Park, Florida in 1976, and then attended the United States Naval Academy, where he graduated first in his class of 976 in 1981, earning a Bachelor of Science degree in aerospace engineering.

After receiving his initial training as a pilot and being designated a Naval Aviator in 1983, he began F-14 Tomcat training at Naval Air Station Oceana, in Virginia Beach, Virginia, while attached to Fighter Squadron 101. After completing his F-14 training, he was deployed to the Mediterranean Sea and to the Indian Ocean while serving in Fighter Squadron 74 aboard the USS *Saratoga* (CV-60). During his time with Fighter Squadron 74, he became an airwing qualified landing signal officer and attended the Navy Fighter Weapons School.

Continuing his education concurrently with his military service, Jett participated in the Navy's Test Pilot School cooperative education program and attended the Naval Postgraduate School in Monterey, California, where he earned a Master of Science degree in aeronautical engineering in 1989.

He also attended the U.S. Naval Test Pilot School, where he was recognized as a distinguished graduate in June 1989.

Serving as a project test pilot in several versions of the F-14 aircraft and the T-45A and A-7E, he worked at the Naval Air Test Center from 1989 to 1991 in the Carrier Stability Department of the Strike Aircraft Test Directorate. In 1991, he returned to Fighter Squadron 74, deployed aboard the USS *Saratoga* (CV-60).

As a pilot and test pilot, Jett has accumulate more than 4,800 hours of flying time in more than 30 different types of aircraft and has made over 450 landings on aircraft carriers. He is a captain in the U.S. Navy.

NASA chose him for astronaut training in March 1992.

Jett's first spaceflight was as pilot of the shuttle Endeavour during STS-72 in January 1996. Launched in the early morning hours of January 11, the flight featured the retrieval of the Japanese Space Flyer Unit (SFU), which Japanese mission specialist Koichi Wakata captured with the shuttle's robotic arm on the third day of the flight. The SFU had been in orbit since March 1995. Wakata also deployed and later retrieved a free flying platform designed by the Office of Aeronautics and Space Technology, the OAST-Flyer, and mission specialists Leroy Chiao and Daniel Barry made two spacewalks totaling nearly 13 hours while testing EVA procedures and equipment for use in the later assembly of the International Space Station (ISS). Endeavour returned to Earth with a nighttime landing at the Kennedy Space Center (KSC) in Florida on January 20, 1996.

Jett flew again in space in January 1997, this time as pilot of the shuttle Atlantis during the 10-day STS-81, the fifth docking of a shuttle and the Russian space station Mir. As part of the ongoing program of cooperation between the United States and Russia that maintained a continuous American presence aboard Mir, Atlantis delivered Jerry Linenger to the space station and brought John Blaha back to Earth. In addition to the crew exchange, the shuttle also delivered more than three tons of food, water, equipment, and supplies to the space station during five days of linked flight. Atlantis also carried the SPACEHAB pressurized module for crew experiments, the second time a SPACEHAB module was carried during a shuttle flight. The mission ended on January 22, 1997, with Atlantis's safe landing at KSC.

In keeping with the goodwill inherent in the joint American-Russian operations of his second spaceflight and the two nations' cooperation in the first phase of development of the ISS, Jett served as NASA's director of operations at the Yuri Gagarin Cosmonaut Training Center in Star City, Russia from June 1997 to February 1998.

On November 30, 2000, Jett began his third space mission, as commander of STS-97 aboard the shuttle Endeavour. In seven days of docked operations at the ISS, the STS-97 crew delivered and installed the solar array structure that generates and converts electrical power for the ISS. At 240 feet long and 38 feet wide, the 17,000-pound P-6 Integrated Truss Segment, which includes the solar arrays and their related electronics, batteries, and cooling radiator, set a new record for the longest human-built object to fly in space. More importantly, it was designed to produce a greater amount of electricity for the ISS than any previous electrical system on past space stations.

In September 2006, Jett commanded STS-115, the third "return to flight" shuttle mission following the loss of the space shuttle Columbia in February 2003. The STS-115 crew of the space shuttle Atlantis successfully delivered and installed the ISS P3/P4 integrated truss segment.

Brent Jett has spent more than 41 days in space during four space shuttle missions.

JING, HAIPENG (JING HAIPENG)

(1966–)

Chinese Taikonaut

Born in Shanxi Province in the People's Republic of China in 1966, Haipeng Jing joined the People's Liberation Army (PLA) in 1985. He has accumulated 1,200 flight hours as a military pilot. In 1998, he was chosen for training as a taikonaut (cosmonaut).

His assignments in the Chinese space program included service as a member of the backup crew for the Shenzhou VI mission, which took place in October 2005.

Jing began his first space mission on September 25, 2008, when he launched in Shenzhou VII with Zhigang Zhai and Buoming Liu from the Jiuquan Satellite Launch Center on China's third manned spaceflight.

In support of his crew mates, who carried out the historic first spacewalk of the Chinese space program, Jing carefully monitored their EVA from inside the spacecraft.

The Shenzhou VII crew also conducted communications experiments and deployed a small satellite before they returned to Earth on September 28, 2008. In his historic first space mission, Haipeng Jing spent 2 days, 20 hours, and 28 minutes in space.

JONES, THOMAS D.

(1955–)

U.S. Astronaut

Thomas Jones was born in Baltimore, Maryland on January 22, 1955. In 1973, he graduated from Kenwood Senior High School in Essex, Maryland, and he then attended the United States Air Force Academy in Colorado Springs, Colorado, where he was recognized as a distinguished graduate and the outstanding graduate in basic sciences in 1977, when he received a Bachelor of Science degree in basic sciences.

He received his initial pilot training in Oklahoma, and was then assigned to Carswell Air Force Base in Texas, where he flew strategic bombers. He led a combat crew of six as pilot and commander of a B-52D Stratofortress aircraft.

During his military career, Jones accumulated more than 2,000 hours of flying time in jet aircraft and had risen to the rank of captain in the U.S. Air Force by the time he left the service in 1983 to pursue a doctoral degree.

As a graduate student at the University of Arizona in Tucson, Jones studied the use of remote sensing technology in space, meteorite spectroscopy, and applications of space resources. He was awarded a NASA graduate student research fellowship in 1987 and became a member of Phi Beta Kappa in 1988, the same year he graduated from the university with a Ph.D. in planetary science.

In 1989, he joined the Central Intelligence Agency's office of development and engineering in Washington, D.C. as a program management engineer. The following year, he became a senior scientist at Science Applications International Corporation, also in Washington, D.C., where he participated in the development of NASA's plans for future robotic missions while working with the agency's Solar System Exploration Division.

Jones became an astronaut in 1991 and first flew in space as a mission specialist during the STS-59 flight of the shuttle Endeavour in April 1994. Coordinating their efforts with teams on Earth at the Jet Propulsion Laboratory (JPL), the STS-59 crew deployed the Space Radar Laboratory (SRL-1), which housed experiments and instruments designed by researchers from 13 countries. The SRL included radar imaging instruments and an experiment that measured the distribution of carbon monoxide in the Earth's troposphere.

Jones began his second spaceflight on September 30, 1994 as payload commander aboard Endeavour during STS-68, the second flight of the Space Radar Laboratory. A part of NASA's Mission to Planet Earth program, STS-68 featured around the clock observations by the crew. Operating in two shifts, the astronauts made studies of locations that were covered during the first flight of the SRL for comparison purposes, and also captured images of areas affected by natural phenomena, including the islands of Japan following a recent earthquake and an erupting volcano in Russia. Endeavour landed on October 11, 1994.

Jones's third flight in space was STS-80, which launched November 19, 1996 and landed 17 days, 15 hours, 53 minutes, and 18 seconds later, on December 7, setting a new record for the longest shuttle flight.

Kenneth Cockrell commanded STS-80, and Kent Rominger served as pilot of the shuttle Columbia during the flight. Mission specialists were Jones, Tamara Jernigan (making her fourth flight in space) and Story Musgrave, who made a then-record sixth space shuttle flight and, at 61, became the oldest human being to fly in space up to that time.

The STS-80 crew launched and retrieved two free-flying spacecraft, including the Orbiting and Retrievable Far and Extreme Ultraviolet Spectrometer-Shuttle Pallet Satellite II (ORFEUS-SPAS II), which made its second shuttle flight during the mission, and the Wake Shield Facility-3 (WSF-3), which was deployed for the third time. In two weeks of observations. the ORFEUS-SPAS made 422 studies of about 150 astronomical bodies, and the Wake Shield Facility successfully cultivated seven thin film semiconductor materials for use in advanced electronics during its three days of free flight. Jones played a central role in the WSF deployment, using the shuttle's Remote Manipulator System (RMS) to release and then retrieve the spacecraft at the start and end of its flight.

On February 7, 2001, Jones launched aboard the space shuttle Atlantis for his fourth trip into space, the STS-98 International Space Station (ISS) assembly mission. As a mission specialist during the 13-day flight, he and Atlantis crew mate Robert Curbeam made three space walks totaling almost 20 hours, coordinating their activities with Marsha Ivins as she operated the shuttle's RMS while they installed the U.S. Destiny Laboratory Module on the ISS. Adding a state-of-the-art 28-foot long cylindrical workspace to the station, the installation of the Destiny module made the ISS the largest space station in history, increasing its mass to about 112 tons. Jones and Curbeam also used a portion of their third EVA to test emergency procedures for rescuing an astronaut in the event that he or she became disabled during a spacewalk.

Atlantis was docked to the ISS for seven days during the flight, while the shuttle crew visited and worked with the ISS Expedition 1 crew. The shuttle landed on February 20, 2001.

Thomas Jones has spent more than 52 days in space, including more than 19 hours in EVA during three spacewalks, during four space shuttle missions.

KADENYUK, LEONID K.

(1951–)

Ukrainian Astronaut; First Representative of the National Space Agency of Ukraine to Fly in Space

The first representative of the National Space Agency of Ukraine (NSAU) to fly in space, Leonid Kadenyuk was born in the Chernivtsi region of the Ukraine on January 28, 1951. He attended the Chernihiv Higher Aviation School in Chernihiv, where he graduated as a pilot-engineer in 1971.

Kadenyuk joined the Air Force of the Soviet Union in the mid-1970s, when the Ukraine was a Soviet republic, and received training as a test pilot at the State Scientific Research Institute of the Russian Air Forces Center.

His remarkable space career began in 1976, when he was selected as a Soviet cosmonaut. He was trained for missions to the Salyut and Mir space stations and as a pilot of the Russian Soyuz spacecraft, and he completed cosmonaut training at the Gagarin Cosmonaut Training Center (GCTC) in Star City, Russia in 1978.

As a test pilot and test cosmonaut, Kadenyuk underwent extensive training and participated in a wide array of tests of experimental technology and spacecraft systems at the GCTC from 1978 to 1983, when he became a test pilot at the Russian State Test Flight Center. He conducted tests of SU-27, SU-27UB, MiG 23, MiG 25, MiG 27, and MiG 31 aircraft during his time at the test flight center and also flew tests related to the development of the Soviet Buran space shuttle. As a test pilot, he played a key role in the design of the cockpit of the SU-27M aircraft.

Continuing his education concurrently with his military career, Kadenyuk attended the Moscow Aviation Institute, where he earned a Master of Science degree in mechanical engineering in 1989.

During his military career, Kadenyuk has accumulated over 2,400 hours of flying time in more than 54 different types of aircraft. He has risen to the rank of colonel and is qualified as a test pilot first class, military pilot second class, and test pilot. As a pilot instructor, he guided 15 students to their graduation from flight training.

After the dissolution of the Soviet Union and the emergence of an independent Ukraine, Kadenyuk was selected to command a planned flight to the Mir space station that was scheduled to include an all-Ukrainian crew. He underwent further extensive training on the Soyuz TM spacecraft and the Buran space shuttle, but the planned Ukrainian flight was subsequently canceled when the Russian space program experienced financial difficulties.

In two decades as a member of the Soviet and Russian cosmonaut corps as a Soviet cosmonaut from the Ukraine, Kadenyuk amassed a great store of expertise in spaceflight and systems, was trained in a wide variety of spacecraft and their attendant equipment and procedures, and witnessed the enormous political, economic, and social changes that shaped the Soviet/Russian space program from a unique perspective. Despite the promise of an opportunity on several occasions, however, he did not fly in space during the first 20 years of his career.

In 1996, he left the cosmonaut corps to become a researcher at the Institute of Botany in the National Academy of Sciences of Ukraine, where he helped to develop the collaborative Ukrainian Experiment (CUE), a joint project of the NSAU and NASA. He was also selected as one of NASA's first group of NSAU astronauts that same year and received training as a payload specialist at the Johnson Space Center (JSC) in Houston, Texas.

On November 19, 1997, Kadenyuk became the first NSAU astronaut to fly in space when he lifted off aboard the space shuttle Columbia from the Kennedy Space Center (KSC) in Florida during STS-87, the fourth flight of the United States Microgravity Payload (USMP-4). As the primary payload specialist assigned to the flight, he was responsible for operating the collaborative Ukrainian-U.S. space biology experiment. After a flight of 6.5 million miles and 252 orbits, the STS-87 crew returned to Earth on December 5, 1997.

Following a wait of 21 years, Kadenyuk had finally flown in space, successfully participating in useful scientific work and representing an independent Ukraine, which, of course, would not have been possible during most of the first two decades of his career, when Ukraine was part of the Soviet state.

Leonid Kadenyuk spent 15 days, 16 hours, and 34 minutes in space during his historic STS-87 flight.

KALERI, ALEXANDER Y.

(1956–)

Russian Cosmonaut

A veteran of four long-duration space missions who has spent more than 600 days in space, Alexander Kaleri was born in Yurmala, Latvia on May 13,

1956. He attended the Moscow Institute of Mechanical Physics, from which he graduated in 1979. He began his space-related career that same year, helping to develop technical documentation for the Mir space station as an employee of the Energia Rocket/Space Corporation (RSC).

He was selected for cosmonaut training in 1984 and completed basic space training by 1986. He was also qualified for test pilot flights the following year, and accumulated 22 hours of flying time in L-39 training aircraft.

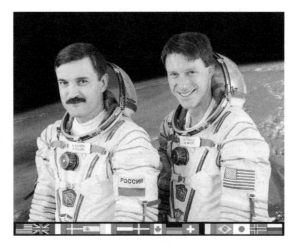

Alexander Kaleri (left) has spent more than 600 days in space during four long-duration space missions. He is seen here with NASA astronaut Michael Foale in 2003. [NASA/courtesy of nasaimages.org]

Kaleri served as a member of the backup crew for the Mir 3 mission and received additional preparation as a member of the Mir 9 backup crew.

He first flew in space during the long-duration Mir 11 mission, from March to August of 1992. With fellow cosmonaut Aleksandr Viktorenko and Klaus-Dietrich Flade of Germany, Kaleri launched aboard Soyuz TM-14 on March 17, 1992. Flade returned to Earth with the Mir 10 crew on March 25, while Kaleri and Viktorenko spent the next five months living and working on the Mir space station. One highlight of their stay came on July 8, when they made a two-hour spacewalk to inspect the station's Kvant module. On their return trip, also aboard Soyuz TM-14, on August 10, 1992, Michael Tognini of France, who had traveled to the station with the Mir 12 crew in late July, accompanied them. From launch to landing, Kaleri and Viktorenko logged 145 days, 14 hours, and 13 minutes in space.

Following that first successful mission, Kaleri next flew in space as a member of the Mir 22 long-duration crew. He flew to Mir with cosmonaut Valery Korzun and French cosmonaut Claudie Andre-Deshays (who was the first French woman to fly in space and is also known by her married name Claudie Haigneré). Andre-Deshays returned to Earth on September 2, with the Mir 21 crew aboard Soyuz TM-23; Kaleri and Korzun lived and worked aboard Mir for more than six months.

Kaleri and Korzun served aboard Mir during a particularly memorable period. As the Mir 22 crew, they played host to two visits by American astronauts aboard the space shuttle Atlantis and shared the station with American Shannon Lucid for the final portion of her epic, record-shattering space endurance flight.

The STS-79 flight of the space shuttle Atlantis culminated with Lucid's return to Earth after 188 days in space, which set a new long-duration mark for U.S. astronauts and a new world record for female astronauts. The mission was also the fourth docking between the space shuttle and the Mir space station, and the first time that one U.S. astronaut, John Blaha, was exchanged for another (Lucid) aboard Mir. The shuttle and the space station were linked together for five days, during which the crews transferred supplies and equipment and conducted joint experiments.

The cosmonauts and Blaha played host to the shuttle again during the Atlantis STS-81 mission, in January 1997. At that time, U.S. astronaut Jerry Linenger replaced Blaha. The visit also marked the largest transfer of equipment and supplies between the shuttle and Mir up to that time.

Kaleri and Korzun made two spacewalks during the Mir 22 mission. The first, on December 2, 1996, was a nearly 6 hour EVA to install cables on solar array panels, and on December 9 they spent more than 6.5 hours outside Mir while installing a docking antenna on the space station's Kurs module.

At the end of their long, productive Mir 22 mission, Kaleri and Korzun returned to Earth on March 2, 1997, aboard Soyuz TM-24, having spent a total of 196 days, 17 hours, and 26 minutes in space. On their return trip, they shared the Soyuz TM-24 with Reinhold Ewald of Germany, who had visited Mir for several weeks after traveling to the station with the Mir 23 crew aboard Soyuz TM-25.

Kaleri's third space mission began on June 16, 2000, with the launch of Soyuz TM-30. The final flight to the Mir space station, Mir 28 was funded by MirCorp, the international consortium that had hoped to transform Mir into a space tourist destination and operate the station on a commercial basis. Kaleri and fellow cosmonaut Sergei Zalyotin spent a little over two months in space during Mir 28. They made a final inspection of the legendary station during a five-hour spacewalk on May 12, and returned to Earth aboard Soyuz TM-30 on June 16, 2000. In the ensuing months, MirCorp's plans to commercialize the station failed to materialize, and Mir was directed to its fiery demise during re-entry into the Earth's atmosphere on March 23, 2001. Kaleri and Zalyotin have the distinction of being the last individuals to visit Mir during its long and illustrious history, bridging the Cold War era to the age of routine international spaceflight and the burgeoning commercialization of space travel.

Following his historic final visit to Mir, Kaleri served as flight engineer of the Expedition 8 crew of the International Space Station (ISS). Launched October 18, 2003 aboard Soyuz TMA-3 with American astronaut Michael Foale and European Space Agency (ESA) astronaut Pedro Duque of Spain, Kaleri added another 194 days, 18 hours, and 33 minutes in space during his six months aboard the ISS. He also added nearly four more hours to his total EVA time when he and Foale ventured outside the station on February 26–27 to retrieve experiments and deploy new ones in their place. At the end of the mission, Kaleri and Foale shared their Soyuz TMA-3 return trip to Earth with ESA astronaut André Kuipers of the Netherlands, who had visited the station aboard TMA-4 with the Expedition 9 crew (Pedro Duque had left the station with the Expedition 7 crew after a brief stay in October 2003).

Alexander Kaleri has spent more than 609 days in space, including 23 hours and 23 minutes in EVA, during his remarkable career as a cosmonaut.

KAVANDI, JANET L.

(1959–)

U.S. Astronaut

Janet Kavandi was born in Springfield, Missouri on July 17, 1959. In 1977, she graduated from Carthage Senior High School in Carthage, Missouri, where she was valedictorian of her graduating class. She was awarded a presidential scholarship by Missouri Southern State College in Joplin, Missouri, from which she graduated magma cum laude in 1980 with a Bachelor of Science degree in chemistry. She then attended the University of Missouri at Rolla, where she received a Master of Science degree in chemistry in 1982.

She began her career as an engineer at Eagle Picher Industries in Joplin, developing new batteries for defense industry applications.

In 1984, she moved to Seattle, Washington to work for Boeing Aerospace Company. As an engineer in the Boeing's Power Systems Technology department, she worked on the company's Short Range Attack Missile II and helped to develop thermal batteries for the Lightweight Exo-Atmospheric Projectile.

She also became involved with the space program for the first time while she was at Boeing, working on the company's plans for future involvement in space station activities and the development of bases on the Moon and Mars, and other space-related vehicles and payloads.

Continuing her education concurrently with her engineering career, she attended the University of Washington in Seattle, where she earned a Doctorate degree in analytical chemistry in 1990. Her doctoral research into the development of a pressure-indicating coating that reveals the amount of pressure on aerodynamic test models in wind tunnels led to her being granted two patents, and her work has been published in scientific journals and presented at technical conferences. In 1991, Boeing honored her with awards for Team Excellence and for Performance Excellence.

NASA selected Kavandi for astronaut training in 1994. Her first flight was as a mission specialist during the June 1998 STS-91 flight of the space shuttle Discovery. The ninth and final docking of a shuttle and the Russian space station Mir, STS-91 marked the successful completion of the first phase of the joint American-Russian cooperative space program. The Mir docking was the first for Discovery.

During four days of docked flight, the STS-91 astronauts worked with the Mir 25 crew to transfer supplies and equipment to the station and to move long-term U.S. experiments and equipment into the shuttle for the return trip to Earth.

The Discovery crew also collected NASA astronaut Andrew Thomas, who had lived aboard Mir for 130 days. Thomas was the last of seven U.S. astronauts to make a long-duration visit to Mir during 907 consecutive days of continual American presence aboard the Russian station. The shuttle landed June 12, 1998, after 154 orbits.

Kavandi's second space mission was STS-99, the Shuttle Radar Topography Mission. Launched on February 11, 2000 aboard the shuttle Endeavour, Kavandi spent 11 days in space, traveling more than four million miles in 181 orbits while using radar instruments to map the surface of the Earth.

STS-99 was conceived and designed by the National Imagery and Mapping Agency, NASA, and the German Aerospace Center (DLR). Using a high resolution radar mapping system specially designed for the flight, the crew was able to capture high quality, detailed, three-dimensional images of the surface of the Earth from 60 degrees north latitude to 56 degrees south latitude, an area populated by about 95 percent of the world's peoples.

The 13-ton radar-mapping instrument was deployed on the first day of the flight at an orbital altitude of 145 miles. Its antennas were mounted in the shuttle's payload bay on a 200-foot mast. It was operated continuously for more than 220 hours while the crew worked around the clock in alternating shifts. Endeavour returned to Earth on February 22, 2000.

Following her second spaceflight, Kavandi received training in using the Remote Manipulator System (RMS) on the shuttle and on the International Space Station (ISS), while attached to NASA's Robotics Branch.

On July 12, 2001, she launched on her third flight in space, STS-104, aboard the shuttle Atlantis. The STS-104 mission was the 10th shuttle flight to travel to the ISS, where its crew delivered and installed the station's Quest Airlock and transferred supplies and equipment to the station's Expedition 2 crew during eight days of docked operations. After a flight of 13 days, Atlantis returned to Earth on July 24, 2001.

Janet Kavandi has spent more than 33 days in space during three space shuttle flights.

She has continued to serve NASA since her third space mission, first as head of the agency's Payloads and Habitability Branch, then as branch chief for the International Space Station, where she oversaw virtually all major aspects of the space station program, including the training and operations of ISS crews, science operations on the ISS, and the involvement of international partners visiting the station. She has subsequently become deputy chief of the Astronaut Office at the Johnson Space Center in Houston, Texas.

KAZAKHSTAN: SPACE PROGRAM.

See Republic of Kazakhstan

KELLY, MARK E.

(1964–)

U.S. Astronaut

Mark Kelly was born in Orange, New Jersey on February 21, 1964. In 1982, he graduated from Mountain High School in West Orange, New Jersey, and he then attended the U.S. Merchant Marine Academy, where in 1986 he received both a Bachelor of Science degree in marine engineering and a Bachelor of Science degree with highest honors in marine transportation. In 1994, he earned a Master of Science degree in aeronautical engineering from the U.S. Naval Postgraduate School.

He served aboard the USS *Midway* while assigned to Attack Squadron 115 and made two deployments to the Persian Gulf as pilot of the A-6E Intruder All-Weather Attack aircraft. His second tour of duty in the Persian Gulf occurred during Operation Desert Storm, during which he flew 39 combat missions.

During his outstanding military career, Kelly has accumulated more than 5,000 hours of flight time in more than 50 different aircraft and he has made more than 375 aircraft carrier landings. Among the many honors he has received for his military service, he has twice been awarded the Defense Superior Service Medal and has also been awarded four Air Medals and several Navy Commendation Medals. He has risen to the rank of captain in the U.S. Navy.

In 1994, he graduated from the U.S. Naval Test Pilot School and was then assigned as a test pilot at the Strike Aircraft Test Squadron and subsequently as an instructor pilot at the Test Pilot School. He was selected for training as an astronaut in April 1996.

Kelly received a U.S. Patent in 2006 for an advanced oxygen mask for use in combat aircraft.

Kelly first flew in space as pilot of the space shuttle Endeavour during STS-108, which launched from the Kennedy Space Center (KSC) on December 5, 2001. The first shuttle flight following the September 11, 2001 terrorist attacks on the United States, STS-108 served as both a forward-looking sign of the nation's will to overcome the fear and sadness of the attacks and as a tribute to the spirit of those who were lost such a short time before the mission's start. The STS-108 crew carried 6,000 small U.S. flags into space with them for later distribution to 9/11 survivors and families of the victims, as part of an initiative known as the "Flags for Heroes and Families" campaign.

The flight was also the 12th shuttle mission to the International Space Station (ISS), and achieved the delivery of the ISS Expedition 4 long-duration crew, the restocking of the station with some three tons of supplies and equipment, some much needed maintenance and repair work, and safe transport back to Earth for the returning Expedition 3 crew. During his first space mission, Kelly spent 11 days, 19 hours, and 36 minutes in space.

He again served as pilot during his second spaceflight, the STS-121 flight of the space shuttle Discovery, which lifted off on July 4, 2006.

NASA's second return-to-flight mission after the loss of the space shuttle Columbia in February 2003, STS-121 was closely monitored for signs of difficulties with the space shuttle in light of the damage done during by foam insulation strikes during the previous shuttle flight, STS-114, in July 2005. Such a strike was deemed the proximate cause of the Columbia accident.

The STS-121 crew launched safely, and once in orbit they continued the test of orbital repair equipment and procedures that had begun with STS-114. They also transported European Space Agency (ESA) astronaut Thomas Reiter of Germany to the ISS, where he became a member of the station's Expedition

13 crew, and they delivered 5,000 pounds of equipment and supplies to the space station. STS-121 crew members conducted a live press conference from space and fielded a congratulatory call from U.S. President George W. Bush. Discovery returned to Earth on July 17, 2006, after a flight of more than 12 days and 18 hours.

On his third spaceflight, Kelly served as commander of STS-124, which was flown aboard the space shuttle Discovery from May 31 to June 14, 2008. The 26th shuttle mission to travel to the ISS, STS-124 brought the pressurized portion of the Japanese Kibo scientific laboratory to the ISS, where it was mated to the previously delivered logistics module and was installed and activated.

STS-124 concluded with Discovery's landing at KSC on June 14, 2008 after a flight of 13 days, 18 hours, and 13 minutes.

During three space shuttle missions, Mark Kelly has spent more than 38 days in space.

KENNEDY, JOHN F.

(1917–1963)

President of the United States, 1961–1963

The 35th president of the United States and a champion of the early efforts of the U.S. space program who set the nation's sights on a manned lunar landing, John

President John F. Kennedy (right), with NASA Deputy Administrator Robert Seamans (left) and Dr. Wernher von Braun (center) at Cape Canaveral in Florida, November 16, 1963. [NASA/courtesy of nasaimages.org]

Fitzgerald Kennedy was born in Brookline, Massachusetts on May 29, 1917. He served in the U.S. Navy during World War II, and was recognized for his heroism when he led his crew to safety despite his own injuries after their boat had been destroyed by enemy fire.

After the war, he pursued a career in politics and was elected as a Democrat to the U.S. House of Representatives and then to the U.S. Senate. In 1960, he was the Democratic candidate for president, and he narrowly defeated sitting vice president Richard M. Nixon to become the 35th president of the United States of America. He was also the first Roman Catholic to be elected to the nation's highest office.

During the presidential campaign, the candidates debated the impact of a perceived missile gap between the United States and the Soviet Union, an issue that attracted much interest in the wake of the Soviet Union's successful Sputnik flights.

As president, Kennedy celebrated the first U.S. manned spaceflight, when Alan Shepard launched aboard Freedom 7 on May 5, 1961. Kennedy contributed to the already intense public interest in the flight by becoming personally involved in marking the event, welcoming Shepard to the White House for a special ceremony.

The success of the nation's first manned spaceflight seemed to invigorate the young president, who in the first year of his administration had already experienced several alarming international setbacks, including U.S. participation in the failed Bay of Pigs invasion of Cuba.

On May 25, 1961, in an address to Congress about urgent national needs, President Kennedy described a remarkable vision that would become the single most important political initiative in the history of the U.S. space program for decades to come. In challenging Congress to fund a long-term effort to place Americans on the lunar surface, he established NASA's primary goal for the ensuing decade, defined the central engine of a vast international competition for supremacy in space and on Earth, and outlined the dream of a great nation that would unite and enthrall all its citizens even years later during a difficult period in the nation's history.

In his address, the Kennedy said, "I believe that this nation should commit itself to achieving the goal, before this decade is out, of landing a man on the Moon and returning him safely to Earth. No single space project in this period will be more impressive to mankind, or more important for the long-term exploration of space, and none will be so difficult or expensive to accomplish."

Kennedy matched the lofty idealism of his space agenda, which ultimately became the goal of both the United States and, surreptitiously, the Soviet Union, with a strong practical challenge to Congress that called for sharp increases in funding for the nation's space activities, adding $148 million immediately to increase the amount devoted in 1962 to the space program's future to $531 million, and candidly predicting that even that figure would be relatively paltry compared to the eventual total cost of the project, which would likely reach as high as $9 billion by the mid-1960s.

Kennedy's Moon landing goal was a politically courageous attempt to elevate the national interest to a new plane of global leadership through the peaceful demonstration of technical and academic superiority. In setting the end of the decade deadline, he admitted the possibility that he might, in fact, be committing the country to a program whose successful completion might well be arrived at after the end of his own presidency, and perhaps even during the administration of a political rival, even though at the time he likely believed and hoped that the landing could be achieved sometime late in his second term.

It can be reasonably argued that Kennedy's fascination with the space program also informed his approach to several of the signature events of his presidency. In any case, he later cited the decision to propose the lunar landing goal as one of his most important initiatives.

With a keen understanding of the historic importance of America's passage into the age of space exploration, President Kennedy on several occasions expressed a desire to attempt to defuse some of the Earth-bound tensions between the United States and the Soviet Union by creating a cooperative space program between the two superpower rivals. Coming as it did at the height of the Cold War and in the midst of the already-intense space race, Kennedy's willingness to cooperate with the Soviets in space was arguably more visionary than even his lunar landing agenda.

Despite the good intent implied by the notion, which Kennedy first tentatively broached during his June 1961 Vienna summit meeting with Soviet leader Nikita Khrushchev, the idea of sharing the first trip to the Moon was clearly not part of the Soviet agenda. While continuing to compete with the United States in space, the Soviets also tested the Kennedy Administration with increasingly provocative challenges on Earth, including the partitioning of Berlin and the test of a 100-megaton nuclear weapon in August 1961. Ultimately, the October 1962 Cuban Missile Crisis brought the two nations to the brink of nuclear confrontation over the Soviets' secret attempt to place nuclear weapons at launch sites in Cuba.

Having surmounted the harrowing confrontation over the possibility of Soviet missiles in Cuba, President Kennedy also presided over all six flights of the pioneering U.S. Mercury program, which concluded with the May 1963 flight of Gordon Cooper in Faith 7. He also lived to see the conclusion of the Soviet Vostok program, which ended with the group flight of Vostok 5 and Vostok 6 in June 1963.

In honor of his memory and his superb commitment to the nation's space program, NASA's Cape Canaveral launch facility was christened the John F. Kennedy Space Center shortly after he was assassinated on November 22, 1963. His vice president, Lyndon B. Johnson, who succeeded him as president, carried on Kennedy's active support of the space program, and although the first lunar landing was eventually achieved during the first administration of Richard Nixon, whom Kennedy had defeated to win the presidency in 1960, a large majority of Americans continue to associate the lunar landings with President Kennedy, who first committed the nation to the lunar missions eight years before Apollo 11 achieved the first landing in July 1969.

KERWIN, JOSEPH P.

(1932–)

U.S. Astronaut

A member of NASA's Group Four selection of scientist astronauts who spent 28 days in space as a member of the first crew of the Skylab space station, Joseph Kerwin was born in Oak Park, Illinois on February 19, 1932. In 1949, he graduated from Fenwick High School in Oak Park and then attended the College of the Holy Cross, where he received a Bachelor of Arts degree in philosophy in 1953.

He went on to attend Northwestern University Medical School, where he earned a Doctor of Medicine degree in 1957, and subsequently completed his internship at the District of Columbia General Hospital in Washington, D.C.

Kerwin joined the U.S. Navy Medical Corps in July 1958 and attended the Navy School of Aviation Medicine in Pensacola, Florida. He was designated a naval flight surgeon later that year, and in 1962 he earned his Navy aviator's wings at Beeville, Texas.

During his outstanding career in the Navy Medical Corps, Kerwin rose to the rank of captain and accumulated more than 4,500 hours of flying time. He retired from the service in 1987.

NASA selected him for training as an astronaut in June 1965, as one of its Group Four selection of scientist-astronauts.

On May 25, 1973, Kerwin lifted off with veteran Gemini and Apollo astronaut—and third person to walk on the Moon—Charles "Pete" Conrad, and Group Five astronaut Paul Weitz on the Skylab 2 mission. Conrad served as commander of the flight and Weitz piloted the Apollo spacecraft that brought the astronauts to the first American space station. Skylab 2 was the first manned flight to the station; the station itself, which launched unmanned on May 14, 1973, had been designated Skylab 1.

Adapted from the third stage of the Saturn V rocket that had launched all previous Apollo flights, the 36 meter-long, 91-metric ton Skylab was damaged during its launch when its meteoroid shield was ripped off. The mishap and its ensuing impact on the station's deployment and operation necessitated a series of complex repairs that were assigned to the Skylab 2 crew.

The accident at launch had caused the loss of one of Skylab's solar arrays and an incorrect deployment of the second array, which deprived the station of much of the electrical power it required for basic operations. At the same time, the loss of the meteoroid shield, which was also intended to function as a sunshade, resulted in much higher than desired temperatures within the station.

Faced with the prospect of losing the station entirely, NASA immediately activated all of its available resources in an effort to devise solutions for each of the problems Skylab faced during its first week and a half in orbit. Teams of astronauts, engineers, scientists, and space agency officials carefully worked through each difficulty, simulating repairs and compiling point-by-point plans for each procedure necessary to salvage the damaged space station.

Although they had originally been scheduled to launch shortly after the station itself, Conrad, Kerwin, and Weitz were immediately preoccupied with the detailed study of the damage and proposed repairs and were thus delayed in their lift-off until May 25.

During their 28-day flight, the Skylab 2 crew managed to implement solutions that had been worked out in the extensive simulations on Earth. They installed a tarp-like shade that sufficiently lowered the temperature in Skylab's workshop area, and repaired the craft's one remaining solar panel, thus providing the station with enough power to allow future crews to continue to do productive work.

Kerwin and Conrad performed the solar array repair during a 3 hour and 25 minute-long spacewalk on June 7, 1973.

In addition to the long hours of difficult repair work, which saved the Skylab space station, Kerwin, Conrad, and Weitz also carried out a good portion of the scientific program that had originally been planned for the Skylab 2 mission. They accumulated a total of 392 hours of work on a variety of experiments, including studies of Earth resources and the Sun, medical studies, and five experiments designed by students.

They also performed the regular regimen of exercise necessary to combat the losses of blood plasma volume and bone density that regularly occur during long stays in space. One of the primary goals of the Skylab program was to test the effects that long-duration stays in space would have on crew members, as NASA medical personnel had only previously been able to collect medical data during the relatively short missions of the Mercury, Gemini, and Apollo programs.

At the end of their successful repair and science mission, Kerwin, Conrad, and Weitz boarded their Apollo command module and returned to Earth, splashing down in the Pacific Ocean on June 22, 1973 after a flight of 28 days and 50 minutes.

After his Skylab 2 mission, Kerwin continued to serve NASA in a variety of positions, initially as head of the on-orbit branch of the Astronaut Office at the Johnson Space Center (JSC) in Houston, and then as NASA's senior science representative in Australia from 1982 to 1983. In 1984, he was named director of Space and Life Sciences at JSC and was responsible for overseeing health care services for astronauts and their families.

Joseph Kerwin spent 28 days and 50 minutes in space, including 3 hours and 25 minutes in EVA, during the Skylab 2 mission.

Kerwin left NASA in 1987 to pursue a private sector career with the Lockheed Corporation, where he and two fellow employees invented the Simplified Aid for EVA Rescue (SAFER) EVA devices that are intended to rescue spacewalkers in the event that they become untethered while conducting extravehicular activities. The SAFER devices were successfully tested in space by NASA space shuttle mission specialists Mark Lee and Carl Meade during STS-64 in September 1994.

Kerwin also served as a member of Lockheed's Assured Crew Return Vehicle team, and participated in the company's involvement in the initial stages of International Space Station (ISS) development. He also continued his association with NASA as a participant in the agency's Human Transportation Study of potential future space transportation systems and as a member of the NASA Advisory Council from 1990 to 1993.

In 1996, Kerwin joined Systems Research Laboratories as manager of the company's medical support activities, and the following year he became president of KRUG Life Sciences, where he oversaw the firm's administration of medical support at the Johnson Space Center. In 1998, Wyle Laboratories of El Segundo, California acquired KRUG.

Kerwin also put his considerable expertise to use as an industry representative member of the board of directors of the National Space Biomedical Research Institute.

KHRUSHCHEV, NIKITA S.

(1894–1971)

First Secretary of the Communist Party of the Soviet Union, 1953–1954; Soviet Premier, 1958–1964

As First Secretary of the Communist Party of the Soviet Union from 1953 to 1964 and Soviet Premier from 1958 to 1964, Nikita Khrushchev played a key role in the early development of his nation's space program and was the driving force behind a string of remarkable milestones the Soviet program achieved under his tenure.

Khrushchev was born in Kalinovka, Russia on April 17, 1894. He joined the Communist Party in 1918, and served in the Red Army for many years, rising to the rank of lieutenant general by the time of his service in World War II. In 1949, he was given the task of overseeing the nation's agricultural reorganization.

When Soviet premier Josef Stalin died in 1953, Khrushchev became the First Secretary of the Communist Party, surprising many who were certain that one of Stalin's more politically powerful competitors would capture the important post.

In 1956, Khrushchev made a remarkable address to the 20th Congress of the Communist Party in which he denounced Stalin and renounced what he called the cult of personality surrounding the late Soviet leader and his policies. That same year he was also responsible for the bloody repression of a popular anti-communist movement in Hungary.

Beginning with the October 4, 1957 launch of Sputnik 1, the first satellite to be placed in Earth orbit, Khrushchev cited Soviet space successes as proof of his nation's technological superiority to its superpower rival, the United States. He derided U.S. launch failures and dismissed the achievement of Explorer 1, the first successful U.S. satellite, which was placed in orbit on January 31, 1958.

Khrushchev consolidated his power on March 27, 1958 when he became Soviet premier. His policies, as well, perhaps, as his flair for the dramatic and willingness to change long-standing norms of behavior between nations, led to the alienation of the People's Republic of China from the Soviet Union and confusion in the West about Soviet intentions on the global stage.

Khrushchev's use of Soviet space achievements for propaganda purposes might have been dismissed as an inevitable by-product of the tense Cold War environment of the time, but his frequent linkage of lofty space goals and provocative Earthly military and political maneuvers eventually led to serious problems in both his international relations campaigns and in his relationship with the leaders of the Soviet space program.

In an effort to win over nonaligned nations and to solidify the unity of Soviet bloc countries, Khrushchev coupled the string of early Soviet successes in space with an increasingly adventurous foreign policy after 1960. Although he initially appeared to present a less intransigent attitude toward the West, his relations with the Eisenhower Administration and with Western European leaders chilled after the Soviets shot down an American U-2 aircraft that was conducting military surveillance over Soviet territory on May 1, 1960.

Later that same year, the Soviet premier made a lasting impression during several meetings of the United Nations General Assembly in New York, when he responded—more than once—to comments by representatives of other nations by removing his shoe and pounding it furiously on the desk in front of him while expressing his displeasure with the comments of other speakers in a loud shouting voice.

With the transformation of the U.S. government that followed the 1960 presidential election, in which Democratic Senator John F. Kennedy of Massachusetts defeated Republican Vice President Richard M. Nixon, who had visited the Soviet Union in 1959 and engaged Khrushchev in an exchange about the advantages of each nation's way of life over that of its rival in what subsequently became known as the "Kitchen Debate," many observers in the United States and around the world cautiously hoped that the Kennedy Administration would enjoy improved relations with the Soviet leader.

The April 1961 U.S.-backed Bay of Pigs invasion, which targeted the Communist regime of Cuba, provided Khrushchev sufficient rationale to rail publicly against the new U.S. president and privately to seek ways to take advantage of the new administration.

By the time Khrushchev and Kennedy met at a summit meeting in Vienna in June 1961, both the Soviet Union and the United States had sent a man into space, and Kennedy had publicly announced his desire for the United States to send a manned mission to the Moon by the end of the 1960s. At the summit, Kennedy broached the possibility of the two nations creating a joint space program to undertake the lunar landing challenge, but Khrushchev did not respond favorably to the idea.

The Soviet premier no doubt felt there was little to be gained by cooperating on space activities when his nation clearly seemed to be out-performing the Americans in that area. As the year wore on, the United States made a second brief sub-orbital spaceflight as the Soviet Union achieved the world's first day-long orbital mission, Vostok 2, in August and, during the same month, began construction on the Berlin Wall and tested a 100-megaton nuclear weapon in the atmosphere, despite a three year moratorium on such testing.

Khrushchev's willingness to engage the West in increasingly tense challenges was matched by the pressure he brought to bear on the Soviet space program in a quest for ever more spectacular spaceflight milestones that could be used for propaganda purposes. Even with an increased risk of failure, and to the potential detriment of the cosmonauts involved in the flights, the pursuit of daring space missions became necessary to support the premier's bold foreign policy.

As a result, the initial propagandized use of space accomplishments was, over time, transformed into a made to order arrangement in which Khrushchev was involved in the conception and design of individual flights. Under his aegis and constant prodding, and in addition to the first satellites in orbit (Sputnik 1 in October 1957 and its successors), the first person in space (Vostok 1, April 1961), and the first day-long spaceflight (Vostok 2, August 1961), the Soviets achieved the first group flight of two spacecraft, with Vostok 3 and Vostok 4 (August 1962), flew the longest single occupant spaceflight (Vostok 5), and the first space mission flown by a woman (Vostok 6) in June 1963. They also achieved the first spaceflight with a crew of more than one person, the three-man Voskhod 1 mission (in October 1964).

The difficulty of translating the achievement of great milestones in space into military superiority on Earth became evident during the October 1962 Cuban Missile Crisis, during which Khrushchev sought to test the will of the U.S. administration by surreptitiously installing nuclear weapons at launch sites in Cuba. During a tense standoff that brought the Soviet Union and the United States to the brink of nuclear confrontation, Khrushchev and Kennedy were ultimately able to arrive at an agreement that resulted in the Soviets removing the weapons.

Khrushchev's close association with the Soviet space program remained a key characteristic of his leadership until the end of his career. As premier, he was instrumental in the creation of the Voskhod 1 mission, which would launch a crew of three into orbit in a hastily redesigned Vostok craft with few safety precautions in order to achieve the first three-person spaceflight before the United States could launch the first manned Gemini spacecraft, which would feature a two-man crew.

He celebrated the launch of Voskhod 1 on October 12, 1964, and placed a congratulatory call to the crew during the flight. When the craft landed the following day, however, he was no longer in control of the country. Replaced by Leonid Brezhnev, Khrushchev was consigned to internal exile within the Soviet Union and was barred from making public appearances or pronouncements for the remainder of his life. He died at his Moscow home on September 11, 1971.

KIZIM, LEONID D.

(1941–)

Soviet Cosmonaut

A veteran of three spaceflights, including two long-duration missions, Leonid Kizim was born in the town of Krasny Liman, Ukraine on August 5, 1941. He attended the Tchernigov Lenin Komsomol Higher Military Aviation School, where he graduated in 1963 after receiving training as a pilot. He then served as a pilot in the Zakavkazski Military District as a member of the Guards Fighter and Bomber Aviation Regiment 168.

Kizim was chosen for training as a cosmonaut as a member of the Air Force Group 3 selection of cosmonaut candidates in October 1965. After intensive training, he became a cosmonaut in December 1967.

Continuing his education concurrently with his cosmonaut career, Kizim attended the Gagarin Higher Military Academy from 1971 to 1975, and received the rating of first-class military pilot in 1974.

His initial assignments as a cosmonaut included service as a member of the backup crew for Soyuz T-2, the first manned test flight of the new Soyuz T-series of spacecraft, which flew in June 1980.

Kizim made his first flight in space as commander of Soyuz T-3, which launched on November 27, 1980. His crew for the flight included flight engineer Oleg Makarov and research engineer Gennady Strekalov; it was the first Soviet mission to feature a three-person crew since the tragic flight of Soyuz 11 in June 1971, which had ended in the death of the Soyuz 11 crew.

During the Soyuz T-3 flight, Kizim, Makarov, and Strekalov visited the Salyut 6 space station, where they spent just under two weeks refurbishing and repairing the station's thermal control system. At the end of their successful mission, they returned to Earth in Soyuz T-3 after a flight of 12 days, 19 hours, 7 minutes, and 42 seconds.

After his first flight, Kizim served as backup commander for the Soyuz T-6 Intercosmos flight, which flew in 1982. The Intercosmos program was designed to provide opportunities for citizens of nations aligned with or friendly to the Soviet Union to fly in space.

Kizim also served on the backup crew for the flight that had originally been designated Soyuz T-10 (and which was later re-named Soyuz T-10–1), which was aborted after a launch vehicle fire on September 26, 1983. The primary crew, Vladimir Titov and Gennady Strekalov, were fortunate to escape injury when their space capsule was hurtled away from the fiery launch site just seconds before the rocket that was to have propelled them into space exploded. They endured a five and a half minute flight, landing a little more than two miles from the launch site. Their intended mission had called for them to visit the Salyut 7 space station; with the aborted flight, their long-duration goal was shifted to the next mission, which was commanded by Kizim.

During that remarkable flight, Kizim and his crew mates, flight engineer Vladimir Solovyov and research engineer Oleg Atkov, lived and worked aboard Salyut 7 for nearly eight months as the station's third long-duration crew.

They quickly settled into a productive routine, conducting scientific experiments and medical and biological investigations, carrying out photography assignments, and participating in a regular program of exercise to lessen the losses in bone density and blood plasma that accumulate over the course of a long-duration stay in space. Their medical and exercise work was made easier by the presence of Atkov, a medical doctor.

In addition to the usual long-duration work, the Soyuz T-10/Salyut 7 crew was also given the extraordinary task of attempting to repair the space station's propulsion system, which had been damaged the previous summer during a mishap with the unmanned Progress 17 refueling and supply spacecraft and subsequently failed.

On April 23, 1984, Kizim and Solovyov began their repair work with a 4 hour and 20-minute spacewalk. They made a second EVA on April 26, adding another 4 hours and 56 minutes to their spacewalking total, and a third on April 29 that lasted another 2 hours and 45 minutes. On May 4, they worked on the damaged system for an additional 2 hours and 45 minutes. During their first forays outside the station, they spent 14 hours and 46 minutes in EVA in just 12 days, and still had more work to do to bring the propulsion system back into working order.

On May 18, Kizim and Solovyov made a fifth spacewalk while adding a solar array to the station. They finished the repair work on August 8 in a final EVA of five hours, and their remarkable efforts made the Salyut 7 propulsion system fully operational again.

During their six Soyuz T-10/Salyut 7 spacewalks, Kizim and Solovyov accumulated a total of 22 hours and 51 minutes in EVA.

Two crews visited Kizim, Solovyov, and Atkov during their long stay on Salyut 7. The first, aboard Soyuz T-11, included Rakesh Sharma, the first citizen of India to fly in space. The Soyuz T-11 crew left the station after a stay of about one week.

When the men received their second set of visitors, the crew of Soyuz T-12, which arrived on July 17, Kizim, Solovyov, and Atkov witnessed an historic moment in the history of space exploration when Svetlana Savitskaya became the first woman to perform a spacewalk.

Commanded by veteran cosmonaut Vladimir Dzhanibekov, the Soyuz T-12 crew also included flight engineer Savitskaya and research engineer Igor Volk. In 1982, Savitskaya had been the second woman to fly in space; during Soyuz T-12 she became the first woman to participate in a spacewalk. She and Dzhanibekov made the historic EVA on July 25, 1984, spending 3 hours and 35 minutes outside Salyut 7. Dzhanibekov, Savitskaya, and Volk left the station and returned to Earth aboard Soyuz T-12 on July 29.

Kizim, Solovyov, and Atkov completed their mission a little more two months later, landing in Soyuz T-11 on October 2, 1984 after a then-record flight of 236 days, 22 hours, and 50 minutes.

Kizim received the rating of cosmonaut, first-class in 1984.

On March 13, 1986, he lifted off on a remarkable third space mission, during which he and Vladimir Solovyov became the first crew to visit two space stations during a single flight. They also served as the first crew of the space station Mir, which had been launched unmanned on February 20, 1986.

Kizim and Solovyov began their Soyuz T-15 mission by traveling to Mir. After docking with the new station, they began their initial inspection of its systems and equipment. They also unloaded supplies delivered by two unmanned Progress supply vehicles and began their assigned program of scientific, medical, and engineering tests.

After nearly two months at Mir, Kizim and Solovyov boarded Soyuz T-15 on May 5 and left the station. They traveled to Salyut 7, where they became the first cosmonauts to enter that station since the hurried exit of the Soyuz T-14 crew in November 1985.

The Soyuz T-14 crew had been forced to leave Salyut 7 because of an illness suffered by commander Vladimir Vasyutin; the illness was subsequently found to be unrelated to his spaceflight or his stay aboard Salyut 7, and as a result, Kizim and Solovyov were given the task of entering Salyut 7 during their Mir mission to finish the work of the previous crew and to retrieve experiments and equipment they had left behind.

During a stay of 51 days aboard the station, Kizim and Solovyov successfully carried out a rigorous program of work that included two experimental spacewalks designed to test procedures for constructing large structures in space.

Kizim and Solovyov made their first EVA of the flight on May 28, 1986, deploying a large truss structure and then folding it for its return to storage within the station. They also retrieved experiment samples during the 3 hour and 50-minute EVA.

On May 31, they tested the truss assembly procedure again and spent another 4 hours and 40 minutes outside Salyut 7.

Kizim and Solovyov returned to Mir on June 25. They were the last cosmonauts to visit Salyut 7, which re-entered Earth's atmosphere in February 1991.

During their second stay aboard Mir, they settled into a productive routine of scientific work. They returned to Earth in Soyuz T-15 on July 16, 1986, after a total flight—to Mir, to Salyut 7, back to Mir, and then back to Earth—of 125 days and 56 seconds.

Leonid Kizim spent more than 374 days in space, including more than 31 hours in EVA, during his cosmonaut career.

He left the cosmonaut corps in 1987 to attend the U.S.S.R. Armed Forces General Staff Military Academy, where he graduated in 1989. Since 1995 he has served as a colonel-general in the Russian Air Force.

KLIMUK, PYOTR I.

(1942–)

Soviet Cosmonaut

A veteran of three spaceflights, including a long-duration mission during which he lived and worked in space for more than two months, Pyotr Klimuk was born in the village of Komarovka, Brest, Byelorussia (now Belarus), on July 10, 1942.

He was educated in Komarovka and then attended the Soviet Air Force military aviation school in Kremenchug until it was disbanded in 1960. In 1964, he graduated from the Chernigov Higher Air Force School with a specialization in the use of aircraft in combat. He received the rating of third-class military pilot in October 1964, and subsequently served in the Leningrad military district as a pilot and senior pilot attached to ADF Guard FAR 57. He rose to the rank of colonel-general in the Russian Air Force by the time of his retirement in 2003.

Klimuk was chosen for training as a cosmonaut as a member of the Air Force Group Three selection of cosmonaut candidates in 1965. After an intensive period of training, he became a cosmonaut in December 1967.

He first flew in space as commander of Soyuz 13, which launched on December 18, 1973 with Klimuk and flight engineer Valentin Lebedev aboard. Although details of the Soyuz 13 flight were not publicly revealed at the time, subsequent revelations indicate that it involved tests of space reconnaissance technology that had originally been scheduled to be carried out aboard the Salyut 2 space station. The flight has officially been referred to as being a scientific mission focused on astrophysics.

Salyut 2 had launched unmanned on April 3, 1973, and quickly developed a fault that caused it to fail in orbit. It was thought to be intended as the first manned military space station, designed to allow Soviet military personnel to carry out high-resolution photo reconnaissance. A closely held secret of the Cold War era, the subsequent Soviet military space stations were known within the Soviet space program as the Almaz series of stations; publicly, they were referred to as Salyuts to give the impression that they were part of the publicly-disclosed civilian Salyut program.

Following the failure of the Salyut 2-Almaz military station, in which the equipment necessary for military spy missions would have been tested, it is thought that the Soyuz 13 flight was used, successfully, to test the involved equipment.

After a flight of 7 days, 20 hours, 55 minutes, and 35 seconds, Klimuk and Lebedev returned to Earth on December 26, 1973.

Continuing his education concurrently with his cosmonaut career, Klimuk also began studies at the Gagarin Red Banner Air Force Academy in 1973, specializing in command staff tactical aviation.

After his first spaceflight, Klimuk served as backup commander of the mission that subsequently became known as Soyuz 18–1, which resulted in the first manned launch abort, on April 5, 1975. The primary crew, Vasili Lazarev and Oleg Makarov, were nearly killed when an electrical fault in their launch vehicle prevented the rocket's stages from separating as planned. Lazarev and Makarov were shot away from the Baikonur Cosmodrome launch site at a harrowing speed, and actually escaped the Earth's atmosphere for a brief time before plummeting to a rough landing about nine hundred miles away from where they had lifted off.

Lazarev and Makarov survived the ordeal despite being injured. Their mission had called for them to occupy the Salyut 4 space station for a period of about two months; after the accident, that task fell to their backup crew, Klimuk and Vitali Sevastyanov.

On May 24, 1975, Klimuk lifted off as commander of Soyuz 18, with flight engineer Sevastyanov. They traveled to Salyut 4 and docked there the next day, beginning their long-duration stay aboard the station.

Over the course of their two months in space, Klimuk and Sevastyanov made extensive studies of the Sun and conducted astronomical observations of several constellations. They also made a comprehensive photographic study of the sprawling landmass that constituted the Soviet Union at the time.

Their intense program of scientific study alternated with the regular routine of exercise and station maintenance typical to long-duration stays in space. In a startling contrast to the experiences of previous long-duration crews, however, Klimuk and Sevastyanov were faced with a frightening health scare when an unidentified green mold began to grow on the walls inside Salyut 4.

Under normal circumstances, the sudden appearance of an unknown bacteria would likely have resulted in a shortening of the cosmonauts' stay aboard the station; in the case of the Soyuz 18/Salyut 4 mission, however, the difficulty developed just as the Soviets were involved in the joint Soviet-American Apollo-Soyuz Test Project (ASTP) flight. The Soviets had previously assured the Americans that the two missions could proceed simultaneously without any conflict, and were strongly opposed to taking any action that might disrupt the cooperative spirit that had developed during the preparations and launch of the ASTP mission.

As a result, Klimuk and Sevastyanov were forced to share their living quarters with the unpleasant green mold for the rest of their flight. Fortunately, the mold was found to be harmless to their health. They returned to Earth on July 26, 1975, after an uncomfortable but productive mission of 62 days, 23 hours, 20 minutes, and 8 seconds.

On June 27, 1978, Klimuk lifted off on his third flight in space, this time as commander of the Soyuz 30 Intercosmos mission. His crew mate for the flight was Miroslav Hermaszewski, the first citizen of Poland to fly in space.

The Intercosmos program was designed to provide citizens of nations aligned with or friendly to the Soviet Union with opportunities to fly in space. Intercosmos missions typically involved short visits to the Salyut 6, Salyut 7 and Mir space stations and frequently served the additional purpose of switching the visiting crew's Soyuz spacecraft for the one at the station, since the systems and equipment aboard the Soyuz were designed to operate safely in space for a period of about 90 days.

In the case of Soyuz 30, Klimuk and Hermaszewski visited Salyut 6, where they spent about a week with long-duration crew members Vladimir Kovalyonok and Alexander Ivanchenkov, who were in the midst of a then-record four and a half month stay aboard the station.

Because another Intercosmos flight was scheduled to visit the station a short time later, Klimuk and Hermaszewski returned to Earth aboard the same spacecraft they had used to travel to the station. They landed in Soyuz 30 on July 5, 1978 after a flight of 7 days, 22 hours, 2 minutes, and 59 seconds.

Klimuk received the rating of first-class cosmonaut in 1978 and that same year was qualified as an instructor in the Air Force Paradrop Training program, after making 65 parachute jumps. In 1979, he began studies at the V. I. Lenin Politico-Military Academy, where he graduated in 1983 with a first class honors degree.

Pyotr Klimuk spent more than 78 days in space during his career as a cosmonaut.

He left the cosmonaut corps in 1982 to begin an administrative career at the Gagarin Cosmonaut Training Center (GCTC), beginning with his appointment as deputy chief of political affairs. In April 1991, he was named deputy chief of the center and chief of the politico-military department, and in September of that year he was named chief of the GCTC.

Klimuk has also continued his academic career since leaving the cosmonaut corps. While attending the V. I. Lenin Politico-Military Academy, he completed refresher courses in staff leadership in 1987, and in 1991 became an academician member of the K. E. Tsiolkovsky Russian Cosmonautics Academy. In 1995, he earned the degree of Candidate of Technical Sciences, and he earned a Doctor of Technology degree in 2000.

Klimuk is the author of *Beside the Stars* (1979) and *Attack on Weightlessness* (1983).

KOMAROV, VLADIMIR M.

(1927–1967)

Soviet Cosmonaut

A pioneering cosmonaut who flew the first manned flight of two separate types of spacecraft, Vladimir Komarov was born in Moscow, Russia on March 16, 1927. In 1941, he graduated from Secondary School Number 235 in Moscow, and then attended the specialized Air Force School Number One in Moscow, where he graduated in 1945. The following year he completed training at the Borisoglebsk Higher Air Force School, and in 1949 he graduated from the Bataysk Higher Air Force School.

Komarov served as a pilot and later as a senior pilot in the Soviet Air Force in the city of Grozny, in the North Caucasus district, from December 1949 to November 1951. He served as a senior pilot until August 1954.

Continuing his education concurrently with his military service, Komarov attended the N. Y. Zhukovsky Air Force Engineering Academy, where he graduated in 1959 with a specialization in air armament. In September 1959, he was assigned to the Central Scientific Research Institute of the Ministry of Defense, where he served as an assistant leading engineer and test pilot in the village of Chkalovsky.

An outstanding pilot and test pilot, Komarov had risen to the rank of colonel in the Soviet Air Force by the time of his death.

He was chosen for training as a cosmonaut in March 1960 as a member of the Soviet Union's first group of cosmonaut candidates. After a period of intensive training, he became a cosmonaut in April 1961.

Komarov's initial assignments as a cosmonaut included service as the backup pilot for the Vostok 4 flight, which was flown by Pavel Popovich in August 1962.

On October 12, 1964, Komarov began his first flight in space when he lifted off as commander of Voskhod 1 from the Baikonur Cosmodrome with flight engineer Konstantin Feoktistov and Boris Yegorov, the first medical doctor to fly in space.

The first manned flight of the Voskhod program, Voskhod 1 was also the first spaceflight to feature a crew of more than one person. The Voskhod spacecraft was derived from the earlier Vostok craft; the former's ejection seat was removed, and a small solid-fuel rocket was attached to slow the spacecraft's descent before the vehicle hit the ground. Where Vostok pilots had ejected from the craft and finished their return to Earth via parachute, the Voskhod crews would land within the spacecraft, a situation which therefore required the braking action of the rocket in order to better their chances for a safe landing.

At 2.3 meters in diameter, the crew compartment of the Voskhod was a tight fit for the crew of three, and left no room for the cosmonauts to wear spacesuits, thus making this flight the first time that a crew would fly in space without spacesuits. Attached to the crew compartment was an instrument module that contained the vehicle's engines and propellant systems.

Without the benefit of spacesuits (except in the case of Alexei Leonov, who required a spacesuit for his Voskhod 2 EVA), in tightly cramped quarters and without a means for the crew to escape in the event of an emergency during launch or landing, the Voskhod flights were risky and required enormous courage on the part of the cosmonauts who flew them. Although many of the basic Voskhod systems had been tested during the Vostok program, since the Voskhod had been derived from the earlier series of spacecraft, there had been just one unmanned test of the Voskhod before the first manned flight. That test, Cosmos 47, had taken place on October 6, 1964, just a week before Voskhod 1 lifted off.

During their Voskhod 1 flight, Komarov, Feoktistov, and Yegorov conducted Earth observation and photography assignments, and Yegorov oversaw a limited program of medical research in the narrow crew cabin. The crew also received a congratulatory call from Soviet leader Nikita Khrushchev, who, during the short period in which the cosmonauts were in space, was replaced by new Soviet leader Leonid Brezhnev. The political change would, in turn, change the future course of the Soviet space program, moving it away from Khrushchev's tendency to design space missions for short-term propaganda gain toward a new policy of focusing on longer-term objectives.

Those changes were to come later, however, after the Voskhod 1 crew returned to Earth. During their flight, Komarov, Feoktistov, and Yegorov continued their scientific and engineering experiments, gauging the spacecraft's performance and proving that a crew of three could work well in space even in the tightly outfitted Voskhod capsule. They made 16 orbits before they returned to Earth, in the first in-vehicle landing of the Soviet space program, on October 13, 1964. Komarov and his crew mates spent 1 day, 17 minutes, and 3 seconds in space during the Voskhod 1 flight.

Komarov received the rating of military pilot, first-class the day he launched aboard Voskhod 1, and was promoted to cosmonaut, third-class two days after landing.

Although neither the crew nor Soviet space officials knew it at the time, the first Voskhod mission would ultimately be one of just two manned flights in the Voskhod program, after which the Soviet space program would be focused on the development of the Soyuz spacecraft, with the intent of trying to reach the Moon with a manned lunar landing mission before the United States.

As the Soyuz program progressed throughout the mid-1960s, anticipation grew in the cosmonaut corps about who would receive the assignment to pilot the first manned Soyuz flights. Political pressures to expedite the launch of the first mission, which was designed as a rendezvous and docking flight between two Soyuz vehicles, gave rise to anxiety within the Soviet space program as administrators and engineering personnel worried that the Soyuz vehicle was not yet ready for manned spaceflights.

The initial mysteries of the first Soyuz mission, the question of who would pilot the first spacecraft and the identity of the crew of the second Soyuz that would serve as a target for rendezvous and docking, were resolved when Komarov received the assignment as commander of Soyuz 1. Yuri Gagarin, the first person ever to fly in space, was assigned as his backup. Valeri Bykovsky, Aleksei Yeliseyev, and Yevgeni Khrunov were assigned to Soyuz 2. The complex mission profile for the two flights called for the three-man crew to follow Komarov into space the day after his launch, with Komarov then docking with Soyuz 2 and Khrunov and Yeliseyev making a spacewalk from their spacecraft to join Komarov aboard Soyuz 1. Even though the two spacecraft would be docked, the spacewalk would still be needed for the crew transfer because an airlock system could not be developed quickly enough for the proposed April 1967 launches.

Komarov and Gagarin underwent intensive training to familiarize themselves with the Soyuz vehicle and its systems, even as Soviet engineers struggled to get the spacecraft ready. Originally intended as part of the Soviet plans to send manned missions to the Moon, the basic Soyuz spacecraft featured a Descent Module, which carried the crew into space and returned them to Earth at the end of a mission, an Orbital Module, in which the crew lived and worked while in space, and an Instrument Module, which housed the equipment necessary to sustain the flight, including the spacecraft's maneuvering engine.

While the cosmonauts designated for the joint Soyuz 1 and Soyuz 2 flights trained for the launch of their rapidly approaching missions, news of a horrific setback to the U.S. Apollo program reached them as they learned of the January 1967 fire that claimed the lives of Apollo 1 crew members Virgil Grissom, Edward White, and Roger Chaffee. Soviet political leaders knew that the inevitable result of the Apollo 1 accident would be a long pause in the Americans' effort to land men on the Moon, but the cosmonauts and Soviet space officials also recognized in the Apollo 1 accident a sobering reminder that space missions required methodical patience and careful attention to detail, lest they end in tragedy.

Unmanned tests of the Soyuz spacecraft provided little comfort to those who worried about the vehicle's readiness: four unmanned test flights all revealed deficiencies in the craft and its systems and equipment, ranging from minor defects to major problems.

The first Soyuz launch, on April 23, 1967, was made amid an atmosphere of concern about the safety of the craft's commander and the resiliency of the spacecraft's systems and equipment.

Komarov experienced difficulties with the vehicle's electrical and control systems shortly after launch. One of the Soyuz 1 solar panels failed to deploy fully, which resulted in substantially less electrical power being available for crucial tasks. The spacecraft's computer system then malfunctioned, making it much more difficult to control the flight.

With the vehicle's systems approaching the point of catastrophic failure, mission controllers decided to abort the complex docking mission and canceled the planned Soyuz 2 launch. They focused all their available resources on returning Komarov safely to Earth, and began preparations for his re-entry during his 16th orbit.

Komarov made his first attempt to orient the spacecraft for re-entry at that point and found that the craft's failing guidance and control systems prevented him from doing so. The veteran cosmonaut kept his focus on the problem at hand despite the severity of his circumstances, and, working with mission controllers advising him from the ground, he was able to properly orient Soyuz 1 after two more trips around the Earth, on his 18th orbit.

Even as he began his return trip through the atmosphere, however, Komarov was quickly presented with another harrowing challenge: his spacecraft's failed electrical systems and malfunctioning computer allowed the Soyuz 1 Descent Module to vibrate wildly. In an attempt to bypass the failing computer system, mission controllers apparently instructed Komarov to place the craft into a ballistic re-entry profile, which caused the vehicle to spin rapidly en route to landing. The maneuver provided a better chance of the spacecraft returning to Earth intact, but also greatly increased the force of the re-entry.

After struggling to control the craft during the final portion of the flight, Komarov is thought to have lost consciousness before Soyuz 1 impacted the Earth, far off course, near Orsk in the Ural Mountains, on April 24, 1967. The craft's parachute lines tangled, and the capsule crashed violently, its retrorockets exploding on impact and engulfing the spacecraft in flames. Komarov was killed in the crash.

The shock of the Soyuz 1 disaster was felt deeply by the Russian people, who knew and revered Komarov as a hero for his participation in the Voskhod 1 mission and knew nothing of the problems that had preceded the Soyuz 1 flight.

Komarov and his fellow cosmonauts had been acutely aware of the shortcomings of the Soyuz craft, of course, and in the years following the tragedy, accounts have emerged of Komarov going through with the flight in spite of his apprehensions because he feared for the life of his backup and close friend, Yuri Gagarin, who would have been assigned to the dangerous Soyuz 1 mission in his place if he had chosen not to continue with the mission.

During his two remarkable spaceflights, Vladimir Komarov accumulated 2 days, 3 hours, 4 minutes, and 55 seconds in space. The married father of two children, he was 40 years old at the time of his death.

KONDAKOVA, YELENA V.

(1957–)

Russian Cosmonaut

Yelena Kondakova was born in Komsomolskna, Russia, on March 30, 1957. In 1980, she graduated from Moscow Bauman High Technical College.

After graduation she began her career at Energia Rocket/Space Corporation (RSC). She was selected for cosmonaut training in 1989, and completed her basic preparation for spaceflight in March 1990, when she was qualified as a test cosmonaut.

Her first flight in space began on October 3, 1994, aboard Soyuz TM-20. As flight engineer for the Mir 17 mission, which was also known as EuroMir '94 in honor of the international cooperation it celebrated, she traveled to the Mir space station with veteran cosmonaut Aleksandr Viktorenko and European Space Agency (ESA) astronaut Ulf Merbold of Germany. Merbold remained aboard Mir for a month, while Kondakova and Viktorenko lived and worked on the station for a six-month stay. They returned to Earth aboard Soyuz TM-20 on March 22, 1995, accompanied by fellow cosmonaut Valeri Polykov, who had flown to Mir on Soyuz TM-18.

During her remarkable first flight, Kondakova accumulated 169 days, 5 hours, and 22 minutes in space.

She journeyed to Mir again during her second space mission, albeit for a far shorter visit. Launched aboard the space shuttle Atlantis for the STS-84 flight on May 15, 1997, she served as a mission specialist during the sixth shuttle-Mir docking. She joined NASA astronauts Charles Precourt, Eileen Collins, Carlos Noriega, Ed Lu, and Michael Foale and ESA astronaut Jean-Francois Clervoy of France on the mission. Highlights of the flight included the transfer of Foale to Mir and the return to Earth of NASA astronaut Jerry Linenger, who had lived aboard Mir for 123 days. The exchange of U.S. crew members continued the ongoing American presence aboard Mir, which had begun with the start of Shannon Lucid's stay at the station on March 22, 1996 and was indicative of the new era of cooperation between the United States and Russia following the end of the Cold War and the long rivalry between the two space superpowers.

The STS-84 mission also saw the transfer of some 7,500 pounds of supplies, equipment, scientific experiments, and water between the space shuttle and the space station. At the conclusion of the successful mission, Atlantis returned to Earth at the Kennedy Space Center in Florida on May 24, 1997, 9 days, 5 hours, and 20 minutes after launch.

Yelena Kondakova has spent more than 178 days in space during her two space missions.

KONONENKO, OLEG D.

(1964–)

Russian Cosmonaut

Oleg Kononenko was born in Chardzhow, Turkmenistan on June 21, 1964. He attended the Zhukovskiy Kharkov Aviation Institute, where he graduated in 1988 with a degree in mechanical engineering. He began his career at the Russian Space Agency shortly after graduation, and in 1996 he was chosen for training as a cosmonaut.

He was named a test cosmonaut in March 1998, and began training in October of that year for missions at the International Space Station (ISS). His initial assignments included service as a member of the backup crew for Soyuz TM-34.

On April 8, 2008, Kononenko launched from the Baikonaur Cosmodrome aboard Soyuz TMA-12 to begin his first spaceflight, a long-duration mission at the ISS as a flight engineer for the station's Expedition 17 crew. He and Expedition 17 commander Sergei Volkov lived and worked at the station for more than six months and made two spacewalks—the first beginning on July 10, 2008 when they spent more than six hours outside the station to make repairs to their Soyuz spacecraft, and the second on July 15, 2008, when they added nearly six more hours in EVA while performing station maintenance and retrieving experiments.

Kononenko and Volkov returned to Earth on October 23, 2008 aboard Soyuz TMA-13.

During his long-duration ISS Expedition 17 mission, Oleg Kononenko spent 198 days, 16 hours, and 20 minutes in space and accumulated more than 12 hours in EVA during two spacewalks.

KOREA: FIRST CITIZEN IN SPACE.

See Yi, So-yeon

KOROLEV, SERGEI P.

(1906–1966)

*Chief Designer of the Soviet Union's Space Program
from its Inception until His Death in 1966*

The guiding spirit behind the many achievements of the Soviet space program during the first era of space exploration, Sergei Korolev was born in Zhtomyr, Ukraine on December 30, 1906 (although his date of birth is sometimes listed as January 12, 1907, in keeping with Russia's change from the Julian calendar, which was in use at the time of Korolev's birth, to the Gregorian calendar, which was introduced in Russia in 1918).

He attended the Odessa Building Trades School, where he was trained as a carpenter, and in 1924 he began courses in aviation at Kiev Polytechnic Institute, where he worked on the construction of a glider aircraft.

His interest in glider construction had begun in 1923, with his membership in the Society of Aviation and Aerial Navigation of Ukraine and the Crimea, where he designed a glider known as the K-5. He also pursued an interest in gymnastics at that time.

Korolev continued his education at the Bauman Higher Technical School in Moscow, where he began special classes in aviation in July 1926. He designed another glider in 1928, and also had his first experience with powered aircraft while attending the Bauman School. In 1929, he flew the glider in a competition and earned his diploma by creating an executable design for an aircraft.

After graduating from Bauman, Korolev began his career as an aircraft designer at the Fourth Experimental Section design bureau. He also designed a new glider that could perform acrobatic flying maneuvers and worked on a bomber aircraft for the Russian military.

Korolev earned his pilot's license in 1930. That same year he co-founded the Soviet Jet Propulsion Research Group to develop liquid rocket engines. He became leader of the group in 1932, and in 1933 he oversaw the successful flight of GIRD-09, the first Russian liquid-fueled rocket launch.

The Soviet government combined Korolev's group with the Gas Dynamics Laboratory in 1933 to create the Jet Propulsion Research Institute, which was led by Ivan Kleimenov. Korolev served as deputy chief of the new organization and was responsible for the design of missiles for military use and for a rocket-powered glider. He subsequently became the institute's chief engineer.

Just as his life-long interest in aviation and his professional achievement as an engineer and manager seemed poised to propel him to the forefront of the burgeoning Soviet rocket program, Korolev fell victim to a campaign of harassment and terror instituted by Soviet leader Josef Stalin. Ultimately known as Stalin's Great Purge, the systematic punishment, imprisonment, and killing of many of the nation's leading scientists, academics, military leaders, and professional people led to the deaths of millions of individuals as Stalin consolidated his authority.

Utilizing the methods employed against many of those targeted during the purge, the Soviet secret police arrested Korolev on June 22, 1938, accused him of subversive activities, and after a severe beating, forced him to sign a contrived confession. He was subsequently sentenced to 10 years of hard labor, a sentence that was later reduced to 8 years. During the early phase of his imprisonment, he spent five months in the Kolyma gulag in Siberia, where he developed a heart condition and suffered severe damage to his health.

Many of Korolev's fellow engineers and scientists from the Jet Propulsion Research Institute also fell victim to Stalin's purge, and the majority were killed. Along with the loss of many of the leaders of the Soviet military, the near-total depletion of those individuals most capable of leading the defense of the nation in the event of war with Germany nearly resulted in the fall of the Soviet Union after the German invasion of 1941. The outbreak of the war forced Stalin to alter his approach toward many of those who had been imprisoned, whose skills he suddenly needed for the defense of the nation.

After his release from the gulag, Korolev remained a prisoner of his own government for much of World War II, working under conditions of extreme duress while helping to develop several major weapons systems for the Soviet military, including the Tupolev Tu-2 bomber aircraft and the Ilyushin II-2 ground attack aircraft. He also worked on rocket technology again, this time as a subordinate to Valentin Glushko, who had previously been a colleague of Korolev's at the Jet Propulsion Research Institute.

His status finally changed with the issuance of a special decree by the Soviet government on June 27, 1944 that cleared him and Glushko and the other purge survivors who had survived the government's pre-war terror campaign.

With dizzying rapidity, Korolov's fortunes changed dramatically with the end of World War II. His was made a member of the Red Army, given the rank of colonel, and his wartime service was recognized with the Badge of Honor. He visited the conquered German territory where the Nazi regime had carried out its V-2 rocket program, and although many of the materials related to the program had been spirited away by the U.S. Army, Korolev and his fellow Soviet scientists were able, over time, to reproduce the V-2 with the assistance of 5,000 German scientists who were captured and moved to Russia.

Korolev became chief designer of the Soviet long-range missile program and was instrumental in developing a series of successively more sophisticated rockets that culminated in the R-7 (which was known in the West as the SS-6 Sapwood), the first Soviet intercontinental ballistic missile (ICBM). Capable of delivering a nuclear warhead to a distance as far as 7,000 kilometers, the two-stage R-7 was successfully tested for the first time in August 1957.

In 1953, in the midst of his work to reconstruct the German V-2, Korolev's career took another turn whose effects at the time were unknown even to him. Even though he had been cleared of criminal activity and promoted at the end of the war, he remained in the service of Soviet premier Josef Stalin, who continued to rule the Soviet Union with an iron hand until his death in 1953; thus, the first phase of Korolev's post-war career was devoted to the benefit of the very regime that had imprisoned and abused him for nearly a decade.

Although Korolev remained very much under the control of the Soviet political and military hierarchy for the rest of his life, the passing of Stalin and his replacement by Nikita Khrushchev as First Secretary of the Communist Party would have a major impact on Korolev's future, as the political change led to the use of rockets for space exploration becoming a national priority for the Soviet Union. Korolev and his fellow engineers soon found themselves transformed from barely-heard proponents of satellite launches and space travel to the central figures in a long, heated international competition for supremacy in space technology, with the results of their work envisioned as key to the future of their country's influence on the rest of the world.

It is likely that Korolev and others who had just barely survived the terrors of the Stalin years were pleased when, in 1956, Khrushchev made a shocking address to the 20th Congress of the Communist Party in which he denounced the excesses of the Stalin years and effectively ended the practice of honoring the former Soviet leader as a heroic figure, a habit he derisively referred to as the "cult of personality."

Initially developed as a weapon of war, the Soviet missile capability was quickly adapted for use as a launch vehicle for satellites. With the advent of the international Geophysical Year of 1957–58, which was conceived as a spur to international scientific development and included a challenge to the nations of the world to develop and launch a scientific satellite into orbit, Korolev personally oversaw the hurried development of Sputnik 1, the first Soviet satellite.

The successful launch of the tiny Sputnik 1 satellite on October 4, 1957 inaugurated both the space age and the intense competition between the Soviet Union and the United States to become the dominant power in the exploration of space.

Achieving and maintaining the leadership position in space was perceived by both sides as a sign of technological superiority and was further seen as proof of the ultimate value of one nation's political and economic philosophy over that of its rival.

Korolev's involvement with the Sputnik development elevated him in Khrushchev's eyes to the preeminent position within the Soviet space program. The Soviet leader made much political and rhetorical use of the success of the first Sputniks, which took place at a time when the United States was struggling in its attempts to orbit a satellite.

With Khrushchev's constant prodding and Korolev's own increasingly lofty ideas for spaceflight, the early Soviet space program achieved a string of impressive—and often, risky—achievements. With his exceptional management and engineering skills, Korolev oversaw virtually every aspect of the Soviet space program from its inception until his untimely death in 1966.

During the decade in which he was given the task of shepherding the Soviet portion of the space race with the United States, Korolev led the teams that designed and carried out the manned Vostok and Voskhod programs, the unmanned Luna Moon explorations, and the Soviets' first attempts to explore Mars and Venus.

Under his leadership, the Soviets placed the first human being in orbit (Yuri Gagarin in Vostok 1, April 12, 1961), made the first day-long orbital flight (Gherman Titov, Vostok 2, August 1961), carried out the first two group flights of two spacecraft simultaneously in orbit (Vostok 3 and Vostok 4, August 1962 and Vostok 5 and Vostok 6, June 1963), conducted the longest flight with a one-person crew (Valeri Bykovsky, 4 days, 23 hours, and 7 minutes in Vostok 5), sent the first woman into space (Valentina Tereshkova in Vostok 6), achieved the first spaceflight with a three-person crew (Voskhod 1, October 1964), and accomplished the first spacewalk (Alexei Leonov, during Voskhod 2, March 1965).

Korolev also oversaw the design and launch of the Luna series, which launched the first vehicles to impact the lunar surface (Luna 2, September 1959) and the first to return photographs of the far side of the Moon (Luna 3, October 1959).

The remarkable string of space milestones was the result of both Korolev's genius for engineering and administration and the unique circumstances in which he lived and worked, when the Soviet Union was intent on besting the United States in space achievements in order to further its political and military aims on Earth. The fevered pitch of the space race took another dramatic turn in 1964, when, in the midst of the Voskhod 1 flight, Nikita Khrushchev was removed from power and replaced by Leonid Brezhnev.

Changes in the Soviet space program inevitably followed the change in political leadership, but for perhaps the first time in his life, Korolev's goals were in harmony with the political climate surrounding him. He had grown tired of the persistent quest for short-term propaganda conquests, and wanted to concentrate on the competition with the United States to be first to land a human being on the surface of the Moon. To that end, he had already begun development of the next generation of Soviet space vehicles, the Soyuz.

The strangeness of Korolev's immense success can be summed up in a single anecdote: for the entire length of his exceptional career as the leader of the Soviet space program, he was publicly identified by the secretive Soviet political regime only as "the chief designer." Through all the years in which he played a key role in the design and launch of the nation's pioneering space vehicles, his identity was a closely held state secret; his name was made public only after his death.

Long suffering ill health as a result of his earlier imprisonment and the difficult circumstances under which he lived and worked for virtually his entire career, Korolev endured a heart attack, a kidney ailment, internal bleeding, and gallbladder difficulties, among other, less threatening health problems. He was diagnosed in late 1965 with a condition requiring surgery that has been clouded by both the contemporary censorship of the government and the passage of time.

In what has in some accounts been described as routine surgery, perhaps to rectify internal bleeding, Korolev underwent an operation in the early days of January 1966, and died in Moscow on January 14, 1966. He was survived by his ex-wife Xenia and their daughter Natasha, and his second wife, Nina.

In honor of his remarkable achievements, Korolev's name has been given to a crater on the far side of the Moon, a crater on Mars, and an asteroid, 1855 Korolev. His home on Ostankinsky Lane in Moscow has been preserved as a museum, and in 1996 Russian President Boris Yeltsin changed the name of the town in which RSC Energia is headquartered to Korolyov in his honor. RSC Energia has also been renamed the S. P. Korolev Rocket and Space Corporation Energia in his honor.

KORZUN, VALERY G.

(1953–)
Russian Cosmonaut

A veteran of long-duration missions to both the Mir space station and the International Space Station (ISS), Valery Korzun was born on March 5, 1953 in Krasny Sulin, Russia. He graduated from the Kachin Military Aviation College in 1974, and then served as a pilot, senior pilot, flight section leader, and squadron commander in the Russian Air Force. He has logged 1,473 hours in various types of aircraft, made 377 parachute jumps, and served as an instructor of parachute training. During his distinguished military career, he has risen to the rank of colonel.

He was selected as commander of the Gagarin Military Air Force Academy in 1987, and that same year was

International Space Station (ISS) Expedition Five commander Valery Korzun (right) with crew mates Sergei Treschev and Peggy Whitson. Korzun has spent more than 380 days in space. [NASA/ courtesy of nasaimages.org]

chosen for cosmonaut training. In 1989, he was qualified as a test cosmonaut after completing two years of basic space training at the Gagarin Cosmonaut Training Center (GCTC). He subsequently received additional preparation for flights aboard the Soyuz TM series of spacecraft and the Mir space station, and served as crew communications supervisor from March 1994 to January 1995.

Korzun first flew in space aboard Soyuz TM-24, as commander of the Mir 22 mission. He launched on August 17, 1996 with fellow Mir 22 cosmonaut Alexander Kaleri and French cosmonaut Claudie Andre-Deshays (who was the first French woman to fly in space and is also known by her married name, Claudie Haigneré). Andre-Deshays returned to Earth on September 2 with the Mir 21 crew, aboard Soyuz TM-23; Korzun and Kaleri lived and worked aboard Mir for more than six months.

As commander during the Mir 22 mission, Korzun played host to two visits by crews of the American space shuttle Atlantis and shared the station with American Shannon Lucid for the final portion of her record long-endurance flight.

After 188 days in space, a new long-duration record for U.S. astronauts and a new world record for female astronauts, Lucid left Mir and returned to Earth during the STS-79 flight of the space shuttle Atlantis. The fourth docking between the space shuttle and the Mir space station, STS-79 also saw the first exchange of U.S. astronauts, when John Blaha replaced Lucid aboard Mir. The shuttle and the space station were linked together for five days, during which the crews transferred supplies and equipment and conducted joint experiments.

Korzun and Kaleri were the host crew aboard Mir during the flight, and served with Blaha until he was replaced by U.S. astronaut Jerry Linenger during the STS-81 Atlantis flight in January of 1997.

Other highlights of the Mir 22 mission included two spacewalks by Korzun and Kaleri. The first, on December 2, 1996, was a nearly 6 hour-long EVA to install cables on solar array panels; and on December 9 the two cosmonauts spent more than 6.5 hours outside Mir while installing a docking antenna on the space station's Kurs module.

At the end of their long, productive Mir 22 mission, Korzun and Kaleri returned to Earth on March 2, 1997 aboard Soyuz TM-24, having spent a total of 196 days, 17 hours, and 26 minutes in space. They also accumulated 12 hours and 33 minutes in EVA during their two spacewalks. On their return trip, they shared the Soyuz TM-24 with Reinhold Ewald of Germany, who had visited Mir for several weeks after traveling to the station with the Mir 23 crew aboard Soyuz TM-25.

Korzun was next named commander of the long-duration Expedition 5 mission to the ISS in 2002. The six-month flight, which he shared with fellow Expedition 5 crew members Peggy Whitson of NASA and Sergei Treschev of Rosaviakosmos, began with the STS-111 launch of space shuttle Endeavour on June 5, 2002. Korzun, Whitson, and Treschev replaced the ISS Expedition 4 crew of Yuri Onufrienko, Dan Bursch, and Carl Walz, who had been aboard the ISS for 196 days. While Korzun and his crew mates settled in at the station, STS-111 mission specialists Franklin Chang-Diaz and Philippe Perrin made three EVAs from the ISS's Quest Airlock to continue the expansion of the station. Korzun and Whitson assisted the spacewalkers by operating the robotic arm of the ISS. Endeavour commander Ken Cockrell used the shuttle's robotic arm, effectively giving Chang-Diaz and Perrin a large extra set of hands to help out while they did their work in space.

At about the halfway point in their six-month stay on the ISS, Korzun and Whitson performed a spacewalk of their own. On August 16, they ventured outside the station for nearly 4.5 hours while installing panels designed to shield the station's Zvezda service module from space debris.

Korzun made a second EVA on August 26, with Sergei Treschev. They attached a frame compartment on the station's Zarya module for the storage of equipment and parts to be used in future construction work at the station, and, on the Zvezda module, they installed hardware for use in tethered EVAs, two ham radio antennas, and two material sampling experiments for the Japanese Space Agency. The two cosmonauts spent 5 hours and 21 minutes outside the ISS, raising Korzun's total spacewalking time during the Expedition 5 mission to 9 hours and 46 minutes.

At the end of their long, productive stay, Korzun, Whitson, and Treschev were replaced by the ISS Expedition 6 crew and returned to Earth during the STS-113 flight of the space shuttle Endeavour, which touched down at the Kennedy Space Center in Florida on December 7, 2002. From lift-off to landing, they had accumulated 184 days, 22 hours, and 14 minutes in space.

Valery Korzun has accumulated 381 days, 15 hours, and 40 minutes in space and 22 hours and 19 minutes in EVA during his two superb long-duration missions.

KOTOV, OLEG V.

(1965–)

Russian Cosmonaut

Born on October 27, 1965 in Simferopol, in what is now the Republic of Crimea in the Ukraine, Oleg Kotov graduated from high school in Moscow in 1982. He then pursued medical studies at the Kirov Military Medical Academy, graduating in 1988.

Kotov put his experience as a medical doctor to good use at the Gagarin Cosmonaut Training Center (GCTC), where he served as deputy lead test doctor and lead test doctor.

In 1996, he was selected as a cosmonaut candidate and subsequently underwent training to prepare for space missions. His initial assignments have included service as a member of the backup crews for the long-duration Mir 26, International Space Station (ISS) Expedition 6 and ISS Expedition 13 missions, and as a Capcom for ISS Expeditions 3 and 4. He also worked at the Johnson Space Center in Houston, Texas for a six-month period in 1999, and in 2004, he became chief of the GCTC Capcom Branch.

Kotov began his first space mission on April 7, 2007 when he lifted off from the Baikonur Cosmodrome in Soyuz TMA-10. During a stay of more than six months at the ISS, he served as flight engineer for ISS Expedition 15.

With Expedition 15 commander Fyodor Yurchikhin, Kotov made two space-walks during his long stay at the space station, venturing outside the ISS on May 30 and again on June 6, 2007.

He returned to Earth on October 21, 2007. During his long-duration Expedition 15 flight, Oleg Kotov spent more than 196 days in space, included more than 11 hours in EVA.

KOVALYONOK, VLADIMIR V.

(1942–)

Soviet Cosmonaut

A veteran of three spaceflights, including two long-duration missions, Vladimir Kovalyonok was born in the village of Beloe, in the Minsk region of Belarus, on March 3, 1942. He received his secondary education at the Zachistensk School, where he graduated in 1959 with silver medal honors, and then attended the Balashov Air Force Higher Military Aviation School, from which he graduated in 1963.

Kovalyonok served as an assistant commander and subsequently as commander in the Soviet Air Force Military Transport, in Division 12 of Military Transport Aviation Regiment 374. Assignments for his unit included work as part of the recovery team for Soviet space missions. In 1967, he received the rating of military pilot, second class.

He was chosen for training as a cosmonaut as a member of the Air Force Group Four selection of cosmonaut candidates in 1967. After intensive preparation, he was qualified as a cosmonaut in August 1969.

His initial assignments as a cosmonaut included service as a member of the backup crew for the Soyuz 18/Salyut 4 long-duration flight.

Continuing his education concurrently with his cosmonaut career, Kovalyonok took correspondence courses from the Gagarin Higher Military Academy, graduating in 1976 with a specialty in command and headquarters training.

Kovalyonok made his first flight in space during Soyuz 25, which launched on October 9, 1977. He served as commander for the flight and was accompanied by flight engineer Valeri Ryumin. They were to have become the first long-duration crew of the Salyut 6 space station, which had been launched unmanned on September 29, 1977.

Although their initial attempts at docking with the station seemed to be successful, the docking mechanism of their Soyuz craft was unable to engage the docking module on Salyut 6 fully. They made several attempts to correct the situation, working closely with mission controllers, but the best efforts of the crew and their support teams on Earth failed to rectify the problem, and the docking—and the long-duration mission—had to be abandoned. Kovalyonok and Ryumin landed on October 9, 1977, after a flight of 2 days, 44 minutes, and 45 seconds.

After his frustrating first spaceflight, Kovalyonok served as backup commander during the Soyuz 26/Salyut 6 mission, which successfully achieved the first long-duration Salyut 6 stay. He also served as a member of the backup crew for Soyuz 27, which was the first Soyuz switching flight, in which one spacecraft was flown to a space station and exchanged for the vehicle that was docked at the station. The short Soyuz replacement flights were made necessary by the limitations of the spacecraft's controls and equipment, which tended to deteriorate after 90 days in space.

On June 15, 1978, Kovalyonok lifted off on his second spaceflight, a remarkable long-duration visit to Salyut 6 in which he and flight engineer Aleksandr Ivanchenkov spent four and a half months in space, setting a new record for the longest stay in space up to that time.

The long flight advanced the cause of space-borne astronomical observation, as Kovalyonok and Ivanchenkov used an advanced telescope to make extensive infrared and ultraviolet observations of the Earth's atmosphere, the Moon, Venus, Mars, and Jupiter. They also conducted a rigorous program of other scientific work, including photography of Earth resources and mapping photography.

During their stay, fuel and supplies were delivered to Salyut 6 three times via unmanned Progress supply spacecraft. They also received several visitors, including the crew of Soyuz 30—which included Miroslav Hermaszewski, the first citizen of Poland to fly in space—and Soyuz 31, which featured Sigmund Jaehn, the first German citizen to fly in space. The two visiting missions were part of the Soviet Intercosmos program, which provided opportunities for the citizens of nations aligned with or friendly to the Soviet Union to fly in space. The second crew left Soyuz 31 at the station after their weeklong visit, so Kovalyonok and Ivanchenkov could have a new Soyuz in which to return to Earth at the end of their stay.

On July 29, Kovalyonok and Ivanchenkov ventured outside of Salyut 6 for a spacewalk. Kovalyonok remained in the station's airlock, tracking his crew mate's progress, while Ivanchenkov retrieved scientific experiment packages from the exterior of the station. The two cosmonauts experienced a scary moment when a meteor passed by, blinding them both for several seconds, but neither was injured.

After a flight of 139 days, 14 hours, 47 minutes, and 32 seconds, Kovalyonok and Ivanchenkov returned to Earth on November 2, 1978, aboard Soyuz 31.

Kovalyonok again visited Salyut 6 during his third space mission, which began on March 12, 1981. He and flight engineer Viktor Savinykh launched aboard Soyuz T-4 and spent two and a half months aboard the station as its fourth, and last, crew.

No doubt benefiting from the veteran commander's previous experience, Kovalyonok and Savinykh settled into a productive routine shortly after their arrival at Salyut 6. They conducted scientific experiments, photography, medical, and biological investigations, and space station maintenance and technology experiments, and also performed the regular program of exercise necessary to combat the losses in bone density and blood plasma volume that routinely occur during long-duration stays in space.

They received two visits from Intercosmos crews. In March, they played host to Soviet cosmonaut Vladimir Dzhanibekov and Jugderdemidiyn Gurragcha, the first citizen of Mongolia to fly in space, and in May they were visited by cosmonaut Leonid Popov and Dumitru Prunariu, the first citizen of Romania to fly in space.

In the usual routine for Intercosmos flights, one of the visiting crews would have left their Soyuz craft at the station, but the switching of spacecraft was not necessary during the Soyuz T-4/Salyut 6 mission, however, because at a planned duration of about 75 days, Kovalyonok and Savinykh would return to Earth before the end of the 90-day Soyuz safe operation limit.

On May 26, 1981, Kovalonok and Savinykh concluded their remarkable stay aboard Salyut 6 when they undocked from the station and returned to Earth after a total flight of 74 days, 17 hours, 37 minutes, and 23 seconds.

Kovalyonok received the rating of cosmonaut first-class in 1981.

Vladimir Kovalyonok spent more than 216 days in space during three space-flights.

He left the cosmonaut corps in 1984 after graduating from the K. E. Voroshilov General Headquarters Military Academy of the U.S.S.R. Armed Forces, where he specialized in command and tactical aviation. He graduated with honors and a gold medal for his superior academic performance.

In 1986, Kovalyonok received a Candidate of Military Sciences degree and became a Doctor of Military Science that same year. He has served as a professor at the Chief Military Aviation Technical University since 1992 and is a colonel-general in the Russian Air Force.

KRAFT, CHRISTOPHER C., JR.

(1924–)

First NASA Spaceflight Director; Director of Johnson Space Center (JSC), 1972–1982

A pioneering NASA official who played a key role in the development of Mission Control, served as director of the Johnson Space Center (JSC) in Houston for a decade, and was involved in every U.S. human spaceflight from Project Mercury to the early years of the space shuttle program, Christopher Kraft was born in Phoebus, Virginia on February 28, 1924.

Kraft attended Virginia Polytechnic University, where he graduated with a Bachelor of Science degree in 1944, and then began his career as an aeronautical engineer with the National Advisory Committee for Aeronautics (NACA) at the Langley Aeronautical Laboratory. After NACA was absorbed into the newly formed National Aeronautics and Space Administration (NASA) on October 1, 1958, Kraft joined the NASA Space Task Group, which was placed in charge of planning and carrying out the first manned U.S. space missions during the Mercury Program.

With his Space Task Group colleagues, Kraft was relocated to the newly constructed Manned Spacecraft Center (MSC) in Houston, Texas in 1962. He played an integral role in the design of Mission Control, the complex of equipment and systems that NASA engineers and support personnel use to guide manned space missions.

Kraft was NASA's first flight director, and served in that capacity during all six manned flights of the Mercury program. He was named operations director for

Project Gemini, for which he also continued to serve as a flight director through the Gemini 7 mission in December 1965. At that time, he became director of flight operations for the Apollo program, a capacity in which he served as a primary planner of each flight of the Apollo lunar landing program. Following the completion of the first lunar landings, Kraft was named deputy director of the MSC, and in 1972 he became director of the center, which was re-named the Johnson Space Center (JSC) a year later in honor of President Lyndon B. Johnson.

His influence on the development and progress of the U.S. space program exceeds even his own long, remarkably successful career at NASA, as he was responsible for the selection and training of those flight directors who succeeded him at Mission Control.

Kraft served as JSC director until his retirement from NASA in 1982. He has since put his experience and expertise in aerospace issues to use as a consultant.

In 2006, he and fellow flight director Eugene Kranz were honored by NASA as Ambassadors of Exploration. The award, which incorporates a Moon rock gathered by the Apollo lunar explorers, has been awarded to a small number of pioneering astronauts and officials.

His autobiography, *Flight: My Life in Mission Control,* was published in 2001.

KRANZ, EUGENE F.

(1933–)

*NASA Flight Director; Played Key Role
   in Rescue of Apollo 13 Crew*

Perhaps the best known of all NASA Mission Control personnel, Eugene Kranz played a significant role in the development of the agency's flight control operations during a career that spanned four decades. Kranz was born in Toledo, Ohio on August 17, 1933. He attended Parks College of Saint Louis University, where he graduated in 1954.

After graduation, Kranz joined the U.S. Air Force Reserve and received his initial training as a pilot at Lackland Air Force Base in Texas. He then served in South Korea, where he flew the F-86 Sabre aircraft during the Korean War.

After completing his military service, Kranz began a private sector career with the McDonnell Aircraft Corporation in 1955 as a flight test engineer.

He began his long association with NASA in 1960 as chief of the Flight Control Operations Branch at the agency's Space Task Group at the Langley Research Center in Hampton, Virginia. He coordinated communications between mission managers and NASA tracking stations for all Mercury missions until the program's fourth manned flight, when he was promoted to assistant flight director.

Kranz first served as a flight director for the Gemini 4 mission in June 1965. In that capacity, he was responsible for overseeing one of the teams of mission controllers that monitored every aspect of a spacecraft's flight during a manned mission. He continued to serve as a flight controller for all remaining Gemini flights.

With the advent of the longer, more complex Apollo missions, which required more preparation and involved increased numbers of support personnel, Kranz served as flight director for alternating flights. His Apollo assignments as flight director included Apollo 7 and Apollo 9, the Apollo 11 lunar landing, the harrowing Apollo 13 flight, and the extended lunar expeditions of Apollo 15 and Apollo 17.

Kranz was named chief of NASA's Flight Control division in 1969.

While his participation in Apollo 11 assured his place in history, as he was the flight director on duty when Neil Armstrong and Buzz Aldrin set down on the lunar surface in the lunar module Eagle, it was Kranz's role in responding to the initial events that followed the Apollo 13 accident that led to his enduring fame as the consummate leader of skilled technicians whose devotion to duty directly contributed to the safe return of the Apollo 13 crew.

The events of the ill-fated April 1970 lunar landing attempt by astronauts James Lovell, Fred Haise, and Jack Swigert were dramatized in the 1995 film "Apollo 13," in which Kranz was portrayed by Ed Harris. Kranz also recounted the details of the flight in his 2000 memoir *Failure is Not an Option*. The film and book together greatly expanded public understanding of the flight director's role in manned space missions and the contributions of Mission Control personnel to every successful spaceflight.

Kranz and his team , along with Lovell, Haise, and Swigert were awarded the Presidential Medal of Freedom in recognition of their exceptional performance during the Apollo 13 mission.

In 1974, Kranz became deputy director of Mission Operations. He served in that capacity throughout the Skylab space station program and into the space shuttle era. He became director of Mission Operations in 1983, and in that role was responsible for the development of space shuttle missions from their planning stages to the control of the flights from launch to landing. Among his many achievements during the shuttle years, Kranz was widely recognized for his key role in developing the STS-61 Hubble Space Telescope repair mission in 1993.

Eugene Kranz retired from NASA on March 3, 1994 after serving 34 years with the space agency. In 2006, he and fellow flight director Christopher Kraft were honored by NASA as Ambassadors of Exploration, and each received an award incorporating a sample from an actual Moon rock, in recognition of their remarkable contributions during the formative years of the U.S. space program.

KREGEL, KEVIN R.

(1956–)

U.S. Astronaut

Kevin Kregel was born on September 16, 1956 and grew up in Amityville, New York. He graduated from Amityville Memorial High School in 1974 and attended the U.S. Air Force Academy, where he received a Bachelor of Science degree in astronautical engineering in 1978.

In 1979, he earned his pilot wings at Williams Air Force Base in Arizona and was subsequently assigned as an exchange officer to serve with the British Royal Air Force (RAF). He later flew A-6E aircraft with the U.S. Navy at Naval Air Station Whidbey Island near Seattle, Washington and made 66 aircraft carrier landings while deployed to the Western Pacific aboard the USS *Kitty Hawk*.

He attended the U.S. Naval Test Pilot School at Patuxent River, Maryland and then served as a test pilot at Eglin Air Force Base in Florida, testing weapons and electronics systems in F-111 and F-15 aircraft.

Continuing his education concurrently with his military service, Kregel attended Troy State University, from which he received a Master's degree in public administration in 1988. In 1990, he resigned from active military service to pursue a career at NASA. During his career as a pilot and test pilot he has accumulated more than 5,000 hours of flying time in 30 different aircraft.

He began his NASA career at Ellington Field in Houston, Texas in April 1990. He served as an instructor pilot in the space shuttle training aircraft and piloted the first flight test of the T-38 avionics upgrade aircraft. In 1992, NASA selected him for training as an astronaut.

Upon completion of his astronaut training, he served the space agency in a variety of roles, including deputy of the space station branch of the Astronaut Office. He has been a member of the astronaut support team at Kennedy Space Center (KSC) in Florida and spacecraft communicator (Capcom) in Mission Control at the Johnson Space Center (JSC) in Houston. He has also worked on the agency's Orbital Space Plane Project.

Kregel first flew in space as pilot of the shuttle Discovery, during STS-70 in July 1995. The STS-70 crew deployed the sixth Tracking and Data Relay Satellite, completing the network of TDRS advanced tracking and communications satellites which began with the deployment of TDRS-1 during STS-6 in April 1983. Discovery made 142 orbits during its mission, landing on July 22, 1995.

During STS-78, which launched aboard Columbia on June 20, 1996, Kregel again served as pilot, during a landmark mission of international cooperation in space.

The primary payload for the flight, the Life and Microgravity Spacelab (LMS), was the work of scientists from 10 countries and the personnel of five space agencies, including NASA, the European Space Agency (ESA), the French space agency Centre National d'Études Spatiales (CNES), the Canadian Space Agency (CSA), and the Italian space agency Agenzia Spaziale Italiana (ASI). The crew conducted more than 40 LMS experiments during the flight, which at nearly 17 days was the longest shuttle flight up to that time.

In addition to the scientific work done by the crew on the shuttle, the mission also achieved a thorough testing of the kind of communications systems necessary to link researchers around the world to astronauts in space, as is necessary for the International Space Station (ISS). Columbia landed July 7, 1996 after 271 orbits.

Kregel next flew in space as commander of Columbia, during STS-87 in 1997. The fourth flight of the U.S. Microgravity Payload, the 16-day mission featured around the clock research by alternating teams of astronauts conducting experiments in the impact of weightlessness and studying the outer atmosphere of the Sun. The crew completed 80 experiments during the flight, and mission specialists Winston Scott and National Space Development Agency (NASDA) astronaut Takao Doi of Japan made a 7 hour and 43-minute spacewalk to manually retrieve the malfunctioning SPARTAN-201–04 free-flying satellite. The SPARTAN craft had been released three days earlier by mission specialist Kalpana Chawla using the shuttle's Remote Manipulator System (RMS). Columbia landed on December 5, 1997, at KSC, after 252 orbits.

On February 11, 2000, Kregel began his fourth space mission as commander of an international crew aboard the shuttle Endeavour. With fellow STS-99 crew mates pilot Dominic Gorie and mission specialists Janice Voss, Janet Kavandi, ESA astronaut Gerhardt Thiele of Germany, and NASDA astronaut Mamoru Mohri of Japan, Kregel spent 11 days in space, traveling more than four million miles in 181 orbits while using radar instruments to map the surface of the Earth.

Also known as the Shuttle Radar Topography Mission, STS-99 was conceived and designed by the National Imagery and Mapping Agency, NASA, and the German Aerospace Center (DLR). Using a high-resolution radar mapping system specially designed for the flight, the crew was able to capture high quality, detailed, three-dimensional images of the surface of the Earth from 60 degrees north latitude to 56 degrees south latitude, an area populated by about 95 percent of the world's peoples.

The 13-ton radar mapping instrument was deployed on the first day of the flight at an orbital altitude of 145 miles. Its antennae were mounted in the shuttle's payload bay on a 200-foot mast. It was operated continuously for more than 220 hours while the crew worked around the clock in alternating shifts. Endeavour returned to Earth on February 22, 2000.

Kevin Kregel spent more than 52 days in space during his four space shuttle missions.

In December 2003 he left NASA to pursue a private sector career.

KRIKALYOV, SERGEI K.

(1958–)

Russian Cosmonaut; Set New Record in 2005
*   for Longest Career Time Spent in Space*

On August 16, 2005, 11 days before his 47th birthday, Sergei Krikalyov became the one person to have spent more total time in space than any other individual.

Krikalyov was born in St. Petersburg, which was then known as Leningrad, Russia on August 27, 1958. He attended the Leningrad Mechanical Institute (since re-named St. Petersburg Technical University), where he received a degree in mechanical engineering in 1981.

He began his career at NPO Energia, where he helped develop and test equipment and techniques for spaceflight and worked in mission control.

Sergei Krikalyov has spent more than 800 days in space—more than any other individual—and has amassed a career total of 36 hours of spacewalking time during six space missions. [NASA/ courtesy of nasaimages.org]

Concurrently with his engineering work, he became a member of the Soviet national aerobatic flying team, winning the title of Champion of Moscow in 1983 and, in 1986, Champion of the Soviet Union.

Krikalyov's first opportunity to demonstrate the exceptional qualities that would prove the hallmark of his long career as a cosmonaut came in 1985, when the Salyut 7 space station suddenly and inexplicably failed. The station was unoccupied at the time when contact was lost, on February 5, so ground controllers were faced with the dual problems of having too little information about the nature of the failure and a pressing need to figure out how to plan a repair mission that would succeed while not endangering its crew.

As a member of the engineering team assigned to devise a salvage mission to the out of control station, which was, by this time, tumbling blindly through space, Krikalyov helped to develop the rendezvous, docking, and repair methods used by Vladimir Dzhanibekov and Viktor Savinykh during their Soyuz T-13 flight in June 1985. The meticulous planning paid off when Dzhanibekov and Savinykh were able to dock with Salyut 7, enter the station, and find the source of the problem, a faulty instrument designed to monitor the amount of power in the station's solar-powered batteries. Re-routing the connection between the batteries and the solar panels to bypass the failed instrument, the cosmonauts were able to revive Salyut 7, which remained in service for the rest of 1985.

In addition to his work with the Salyut 7 rescue team, Krikalyov was that same year selected for cosmonaut training. He was briefly assigned to the Buran space shuttle program, which was later canceled because of cost considerations.

Krikalyov first flew in space aboard Soyuz TM-7, launched on November 26, 1988 with fellow cosmonaut and mission commander Alexander Volkov and Jean-Loup Chrétien of France. In 1982, Chrétien had been the first French citizen in space; during his stay aboard Mir with Volkov and Krikalyov, he made the first spacewalk by a citizen from a country other than the United States or the Soviet Union (he and Volkov spent six hours in EVA while deploying an experimental structure on the outside of the station).

For the first 25 days of their long-duration stay aboard Mir, Krikalyov, Volkov, and Chrétien were accompanied by the previous long-duration crew of Vladimir Titov, Musa Manarov, and Valeri Polyakov. The nearly month-long overlap set a record for the longest time in orbit for a six-person crew.

Chrétien, Titov, and Manarov returned to Earth on December 21; the two Soviet cosmonauts had been the first people to spend an entire year in space. Krikalyov and Volkov remained aboard Mir with Polyakov, conducting scientific research while living and working on the station for the next five months, until their return to Earth aboard Soyuz TM-7 on April 27, 1989.

When they left Mir at the end of their mission, Krikalyov, Volkov, and Polyakov were not greeted by a replacement crew, as had been the norm during the station's first three years of operation; and when they returned to Earth, they found a rapidly changing Soviet Union that would soon pass into history.

During their time in space, nascent democracy had taken hold in the country, under the aegis of Communist Party General Secretary Mikhail Gorbachev. The resulting political and economic changes had a vast and lasting positive impact on Russian society and on the former Soviet bloc nations, but in the far narrower case of the Russian space program, uncertainty born of the changes caused a series of cutbacks in funding and ambition. Thus began a long period of unoccupied flight for Mir, as future missions were reconsidered.

The contrast between the heroic first era of the Soviet program and the disarray inherent in the late 1980s and early 1990s was evident in a humorous sidelight of Krikalyov's first trip to Mir. As a member of the Soviet Army reserve, Krikalyov was subject to be called to duty at any time—which he was, during his long stay aboard the space station. The army persisted with its summons until higher officials intervened, saving both the cosmonaut and the military from further embarrassment. Still, the incident made it obvious that, at a time when virtually every citizen in the nation knew who he was and what he was doing, there were still government officials who were entirely unaware of the activities of one of their own space heroes.

Earthly concerns aside, Krikalyov remained focused on his cosmonaut duties and achieved a brilliant second flight as a member of the Mir 9 crew. Launched on May 19, 1991 aboard Soyuz TM-12 with mission commander Anatoli Artsebarsky and Helen Sharman, the first British citizen to fly in space, Krikalyov left the Soviet Union for a mission of more than 300 days in space. Sharman returned to Earth a week later with cosmonauts Viktor Afanasyev and Musa Manarov aboard Soyuz TM-11, while Krikalyov and Artsebarsky began their epic stay.

The two cosmonauts made six long spacewalks during their Mir 9 mission. They replaced an antenna during their first EVA, which took 4 hours and 48 minutes, on June 24, and on June 28 they deployed an experiment on the outside of the station during a 3.5-hour spacewalk. Their next four EVAs were devoted to construction tasks; each of the first three, on July 15, July 19, and July 23, lasted for more than 5 hours, and their final EVA, on July 27, 1991, was 6 hours and 49 minutes.

In their six trips outside the space station, Krikalyov and Artsebarsky amassed a total of 31 hours and 48 minutes in EVA.

The ongoing sea change in Soviet society contributed to impact the future course of the space program, even as the cosmonauts continued their work in orbit. Economic constraints led mission planners to alter flight plans to accommodate guests from other countries, and in the case of the next-planned visits to Mir, Soviet space officials decided to send two paying customers with one cosmonaut.

The result for Krikalyov was an opportunity to extend his stay aboard Mir for an additional six months and to continue to serve as flight engineer for a new commander. Having accepted the assignment, he welcomed Alexander Volkov, who had traveled to the station aboard Soyuz TM-13 with Franz Viehbok of Austria and Toktar Aubakirov of Kazakhstan on October 4. A week later, he bid farewell to Artsebarsky, who returned to Earth aboard Soyuz TM-12 with Viehbok and Aubakirov.

As a member of the Mir 10 crew, Krikalyov made another EVA (with Volkov, on February 20, 1992), to perform maintenance tasks on the station, adding another 4 hours and 12 minutes to his spacewalking total.

Krikalyov and Volkov remained on Mir until their return to Earth in March 1992, aboard Soyuz TM-13. They were accompanied on the trip home by Klaus-Dietrich Flade of Germany.

When he touched down on March 25, 1992, Krikalyov emerged from his Soyuz berth to take his first steps in the independent state of Kazakhstan, the Soviet Union having finally disintegrated during the 311 days, 20 hours, and 1 minute he had been in space.

Although the initial uncertainties surrounding the collapse of the totalitarian Soviet Union would continue to impact the immediate future of the Mir space station and near-term planning of Soyuz flights, the Russian space program would soon reemerge as a source of pride and inspiration for the Russian people. Krikalyov continued at the forefront of the program's progress as a member of the STS-60 flight of the space shuttle Discovery, the first joint shuttle mission of the United States and Russia.

The STS-60 mission began on February 3, 1994. Krikalyov made extensive use of the shuttle's robotic arm, the Remote Manipulator System (RMS), as a mission specialist during the flight, which came to a close with Discovery's landing at the Kennedy Space Center (KSC) in Florida on February 11, 1994. He added 8 days, 7 hours, and 9 minutes to his total time in space during the historic mission.

For several years after his STS-60 flight, Krikalyov split his time between Russia and the United States as the early days of the new era of cooperative missions began to take hold between the former enemies. He played an instrumental role in supporting the two nations' joint space missions.

In December 1998, Krikalyov returned to spaceflight, lifting off aboard the space shuttle Endeavour on December 4 for STS-88, the first assembly mission for the International Space Station (ISS). The Endeavour crew linked the ISS Unity and Zarya modules during the flight, thereby completing a crucial construction task for the station. Endeavour returned to Earth on December 15, 1998, after 11 days, 19 hours, and 18 minutes in space.

Krikalyov's next space mission brought him back to the ISS, this time as a member of the long-duration ISS Expedition 1 crew. Launched from the Baikonur Cosmodrome in Kazakhstan on October 31, 2000 with fellow cosmonaut Yuri Gidzenko and Expedition 1 commander William Shepherd of the United States aboard Soyuz TM-31, Krikalyov logged another 140 days, 23 hours, and 40 minutes in space during his first stay on the ISS. The Expedition 1 crew returned to

Earth aboard the space shuttle Discovery at the end of its STS-102 flight, which landed at KSC in Florida on March 21, 2001.

The veteran cosmonaut whose career has mirrored the remarkable changes of his time began his sixth mission in space on April 15, 2005 as commander of ISS Expedition 11. Lifting off aboard Soyuz TMA-6 with fellow Expedition 11 crew mate John Phillips of NASA and European Space Agency (ESA) astronaut Roberto Vittori of Italy (who returned to Earth on April 24 with the Expedition 10 crew), Krikalyov spent six months on the space station.

On August 16, 2005 Krikalyov's career-long total time in space surpassed the previous high mark of 747 days, 14 hours, 14 minutes, and 11 seconds held by cosmonaut Sergei Avdeyev. When his ISS Expedition 11 flight came to a close two months later on October 11, 2005, Krikalyov had amassed a career total time in space of 803 days, 9 hours, and 40 minutes, including 36 hours in EVA.

Sergei Krikalyov has lived in space for more than two years of his life. His remarkable career is a living testament to the ability of both nations and individuals to surmount the tides of historic change to achieve feats of lasting significance.

KUBASOV, VALERI N.

(1935–)
Soviet Cosmonaut

A veteran of three spaceflights, including the first cooperative space mission undertaken by the Soviet Union and the United States, Valeri Kubasov was born in Vyazniki, Russia on January 7, 1935. He attended the Moscow Aviation Institute, where he earned a Candidate of Technical Sciences degree in aeronautics and astronautics in 1958.

He was chosen for training as a cosmonaut as a member of the civilian specialist group two selection of cosmonaut candidates in 1966. His initial assignments as a cosmonaut included service as a member of the backup crew for the Soyuz 5 flight, which took place in January 1969.

Kubasov first flew in space as flight engineer for Soyuz 6, which lifted off from the Baikonur Cosmodrome on October 11, 1969. Launched with commander Georgi Shonin during the first of three Soyuz launchings on consecutive days, Kubasov carried out innovative welding experiments with a Vulkan smelting furnace during his first flight in space.

An impressive achievement for the Soviet space program, the group flight of Soyuz 6, Soyuz 7, and Soyuz 8 simultaneously placed seven cosmonauts in orbit at a time when the Soviets were searching for ways to boost morale after the United States landed the first astronauts on the Moon in July 1969. As the three Soyuz vehicles launched in October, the Americans were preparing Apollo 12, which would carry out the second manned lunar landing in November.

The simultaneous launch and control of three spacecraft also contributed to the Soviets' later goals in space, which included the orbiting of the Salyut series of space stations.

On October 15, Kubasov carried out his assigned welding experiments, and he and Shonin then maneuvered their craft into position to observe rendezvous exercises carried out by Soyuz 7 and Soyuz 8 before returning to Earth on October 16, 1969.

During his first flight, Kubasov spent 4 days, 22 hours, and 43 minutes in space.

He next served as a member of the crew that was originally assigned to the Soyuz 11/Salyut 1 mission. With crew mates Aleksei Leonov and Pyotr Kolodin, he underwent extensive training for the flight, and after the Soyuz 10 crew was unable to board Salyut 1 during their attempt in April 1971, Kubasov and his crew mates looked forward to being the first individuals in history to occupy a space station in orbit.

While he was undergoing the extensive medical examinations that routinely precede any planned spaceflight, however, Kubasov was found to be suffering from a lung ailment that was considered serious enough for him to be removed from the Soyuz 11 prime crew. Faced with the sudden loss of their primary flight engineer, Soviet mission planners initially considered replacing him with backup crew member Vladislav Volkov and letting the flight proceed, but then, in a decision whose tragic consequences they could not at the time have foreseen, they instead replaced the entire crew with backup crew members Georgy Dobrovolsky, flight engineer Volkov, and research engineer Viktor Patsayev.

The three backup crew members performed expertly during their flight, successfully occupying Salyut 1 for more than three weeks and achieving the historic milestone of being the first crew to occupy a space station in orbit. Sadly, Dobrovolsky, Volkov, and Patsayev were killed when their Soyuz spacecraft malfunctioned during their return to Earth.

After his tumultuous assignment to and removal from the Soyuz 11/Salyut 1 mission, Kubasov received medical treatment for the ailment that had forced him off the flight and returned to active duty status as a cosmonaut.

Kubasov made his second flight in space as flight engineer for the milestone Soyuz 19 flight—the Soviet half of the joint Soviet-American Apollo-Soyuz Test Project—in July 1975.

Launched on July 15 with commander Alexei Leonov, Soyuz 19 met and docked with an Apollo craft carrying U.S. ASTP crew members Tom Stafford, Donald "Deke" Slayton, and Vance Brand on July 17.

Following years of intense competition between the Soviet and U.S. space programs, the ASTP mission had begun with an agreement signed by Soviet leader Leonid Brezhnev and U.S. President Richard Nixon in 1972. The two crews trained together in the United States and the Soviet Union during their long preparations for the flight, and the engineers, scientists, and administrators of each nation's space program shared information and expertise with each other in an unprecedented fashion while preparing the systems and equipment that would be used by the crew members while in orbit.

Both the Soyuz 19 and the American Apollo launched on July 15. Kubasov and Leonov spent the early portion of their flight making repairs to a television camera, which would be used to televise much of the historic flight live.

After successfully linking the two craft on July 17, the crew members visited each other's quarters and exchanged gifts that exemplified the hopeful nature of the mission, such as native seeds that each crew returned to their own country for later planting. Each crew also received congratulatory calls from their respective political leaders, and all five crew members participated in a news conference that was aired live in both the Soviet Union and the United States. The two spacecraft remained docked for a total of 47 hours.

Kubasov and Leonov returned to Earth on July 21, 1975, after a flight of 5 days, 22 hours, and 31 minutes in space.

After his historic second flight, Kubasov served as the backup commander of the Soyuz 30 Intercosmos flight. A result of the need to replace Soyuz spacecraft docked at a space station for an extended period of time with a fresh Soyuz before the docked vehicle exceeded its safe operations limit, the Intercosmos Program made use of the short Soyuz switching flights to provide spaceflight opportunities for citizens of nations aligned with or friendly to the Soviet Union. In the case of Soyuz 30, which was commanded by Pyotr Klimuk, Miroslav Heremaszewski became the first citizen of Poland to fly in space. Kubasov and Zenon Jankowski served as the backup crew for the flight.

On May 26, 1980, Kubasov began his third flight in space as commander of the Soyuz 36 Intercosmos mission. Bertalan Farkas, the first citizen of Hungary to fly in space, accompanied him.

Kubasov and Farkas traveled to the Salyut 6 space station, where they visited with the station's long-duration crew members Leonid Popov and Valeri Ryumin, who were in the midst of a stay of 184 days at Salyut 6. After a visit of about a week, Kubasov and Farkas accomplished the Soyuz switching objective of their mission profile when they returned to Earth on June 3, 1980 in Soyuz 35, leaving their fresh Soyuz 36 vehicle at Salyut 6 for Popov and Ryumin.

Valeri Kubasov spent more than 18 days in space during his three space missions.

He continued to serve in the Soviet and Russian space programs after his third spaceflight, and helped to develop the Mir space station. He retired from the cosmonaut corps in 1993, but continued to work for NPO Energia, where he subsequently served as deputy director.

L

LAVEYKIN, ALEKSANDR I.

(1951–)

Soviet cosmonaut

A veteran of a long-duration spaceflight during which he spent more than five months in space aboard the Mir space station, Aleksandr Laveykin was born in Moscow, Russia on April 21, 1951. He attended the Bauman Higher Technical School in Moscow, where he graduated in 1974, and then began his career at NPO Energia.

Laveykin was chosen for training as a cosmonaut in 1978 as a member of the civilian specialist group six selection of cosmonaut candidates.

On February 5, 1987, he launched with Commander Yuri Romanenko from the Baikonur Cosmodrome as flight engineer for the Soyuz TM-2/Mir 2 long-duration mission. Laveykin and Romanenko had originally been assigned as the backup crew for the flight, but primary crew member Aleksandr Serebrov had developed a medical problem during training, and he and his crew mate Vladimir Titov had been forced to give up the assignment.

Aboard Mir, Laveykin and Romanenko settled into a productive routine, working on their assigned program of scientific experiments, carrying out engineering and space station maintenance tasks, and regularly exercising to combat the losses in bone density and blood plasma volume that routinely occur during long stays in space.

About two months into their stay aboard the station, the cosmonauts faced their first major challenge when the Kvant 1 module arrived at Mir. Launched unmanned on March 31, 1987, Kvant 1 was equipped with X-ray telescopes that had been designed and built by the European Space Agency (ESA), Great Britain, The Netherlands, and West Germany. Although the module was designed to be automatically docked with the station under the control of mission managers on Earth, two attempts at docking failed, and Mission Control decided that their best chance for fixing the problem would be a hands-on approach, with Laveykin and Romanenko making a spacewalk to inspect the docking mechanism and remedy any difficulty they might find.

The cosmonauts carried out their impromptu repair mission on April 11. During a 3 hour and 35-minute excursion outside the station, they found a most unlikely source for the docking failures: a piece of plastic that had been wedged into Mir's rear docking port. Later investigation revealed that the plastic had been inadvertently left during a visit by an unmanned Progress resupply spacecraft that had been improperly closed out before it undocked.

Laveykin and his crew mate were able to strip away the plastic, and Kvant 1 docked successfully on April 12. The cosmonauts also made two other EVAs during their stay to install a third solar panel on Mir's solar arrays. They accomplished this task on June 12 and June 16, adding another hour and 53 minutes to their EVA total during the first excursion and another 3 hours and 15 minutes during their second.

Just as things seemed to settle down after the Kvant 1 installation and the two solar panel EVAs, Laveykin was the recipient of bad news from the ground: Soviet medical personnel monitoring his health found indications that he had developed a heart ailment. After due consideration, the officials responsible for the safety of the crew decided that their wisest course of action would be to bring Laveykin back to Earth as quickly as possible, and they arranged for him to return with the Soyuz TM-3 crew, which was scheduled to visit Mir in late July.

Soviet cosmonauts Aleksandr Viktorenko and Aleksandr Aleksandrov arrived at the station in Soyuz TM-3 with guest cosmonaut Muhammad Faris, the first citizen of Syria to fly in space. Laveykin joined Viktorenko and Faris aboard Soyuz TM-2 for his return to Earth, while Aleksandrov transferred to Mir to replace him and to serve out the rest of the long-duration flight with Romanenko.

Fortunately, Laveykin was subsequently found to be free of serious heart difficulties.

Aleksandr Laveykin spent more than 174 days in space, including over 8 hours in EVA during three spacewalks, during his long-duration flight.

In 1989, Laveykin became deputy commander of civilian specialist cosmonauts at NPO Energia, a position in which he served until his retirement in 1994.

LAWRENCE, ROBERT H., JR.

(1935–1967)

U.S. Astronaut

One of the pioneering astronauts chosen for training in the U.S. Air Force Manned Orbiting Laboratory (MOL) program and the first African American astronaut,

Robert Lawrence was born in Chicago, Illinois on October 2, 1935. In 1952, he graduated in the top 10 percent of his class at Englewood High School in Chicago when he was only 16, and he then attended Bradley University, where he earned a Bachelor of Science degree in chemistry in 1956.

During his undergraduate years, Lawrence participated in the U.S. Air Force ROTC program at Bradley and served as a cadet commander in the program. Upon graduation, he was commissioned in the Air Force reserve as a second lieutenant.

He received his initial training as a pilot at Malden Air Force Base and subsequently served as an instructor pilot and as a test pilot.

Continuing his education concurrently with his military service, he attended Ohio State University, where he earned a Doctorate degree in physical chemistry in 1965.

Lawrence was next chosen to attend the Air Force Test Pilot School at Edwards Air Force Base in California, where he graduated in 1967. Upon graduation, he was selected by the Air Force as a member of its third group of MOL astronauts.

During his outstanding military career, Lawrence accumulated more than 2,500 hours of flying time, including 2,000 hours in jet aircraft. He rose to the rank of major in the United States Air Force, and had earned the Air Force Commendation Medal and the Air Force Outstanding Unit Citation by the time of his tragic death in an airplane crash in December 1967.

Air Force officials had developed the MOL program as a series of small space stations that would enable specially trained members of the U.S. military to conduct surveillance from space. Program managers envisioned MOL missions as 30-day flights with crews of two astronauts traveling into orbit in a modified Gemini spacecraft outfitted with advanced photographic and radar equipment for conducting surveillance.

First announced in December 1963, the MOL program was in development for most of the 1960s. Three groups of astronauts were selected for the proposed spaceflights, and a single, unmanned test flight was carried out on November 3, 1966.

Robert Lawrence was killed on December 8, 1967 in the crash of an F-104 Starfighter aircraft, in which he was serving as an instructor pilot for a pilot trainee. He had previously flown many tests in the F-104 Starfighter himself, honing the gliding technique he was teaching at the time of the crash. He was 32 years old at the time of his death.

The MOL program ultimately ran into financial and technical difficulties and was canceled in June 1969, but the technology originally developed for the project was later adapted for use in unmanned military spy satellites. NASA invited those Air Force astronauts who were under the age of 35 to transfer to the civilian agency, and in August 1969 announced its Group Seven selection of seven MOL astronauts. Given his outstanding service in the Air Force and his age eligibility, Lawrence would have been eligible to join NASA with his fellow MOL astronauts, had he lived.

Robert Lawrence was survived by his wife, Barbara. In 1997, NASA honored him by adding his name to the Kennedy Space Center Astronaut Space Mirror Memorial.

LAWRENCE, WENDY B.

(1959–)

U.S. Astronaut

Wendy Lawrence was born in Jacksonville, Florida on July 2, 1959. In 1977, she graduated from Fort Hunt High School in Alexandria, Virginia and then attended the United States Naval Academy, where she received a Bachelor of Science degree in ocean engineering in 1981.

She was recognized as a distinguished graduate of her class when she completed her initial flight training and was designated a naval aviator in 1982. She subsequently served as one of the first two female helicopter pilots to be deployed to the Indian Ocean for an extended period as part of a carrier battle group, while stationed at Helicopter Combat Support Squadron Six (HC-6). In 1986, she received the National Navy League's Captain Winifred Collins Award for inspirational leadership.

Continuing her education concurrently with her military career, Lawrence pursued her graduate degree in the joint program of the Massachusetts Institute of Technology and the Woods Hole Oceanographic Institution. She received a Master of Science degree in ocean engineering in 1988, and was subsequently assigned as officer-in-charge of Detachment ALFA of the Navy's Helicopter Anti-Submarine Squadron Light Thirty (HSL-30).

In 1990, Lawrence returned to the U.S. Naval Academy as a physics instructor. She also served as the novice women's crew coach, and remained at the academy until NASA selected her for astronaut training in March 1992.

During her exceptional military career, Lawrence has accumulated over 1,500 hours of flying time in six different types of helicopters, and she has made more than 800 landings on ships. She is a captain in the U.S. Navy and a member of Women Military Aviators and the Naval Helicopter Association.

Her NASA assignments have included service at the Gagarin Cosmonaut Training Center (GCTC) in Star City, Russia, where she served as assistant training officer, and she has served as the Astronaut Office representative for space station training and crew support.

She first flew in space as a mission specialist during STS-67 aboard the space shuttle Endeavour. The second flight of the Astro observatory, STS-67 launched from the Kennedy Space Center (KSC) in Florida in the early morning hours of March 2, 1995. During the flight, the crew used the three Astro-2 telescopes to make around the clock ultraviolet observations while making studies of the far ultraviolet spectra of faint astronomical objects and the ultraviolet light emitted by hot stars and distant galaxies. Endeavour returned to Earth with a landing at Edwards Air Force Base in California on March 18, 1995, after a flight of 16 days, the longest shuttle mission up to that time.

Lawrence made her second flight in space in 1997. Launched on September 25 aboard the space shuttle Atlantis, the STS-86 mission was the seventh docking of a shuttle and the Russian space station Mir.

The STS-86 crew delivered NASA astronaut David Wolf to Mir, where he replaced Michael Foale to become the sixth American astronaut to make a long-duration visit to the station as part of a U.S.-Russian cooperative program of continual presence aboard Mir. Foale returned to Earth with the STS-86 crew after having spent 134 days aboard the station.

Atlantis docked with the station on September 27. Mission specialists Scott Parazynski of NASA and Vladimir Titov of the Russian Space Agency conducted the first joint American-Russian EVA during a shuttle mission during a repair to the station's Spektr module, which had been damaged during an earlier accident with a Progress cargo ship.

In six days of linked flight, the STS-86 and Mir 24 crews transferred more than four tons of supplies and equipment from Atlantis to Mir. When the two spacecraft undocked, the shuttle conducted a flyaround of the station while they inspected Mir for damage. Atlantis returned to Earth on October 6, 1997.

Lawrence returned to Mir the following year, this time as a mission specialist aboard the space shuttle Discovery during STS-91. The ninth and final docking of a shuttle and the Russian space station, the mission marked the successful completion of the first phase of the joint American-Russian cooperative space program. The Mir docking was the first for Discovery.

During four days of docked flight, the STS-91 astronauts worked with the Mir 25 crew to transfer supplies and equipment to the station and to move long-term U.S. experiments and equipment into the shuttle for the return trip to Earth.

The Discovery crew also collected NASA astronaut Andy Thomas, who had lived aboard Mir for 130 days. Thomas was the last of seven U.S. astronauts to make a long-duration visit to Mir during 907 consecutive days of continual American presence aboard the Russian station. Discovery returned to Earth on June 12, 1998 after 154 orbits.

On July 26, 2005, Lawrence began a remarkable fourth spaceflight, as a mission specialist aboard Discovery for the STS-114 return-to-flight mission.

Following the tragic loss of the shuttle Columbia in February 2003 and the long, painful recovery period after the accident, the STS-114 flight was intended to prove that NASA had solved the problems that had resulted in the loss of the Columbia crew. As the first flight after the accident, and in light of subsequent events, STS-114 turned out to be a remarkable testament to the courage of its crew and a good example of just how difficult it is to eliminate the risks involved in spaceflight.

The commander of STS-114 was Eileen Collins, and James Kelly served as pilot of Discovery during the flight. Lawrence's fellow mission specialists were Charles Camarda, Steve Robinson, Andy Thomas, and Japan Aerospace Exploration Agency (JAXA) astronaut Soichi Noguchi. The crew visited the International Space Station (ISS), where they worked with the ISS Expedition 11 crew, Sergei Krikalyov and John Phillips, and used the Raffaello Multi-Purpose Logistics Module (MPLM) to deliver more than 11,000 pounds of equipment and supplies to the station during nine days of docked operations. Robinson and Noguchi performed three EVAs while testing repair techniques on the outside of the shuttle and making long-delayed repairs to the ISS.

Magnificent in its achievement, STS-114, would in normal times, have been lauded as simply another successful flight in the long string of NASA successes; coming as it did as the first flight after the two and a half year recovery from the Columbia disaster, however, it quickly turned into a national drama. The crew's two weeks in space were watched with nervous tension by observers on Earth who feared the worst, that the same damage that had doomed Columbia might have befallen Discovery during its lift-off on July 26.

The elaborate system of cameras and data acquisition instruments that re-corded the launch in great detail plainly showed a chunk of foam insulation flying past the shuttle as it left the launch pad, providing clear evidence that more than two years' worth of work to eliminate the chance of errant insulation hitting the shuttle, as had occurred with Columbia, had not solved the problem. Although there seemed to be no direct hit, and therefore no damage to Discovery, the failure was enough for NASA officials to decide to ground all future shuttle flights until a better solution could be found.

Poor weather in Florida forced the scheduled landing to be postponed for one day, adding to the apprehension surrounding the end of the flight. The landing then had to be diverted to Edwards Air Force Base in California, which raised a multitude of new concerns on the part of a wary public. In response to worries about the shuttle's flight path, NASA officials altered their plans so Discovery would not fly directly over Los Angeles; and where past landings at Edwards were open to the public, post-9/11 security procedures sharply limited public access to the base during the STS-114 landing.

In spite of all the worries, Discovery landed safely in the early morning hours of August 9, 2005, and the nation breathed a sigh of relief as the superb return-to-flight mission demonstrated the astronauts' courage and skill under enormous pressure.

Wendy Lawrence has spent more than 1,225 hours in space during four space missions.

LAZAREV, VASSILY G.

(1928–1990)

Soviet Cosmonaut

A veteran of two space missions, including the first manned launch abort in the history of space exploration, Vassily Lazarev was born in the village of Poroshino, Russia on February 23, 1928. He attended the Sverdlovsk Medical Institute, where he graduated in 1951 with a specialization in surgery.

In 1952, he completed further medical studies at the Saratov Medical Institute and was then assigned to Detached Air Base Service Battalion 336 of the Soviet Air

Force Army 30, as a medical specialist. Later in 1952, he became a Hospital Master in Service Battalion 343.

Lazarev then attended the Kharkov Higher Air Force School, where he graduated in 1954 after being trained as a fighter pilot. He was an instructor pilot in Air Regiment 810 from December 1954 to January 1956, and then became a test pilot at the Air Force State Red Banner Scientific Research Institute.

He became a senior test pilot in 1957, and subsequently also served as a hygienist and senior test pilot doctor. From 1964 to 1966, he was a senior research assistant and pilot doctor at the State Scientific Research Test Institute for Aviation and Space Medicine.

Lazarev was initially chosen for training as a cosmonaut as a member of the Voskhod Medical Group One selection of cosmonaut candidates in May 1964, with the intention of his serving as a flight surgeon for a Voskhod crew.

His initial duties as a cosmonaut included service as the backup cosmonaut to Boris Yegorov, who served as flight surgeon for the Voskhod 1 flight in October 1964.

After the final Voskhod flight in March 1965, Lazarev was transferred to the Air Force Group Three selection of cosmonaut candidates in 1966 and underwent additional training.

Lazarev first flew in space as commander of Soyuz 12, the first manned test of the Soyuz spacecraft after the tragic Soyuz 11 accident, which had resulted in the deaths of three cosmonauts in June 1971.

Launched on September 27, 1973 with Lazarev and with Oleg Makarov serving as flight engineer, the Soyuz 12 flight was a test of the Soyuz systems and equipment that had been thoroughly redesigned in the aftermath of the Soyuz 11 accident. The spacecraft and its systems, equipment, and procedures had been minutely studied and re-worked to better ensure the safety of future crews and the success of future missions, which would soon include flights to the Salyut series of space stations.

The short Soyuz 12 flight successfully returned the Soviet space program to manned spaceflight. Lazarev and Makarov landed safely on September 29, 1973, after a flight of 1 day, 23 hours, 15 minutes, and 32 seconds.

Lazarev received the rating of cosmonaut, third class in 1973.

After completing his first space mission, Lazarev served as backup commander for the Soyuz 17/Salyut 4 flight in 1975.

He was to have made his second flight in space, again with Oleg Makarov, on April 5, 1975. Lazarev and Makarov were set to launch that day aboard what was then known as Soyuz 18 (and which was later designated Soyuz 18-1) to travel to the Salyut 4 station for a stay of about two months.

An electrical fault in their launch vehicle prevented the stages of the rocket that were to have propelled them into space from separating properly, and mission controllers were forced to engage the launch escape system, which separated the capsule containing the cosmonauts from the rest of the launch vehicle. Lazarev and Makarov were shot away from their launch vehicle in a long, fast arc that rose to an altitude of 180 kilometers. They traveled about 900 miles from the launch site at the Baikonur

Cosmodrome and landed near the border between the Soviet Union and China. The brief, harrowing flight ended with the cosmonauts' space capsule wedged in the branches of a tree on the side of a mountain; Lazarev and Makarov were both injured during the rough 20-minute flight, and recovering them was a difficult task.

As the first individuals ever to experience the abort of a manned spaceflight, Lazarev and Makarov were fortunate to be alive.

Vassily Lazarev spent just under two days in space during his two space missions.

Lazarev remained a member of the cosmonaut corps for more than a decade after the 1975 launch abort, but did not fly in space again. He was trained for a mission in the Soyuz T-series of spacecraft and served as a backup crew member for several missions before leaving the cosmonaut corps in 1985 because of illness. Lazarev was a colonel in the reserve at the time. He died on December 31, 1990.

LAZUTKIN, ALEKSANDR I.

(1957–)

Russian Cosmonaut

A long-time cosmonaut who spent more than six months in space during the long-duration Mir 23 mission, Aleksandr Lazutkin was born in Moscow, Russia on October 30, 1957. He attended the Moscow Aviation Institute, where he graduated with a degree in mechanical engineering in 1981 and then remained at the institute to begin his career as a research scientist. He joined NPO Energia in November 1984.

Lazutkin was chosen for training as a cosmonaut as a member of the civilian specialist group 12 selection of cosmonaut candidates in March 1992. After a period of intensive training, he became a cosmonaut in February 1994.

His initial duties as a cosmonaut included service as a member of the backup crew for the Soyuz TM-23/Mir 12 long-duration mission, which launched in February 1996, and for the Soyuz TM-24/Mir 22 long-duration mission, which began on August 17, 1996.

On February 10, 1997, Lazutkin lifted off from the Baikonur Cosmodrome as flight engineer for the Soyuz TM-25/Mir 23 long-duration mission, with commander Vasili Tsibliyev and guest cosmonaut Rehinhod Ewald. The three cosmonauts were met at Mir by cosmonauts Valery Korzun and Aleksander Kaleri, and NASA astronaut Jerry Linenger, who was in the midst of a long stay at the station as part of a program of cooperation in space between Russia and the United States that served as the first phase of the two nations' joint involvement in the development of the International Space Station (ISS).

A fire broke out on the space station on February 24 when an oxygen-generating device malfunctioned in the Mir Kvant 1 module. The six crew members aboard at the time rushed to put on protective masks and to gather fire-fighting gear, and they were able to quickly put out the blaze, but the after-effects of the incident, including heavy smoke and damage to electric cables and equipment, lingered for some time. Linenger, a medical doctor, examined each of his fellow crew mates to confirm that none had suffered lasting damage to his health.

Mir 22 crew members Korzun and Kaleri left the station as scheduled on March 2, returning to Earth with Ewald aboard Soyuz TM-24, while Lazutkin and Tsibliyev remained at the station with Linenger and tried to settle into a productive routine after the chaotic start of their long-duration stay. The station suffered a number of other mishaps and malfunctions during their six-month stay, including a coolant system leak that threatened the crew's breathable atmosphere, an electrical outage, and a failure of the station's attitude control system. Fortunately, none of the problems resulted in injury to the crew.

Lazutkin witnessed a bit of space history on April 29, when Tsibliyev and Linenger made the first spacewalk involving a Russian cosmonaut and an American astronaut. On May 16, he and his crew mates welcomed the STS-84 crew of the U.S. space shuttle Atlantis, which arrived at Mir for the sixth docking of a U.S. shuttle and the Russian space station. Arriving with the shuttle crew, NASA astronaut Michael Foale joined Lazutkin and Tsibliyev aboard Mir, replacing Linenger, who returned to Earth as scheduled with the STS-84 crew on May 24.

After Atlantis undocked and returned to Earth, Lazutkin, Tsibliyev, and Foale turned their attention to their assigned program of scientific experiments and attempted to resume a productive routine. They suffered further difficulties, however, when an unmanned Progress supply craft collided with the Mir Spektr module on June 25. The life-sustaining atmosphere within the Spektr module vented into space during the incident, and only the quick response of Lazutkin, Tsibliyev, and Foale prevented a wider catastrophe.

In the aftermath of the accident, mission controllers suggested that the cosmonauts make an internal spacewalk—that is, an EVA into the depressurized Spektr module—to assess the damage caused by the collision and to try to devise repairs. While they were preparing for the exercise, however, a further mishap caused the station to lose electrical power for a brief period, and the proposed EVA was, as a result, assigned to the next crew, who were scheduled to arrive in early August.

Lazutkin and Tsibliyev returned to Earth on August 14, 1997 aboard Soyuz TM-25.

Aleksandr Lazutkin spent more than 184 days in space during the Soyuz TM-25/Mir 23 flight.

After his first space mission, Lazutkin served as a member of the backup crew for the Soyuz TMA-1 mission, whose prime crew traveled to the ISS in 2002.

LEBEDEV, VALENTIN V.

(1942–)

Soviet Cosmonaut

A veteran of two spaceflights, including one long-duration mission during which he spent more than 210 days in space, Valentin Lebedev was born in Moscow, Russia on April 14, 1942.

He attended the Higher Air Force School in Orenburg, where he graduated in 1960, and worked as an aircraft designer prior to his selection as a cosmonaut.

Lebedev was chosen for training as a cosmonaut as a member of the civilian specialist group four selection of cosmonaut candidates in 1972.

Lebedev began his first space mission as a flight engineer for Soyuz 13, which was commanded by Pyotr Klimuk. Launched on December 18, 1973, the Soyuz 13 flight apparently involved tests of space reconnaissance technology that were originally planned for the Salyut 2 Almaz space station ("Almaz" was the designation given to those space stations known to have been part of the Soviet military series, which was similar in design and intent to the U.S. Manned Orbiting Laboratory program of the 1960s).

Salyut 2 had been launched unmanned on April 3, 1973, and quickly developed serious difficulties that resulted in its being de-orbited before it could be occupied. It has long been assumed that Soyuz 13 tested the high-resolution photographic reconnaissance equipment that had apparently been planned for the failed space station. Whatever the details of their closely-held mission profile, Lebedev and Klimuk were known to have landed on December 26, 1973, after a flight of 7 days, 20 hours, and 56 minutes.

After his first spaceflight, Lebedev served as a member of the backup crew for the Soyuz 32/Salyut 6 long-duration mission, which launched in February 1979.

He was then assigned as flight engineer for the Soyuz 35/Salyut 6 long-duration mission in 1980, but suffered an injury while training for the flight and was forced to give up the assignment.

On May 13, 1982, Lebedev lifted off with commander Anatoly Berezovoy aboard Soyuz T-5 on a remarkable second space mission, in which he served as flight engineer of the first long-duration crew of the Salyut 7 space station.

Lebedev and Berezovoy would spend seven months in orbit during the flight, setting a new record for the longest stay in space up to that time. They carried out a program of scientific experiments and performed engineering and space station maintenance tasks during the mission, and received two historic visits from crews that included Jean-Loup J. M. Chrétien, the first citizen of France to fly in space, and Soviet cosmonaut Svetlana Savitskaya, the second woman to fly in space (Valentina Tereshkova had been first, during Vostok 6 in June 1963).

Lebedev also made a spacewalk during his Salyut 7 mission, venturing outside the station on July 30, 1982 for 2 hours and 33 minutes to retrieve scientific experiments attached to Salyut 7's exterior while Berezovoy monitored his progress from the station's airlock.

At the end of their long stay, Lebedev and Berezovoy returned to Earth aboard Soyuz T-7 on December 10, 1982, after a total flight of 211 days, 8 hours, and 5 minutes in space.

Continuing his education concurrently with his career as a cosmonaut, Lebedev returned to the Higher Air Force School after his first space mission, and in 1975 he earned the degree of Candidate of Technical Sciences. In 1985, after his second spaceflight, he earned a Doctorate of Technical Sciences degree.

He was assigned to fly aboard the Buran space shuttle, but his planned flight was canceled when the program was abandoned because of financial difficulties in the initial post-Soviet period.

Valentin Lebedev spent more than 219 days in space, including more than two hours in EVA, during his two space missions.

After leaving the cosmonaut corps, Lebedev joined the Russian Academy of Sciences, where he served as director of the Institute of Geosciences until his retirement. He is the author of *Diary of a Cosmonaut: 211 Days in Space,* an account of his long-duration stay aboard Salyut 1.

LEESTMA, DAVID C.

(1949–)

U.S. Astronaut

David Leestma was born in Muskegon, Michigan on May 6, 1949. In 1967, he graduated from Tustin High School in Tustin, California, and then attended the United State Naval Academy, where he graduated first in his class in 1971 and received a Bachelor of Science degree in aeronautical engineering.

He initially served aboard the USS *Hepburn* (DE-1055) in Long Beach, California, and then attended the U.S. Naval Postgraduate School, where he earned a Master of Science in aeronautical engineering in 1972.

After receiving his initial flight training in 1973, he was trained in the F-14A Tomcat aircraft while assigned to VF-124 in San Diego, California, and then served in VF-32 at Virginia Beach, Virginia.

While assigned to the USS *John F. Kennedy,* he made three deployments overseas. In 1977, Leestma became an operational test director for the F-14A while assigned to Air Test and Evaluation Squadron Four (VX-4) at Naval Air Station Point Mugu, California. In that capacity, he made the first operational test of new tactical software for the aircraft and completed the testing of a new programmable signal processor for use in the F-14A.

During his outstanding military career, Leestma accumulated more than 3,500 hours of flying time, including almost 1,500 hours in the F-14A. He had risen to the rank of captain in the U.S. Navy by the time of his retirement from the service.

NASA selected Leestma for astronaut training in 1980, and he first flew in space as a mission specialist aboard the space shuttle Challenger during STS-41G, which lifted off on October 5, 1984. Robert Crippen was commander of the flight, and Leestma's other crew mates included pilot Jon McBride, mission specialists Kathryn Sullivan and Sally Ride, and payload specialists Marc Garneau (the first Canadian citizen to fly in space) and Paul Scully-Power.

On October 11, 1984, Leestma and Sullivan made a remarkable spacewalk of 3 hours and 29 minutes. Using hydrazine fuel and the Orbital Refueling System (ORS), they performed a test that proved that satellites can be refueled in orbit. Sullivan was the first American woman to perform a spacewalk.

The crew also deployed the Earth Radiation Budget Satellite (ERBS) and conducted three experiments for NASA's Office of Space and Terrestrial Applications (OSTA-3), and Garneau conducted a series of experiments dubbed "CANEX," in honor of their having been designed by scientists in Canada. As a representative of the U.S. Naval Research Laboratory, Scully-Power conducted oceanographic research. At the end of the eight-day flight, Challenger returned to Earth on October 13, 1984.

Following his first flight, Leestma served as chief of the Mission Development Branch of the Astronaut Office, and then as deputy director of Flight Crew Operations.

In August 1989 Leestma made his second spaceflight, as a mission specialist during STS-28, the fourth shuttle mission devoted to the classified activities of the U.S. Department of Defense. Launched on August 8, Leestma joined commander Brewster Shaw, Jr., pilot Richard Richards and fellow mission specialist James Adamson and Mark Brown on the short mission. Columbia returned to Earth at Edwards Air Force Base in California on August 13, 1989, after 81 orbits, in 5 days, 1 hour, and eight seconds.

Leestma next served as deputy chief and then acting chief of the Astronaut Office before making his third flight in space, STS-45, in March 1992. Launched aboard the shuttle Atlantis on March 24, the STS-45 crew successfully deployed the Atmospheric Laboratory for Applications and Science (ATLAS-1). The flight was also notable for the presence of payload specialist Dirk Frimout of Belgium, the first Belgian citizen to fly in space. The mission was extended for one day to allow the crew time to complete their Atlas-1 experiments; Atlantis landed at the Kennedy Space Center (KSC) in Florida on April 2, 1992.

David Leestma has spent more than 532 hours in space, including 3 hours and 29 minutes in EVA, during his astronaut career.

Since his third flight, Leestma has continued to serve NASA at the Johnson Space Center (JSC) in Houston, initially as director of the Flight Crew Operations Directorate, overseeing 41 shuttle flights and the selection of NASA astronaut groups 15, 16, and 17, and then as deputy director of Engineering. In 2001, he was named JSC project manager for the Space Launch Initiative and also served as assistant program manger for the Orbital Space Plane, a new vehicle intended to transport crews to and from the International Space Station (ISS).

He has also served as manager of the JSC Exploration Programs Office, and, subsequently, as manager of the JSC Advanced Planning Office, which is responsible for developing future space missions to the Moon and to Mars.

LEONOV, ALEXEI A.

(1934–)

Soviet Cosmonaut; First Human Being to Conduct a Spacewalk

The first person ever to conduct a spacewalk and a veteran of the first cooperative spaceflight undertaken by the Soviet Union and the United States, Alexei Leonov

On March 18, 1965, Alexei Leonov became the first human being ever to conduct a spacewalk. He is seen here (left) in training for the historic Apollo-Soyuz Test Project (ASTP) flight, with NASA astronaut Thomas Stafford. [NASA/courtesy of nasaimages.org]

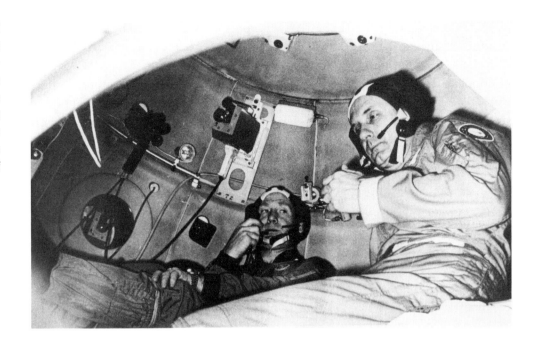

was born in the village of Listvianka, in the Kemerovo region of Russia, on May 30, 1934. In 1953, he graduated from Konigsberg Secondary School Number 21, and then attended the Kremenchough Military Aviation Pilot Basic Training School, from which he graduated in 1955.

Leonov served as a pilot in Air Army 69, in the Kiev military district, while attached to parachute aviation regiment 113 of the Guards engineering aviation division. In December 1959 he became a senior pilot in Scouting Air Regiment 294 of Air Army 24, stationed in East Germany.

He also graduated from the Tchuguev Higher Air Force Pilot School, and began engineering studies at the Zhukovsky Higher Military Engineering Academy in September 1961. He completed the academy's academic program in 1968.

Leonov was chosen for training as a cosmonaut as a member of the Soviet space program's first selection of cosmonaut candidates, on March 7, 1960. His initial duties included service as a backup crew member during the Vostok program. He had been scheduled to fly in space during the proposed Vostok 11 flight, which had been scheduled for June 1965, but the rise of the Voskhod program resulted in the cancellation of all future Vostok flights after Vostok 6.

On March 18, 1965, Leonov lifted off on his first flight in space, as co-pilot of Voskhod 2 with pilot Pavel Belyayev. His mission for the flight was to attempt the world's first spacewalk.

Shortly after reaching orbit, Leonov and Belyayev began their preparations for the EVA. For his remarkable spacewalk, Leonov wore a pressurized spacesuit identified as the Berkut, or Golden Eagle, suit. It was a modified version of the standard Vostok Sokol-1 intravehicular spacesuit, specially adapted for the task at hand. A life support backpack capable of supplying as much as 45 minutes' worth of oxygen was attached to the back of the suit; as it was expended, the oxygen would be vented into space, along with exhaled carbon dioxide.

A special passageway known as the Volga Airlock had been attached to Voskhod 2 for the flight. The inflatable airlock, which was made rigid by the activation of a series of air pumps, was designed to fit over the spacecraft's hatch, which opened inward. That configuration allowed for the airlock to be used without the necessity of depressurizing the spacecraft itself. At its fully extended length, the Volga airlock deployed to a length of 8 feet, 3 inches (2.5 meters).

When Leonov was ready to leave the Voskhod 2 cabin and the airlock had been inflated to its full length, he opened the spacecraft's hatch and ventured into the long, narrow tube. Belyayev closed the hatch behind him and activated the airlock's venting system. Gently letting out the slack of the 50 foot and seven inch-long (15.35 meter) umbilical that kept him tethered to Voskhod 2, Leonov floated freely and began the world's first extra-vehicular activity (EVA).

Soviet officials, who sought to present the event as evidence that the nation's space effort was outpacing that of its superpower rival, the United States, carefully crafted contemporaneous reports of Leonov's EVA experience. The original official account of the event made no mention of several difficulties he experienced, beginning with the unintended over-inflation of his pressure suit, which severely hampered his return to the inside of the spacecraft.

Leonov was able to make some of the Earth observations called for in his mission profile, but his billowing spacesuit made it difficult for him to carry out his photography assignments.

Subsequent revelations about the flight indicate that he was actively engaged in the objectives of the EVA for about half of the total time he was spacewalking, with the rest of the 24-minute excursion taken up by the struggle to re-enter Voskhod 2.

After a difficult struggle during which he apparently became wedged in the airlock opening while in his expanded suit, Leonov decided to cut the pressure in the suit, a risky maneuver, considering that the rapid drop in pressure might have subjected him to an experience similar to what undersea divers routinely refer to as the "bends," or decompression sickness. Fortunately, Leonov's quick thinking proved to be the proper solution to his dilemma, and he was finally able to propel himself into the Volga airlock and seal it behind him.

Belyayev repressurized the airlock and opened the Voskhod 2 hatch to allow Leonov to re-enter the spacecraft. The cosmonauts closed the hatch, jettisoned the airlock, and continued their flight. They made a total of 16 orbits, making observations of the Earth and photographing its surface and cloud formations in the atmosphere.

As they prepared to return to Earth, they were met by an unpleasant surprise. The heavily automated control systems of the Voskhod 2 craft suffered a malfunction, and Belyayev had to assume manual control of the vehicle to orient it for re-entry. Fortunately, the cosmonauts were able to re-enter safely, and survived a landing, on March 19, 1965, that deposited them in a snowy forest between the villages of Sorokovaya and Schuchino, more than 3,000 kilometers from the planned landing site.

They endured a harrowing ordeal after landing, as recovery crews searched for them in the wooded area without the benefit of the antenna that would normally have transmitted their location but had been broken off during the rough landing. They were located within hours, but extracting them from the area took as long as two days. Some accounts of the cosmonauts' long wait for rescue recount their barring themselves inside the spacecraft to avoid marauding timber wolves; in any case, they spent at least one long, difficult night at the landing site before being evacuated.

Leonov was widely celebrated for his remarkable EVA, even when most of the world was unaware of the true, heroic nature of his struggle during the exercise.

He was given the ratings of Air Force parachute training instructor and first-class military pilot in 1965.

His spacewalk marked the last highlight of the first phase of the Soviet space program. The Voskhod 2 mission was the last flight of the Voskhod program, and the Soviet space program entered a long period during which no manned flights were made, while preparations for the first flight of the new Soyuz series of spacecraft progressed.

As the United States edged closer to achieving a manned lunar landing with its Apollo program, the Soviets planned a series of flights that would keep them on par with the Americans' progress. The unmanned Zond missions of the late 1960s were the first of the Soviet efforts directly related to manned lunar missions, and a series of three manned lunar orbit flights were reportedly planned and then abandoned after the American Apollo 8 lunar orbit flight of December 1968. Leonov had been proposed to serve as commander of the first planned Soviet lunar orbit flight, which had apparently been scheduled for March 1969.

Leonov served as the backup commander for the Soyuz 10/Salyut 1 mission and for Soyuz 11/Salyut 1, which, in June 1971, resulted in the first occupation of an orbiting space station for a long-duration stay. Sadly, the primary Soyuz 11/Salyut 1 crew was killed during their return to Earth because of a malfunction in their Soyuz capsule.

Leonov's second flight in space proved to be another historic milestone in the history of spaceflight, as he served as commander of the Soviet portion of the 1975 Apollo-Soyuz Test Project (ASTP), the first cooperative spaceflight undertaken by the Soviet Union and the United States.

In a mission that seemed at best unlikely even just a few years previously, the ASTP flight featured shared training, cooperative engineering, and prolonged, intensive communication between the Soviet and American space programs.

With the goal of achieving a rendezvous and docking in orbit with their American counterparts, Leonov and his flight engineer, Valeri Kubasov, launched in Soyuz 19 on July 15, 1975. Fewer than eight hours later, the U.S. crew, made up of commander Thomas Stafford, Donald "Deke" Slayton, and Vance Brand, lifted off aboard the last Apollo vehicle to fly in space. The launches were both broadcast live in the Soviet Union and in the United States.

On July 17, the two vehicles achieved a smooth docking, and a short while later, the two commanders met at the Soyuz end of the customized docking module. The two veteran space commanders shook hands, and Stafford marked the historic moment with the Russian word for comrade, "*tovarish*." Leonov replied, in English, "Very happy to see you."

The Soviet and American ASTP crews visited each other's spacecraft during their docked flight, which encompassed a total of 47 hours. Crew members shared several meals together, and exchanged symbolic gifts that spoke to the hopeful nature of their cooperative mission, including seeds that each crew had brought from their own nation for later planting in their counterparts' country.

Soviet President Leonid Brezhnev and U.S. President Gerald Ford each made congratulatory calls to their respective crews, and the crew members also took part in a news conference that was aired live in both the Soviet Union and the United States.

Although the initial docking had been achieved with the ASTP Apollo vehicle acting as the active craft and Soyuz 19 serving as the passive target, the two vehicles undocked on July 19 to conduct a series of rendezvous experiments, and Leonov then achieved a second docking with the Soyuz craft as the active vehicle.

After the final undocking, on July 21, 1975, Leonov and Kubasov returned to Earth in Soyuz 19, after 5 days, 22 hours, 30 minutes, and 51 seconds in space.

Continuing his education concurrently with his cosmonaut career, Leonov was granted a degree of Candidate of Technology in 1975. He also received the rating of cosmonaut second class that same year.

During his remarkable career as a cosmonaut, Alexei Leonov spent more than 7 days in space, including 23 minutes and 41 seconds in EVA during the world's first spacewalk.

Leonov was widely celebrated for both of his historic spaceflights. Among many other awards and international recognition for his two landmark spaceflights, he was awarded the Harmon International Aviation Prize, and received both the Gold Star Hero of the U.S.S.R. medal and the Order of Lenin on two separate occasions, for his first flight in 1965 and his second, 10 years later. He has also been named an honorary citizen of 12 cities.

He left the cosmonaut corps in January 1982 to become the first deputy for flight and space training at the Gagarin Cosmonaut Training Center (GCTC), a capacity in which he served until 1991. He left active duty military service in March 1992, and has served as a major-general the Air Force reserve since that time.

LIND, DON L.

(1930–)

U.S. Astronaut

Don Lind was born in Midvale, Utah on May 18, 1930. He graduated from Jordan High School in Sandy, Utah, and then attended the University of Utah, where he received a Bachelor of Science degree with high honors in physics in 1953.

He entered active duty with the U.S. Navy, receiving his aviator's wings in 1957, and was deployed on the USS *Hancock*. After four years of active duty service, he entered the U.S. Naval Reserve and then attended the University of California at Berkeley, where he earned a Ph.D. in high energy nuclear physics in 1964.

Lind worked as a scientist at the Lawrence Radiation Laboratory in Berkeley, and then began his association with NASA in 1964, as a space physicist at the agency's Goddard Space Flight Center. His research at Goddard included study of low-energy particles in the magnetosphere and in interplanetary space.

A commander in the U.S. Naval Reserve, Lind has accumulated over 4,500 hours of flying time, including 4,000 hours in jet aircraft.

NASA selected him for training as an astronaut in April 1966, as one of its Group Five selection of pilot astronauts. He was a member of the backup crew for the Skylab 3 and Skylab 4 missions, and served as a member of the missions development group in the Astronaut Office at the Johnson Space Center (JSC) in Houston.

In 1975, he took a one-year leave from his duties at NASA to conduct academic research at the University of Alaska's Geophysical Institute.

On April 29, 1985, Lind lifted off as a mission specialist aboard the space shuttle Challenger, during the STS-51B Spacelab-3 mission.

The STS-51B crew conducted a variety of experiments in atmospheric physics and astronomy, fluid mechanics, life sciences, and materials sciences using the Spacelab-3 orbital laboratory developed by the European Space Agency (ESA).

Lind's role in the mission included the operation of an experiment he designed to make three-dimensional video recordings of the Earth's aurora.

Spacelab-3 had originally been intended to fly on mission 51-E, but that flight was canceled during the long delays prior to the shuttle Discovery's first flight (STS-41D). As a result, Spacelab-3 actually flew before Spacelab-2, which flew aboard STS-51F.

STS-51B was the first operational flight for Spacelab, which was designed to prove that the shuttle could serve as a stable microgravity environment in which to carry out the sort of experiments in materials processing and fluid mechanics that the ESA scientists had envisioned when they first designed the Spacelab module. The results were encouraging: 14 of the 15 primary Spacelab experiments were deemed successful. At the end of the productive flight, Challenger landed at Edwards Air Force Base in California, on May 6, 1985.

Don Lind spent more than seven days in space during STS-51B. He left NASA in 1986.

LINDSEY, STEVEN W.

(1960–)

U.S. Astronaut

Steven Lindsey was born in Arcadia, California on August 24, 1960. In 1978, he graduated from Temple City High School and then attended the United States Air Force Academy, where he received a Bachelor of Science degree in engineering sciences in 1982.

Upon graduation, he was commissioned as a second lieutenant in the Air Force, and received his undergraduate pilot training at Reese Air Force Base in Texas, where he was recognized as a distinguished graduate in 1983. He received further training as a pilot in the RF-4C Phantom II aircraft and was then assigned to Bergstrom Air Force Base in Texas as a member of the 12th Tactical Reconnaissance Squadron.

Lindsey served as a combat-ready pilot, instructor pilot, and as an academic instructor from 1984 until 1987, when he was selected to attend the Air Force Institute of Technology at Wright-Patterson Air Force Base in Ohio.

Concurrently with his graduate education, he also attended the U.S. Air Force Test Pilot School at Edwards Air Force Base in California, where he was recognized as a distinguished graduate and received the Liethen-Tittle Award as the outstanding test pilot of his class when he graduated in 1989.

In 1990, he earned a Master of Science degree in aeronautical engineering from the Air Force Institute of Technology.

He next served at Eglin Air Force Base in Florida, where he flew F-16 and F-4 aircraft while conducting weapons and systems tests, and he served as deputy director of the Advanced Tactical Air Reconnaissance System Joint Test Force while attached to the 3247th Test Squadron.

Lindsey graduated from the Air Force Air Command and Staff College at Maxwell Air Force Base in Alabama in 1994 and returned to Eglin Air Force Base as an Integrated Product Team leader responsible for weapons certification for a variety of aircraft. NASA selected him for astronaut training in 1995.

During his outstanding military career, Lindsey has accumulated more than 5,000 hours of flying time in more than 50 different types of aircraft. He is a colonel in the U.S. Air Force.

Lindsey's initial technical assignments at NASA included work on the Multifunction Electronic Display System (MEDS) upgrade program for the space shuttle cockpit.

He first flew in space as pilot of the space shuttle Columbia, during STS-87 in 1997. The fourth flight of the U.S. Microgravity Payload, the 16-day mission featured around the clock research by alternating teams of astronauts conducting experiments in the impact of weightlessness and studying the outer atmosphere of the Sun.

Lindsey piloted the first flight of the AERCam Sprint, a free-flying robotic camera, during STS-87.

The crew completed 80 experiments during the flight, and mission specialists Winston Scott and National Space Development Agency (NASDA) astronaut Takao Doi of Japan made a 7 hour and 43-minute spacewalk to manually retrieve the malfunctioning SPARTAN-201–04 free-flying satellite. Mission Specialist Kalpana Chawla had released the SPARTAN craft three days earlier, using the shuttle's Remote Manipulator System (RMS). Columbia landed on December 5, 1997 at the Kennedy Space Center, after 252 orbits.

Lindsey's second spaceflight, STS-95 in 1998, featured the return to space of Mercury program pioneer John Glenn. Launched October 29 aboard the shuttle

Discovery, the STS-95 crew was commanded by Curtis Brown and also included payload specialist Stephen Robinson, mission specialists Scott Parazynski, European Space Agency (ESA) astronaut Pedro Duque, and payload specialist Chiaki Mukai of the National Space Development Agency (NASDA) of Japan.

The crew launched the SPARTAN 201 free flyer and retrieved it after two days of independent flight, during which it gathered data about the solar wind. Discovery also carried a SPACEHAB pressurized module in its payload bay to facilitate the crew's scientific experiments, and hardware that was being tested for its flight readiness prior to its use on a later mission in which it would be used in the servicing of the Hubble Space Telescope (HST). The nine-day flight came to a close with Discovery's return to Earth on November 7, 1998.

Following his second flight, Lindsey served as deputy for Shuttle Operations and as co-chair of the Space Shuttle Cockpit Council, which was responsible for the $400 million avionics upgrade of the shuttle's cockpit.

On July 12, 2001, Lindsey launched on his third space mission, STS-104, aboard the shuttle Atlantis. The STS-104 flight was the 10th shuttle mission to travel to the ISS; its crew delivered and installed the station's Quest Airlock and transferred supplies and equipment to the station's Expedition 2 crew. Michael Gernhardt and James Reilly made three spacewalks while installing the Quest Airlock, including the first from the ISS itself. Atlantis was docked with the ISS for eight days before it returned to Earth on July 24, 2001.

Steven Lindsey has spent more than 896 hours in space during his career as an astronaut.

Since his third flight he has continued to serve in the Astronaut Office, initially as chief of International Space Station Operations, and as an active astronaut available for assignment to future missions.

LINENGER, JERRY M.

(1955–)

U.S. Astronaut

Jerry Linenger was born on January 16, 1955 and was raised in Eastpointe, Michigan. In 1973, he graduated from East Detroit High School in Eastpointe and then attended the United States Naval Academy, where he received a Bachelor of Science degree in bioscience in 1977.

He went on to attend medical school at Wayne State University, where he received a Doctor of Medicine degree in 1981, and subsequently served a surgical internship at Balboa Naval Hospital in San Diego, California. He received aerospace medicine training at the Naval Aerospace Medical Institute in Pensacola, Florida and was the top graduate in his class in Naval Flight Surgeon Training and Naval Safety Officer's School. Upon completion of his training, Linenger was assigned to Cubi Point in the Republic of the Philippines, where he served as a naval flight surgeon.

Linenger returned to San Diego to serve as medical advisor to the commander of Naval Air Forces of the U.S. Pacific Fleet and was then assigned to the Naval

Health Research Center in San Diego after completing doctoral-level training in epidemiology. During his assignment to the Center, he also taught sports medicine as a member of the faculty of the University of California-San Diego School of Medicine.

Continuing his academic career concurrently with his military service, he attended the University of Southern California, where he earned a Master of Science degree in systems management in 1988, and the University of North Carolina, where he received a Master of Public Health degree in health policy in 1989 and a Ph.D. in epidemiology the same year.

Linenger is a member of the Association of Naval Aviation, the U.S. Navy Flight Surgeons Association, the American Medical Association, the Aerospace Medicine Association, the American College of Preventive Medicine, the Society of U.S. Navy Preventive Medicine Officers, and the American College of Sports Medicine. He is board certified in preventive medicine. Linenger retired from the U.S. Navy in 1998, with the rank of captain.

NASA selected him for astronaut training in 1992, and he made his first flight into space aboard the space shuttle Discovery during STS-64 in September 1994.

Launched on September 9, STS-64 featured the successful test of an innovative optical radar system, the Lidar in Space Technology Experiment (LITE), and the crew deployed and retrieved a SPARTAN free-flyer satellite. The mission also featured the first untethered spacewalk by American astronauts in 10 years.

The radar exercise was part of NASA's "Mission to Planet Earth" initiative. An experimental system that utilized laser pulses instead of radio waves, the LITE instrument was trained on a variety of targets, including cloud structures, dust clouds, and storm systems, among others. Groups in 20 countries around the globe collected data using ground- and aircraft-based radar instruments to help verify the data collected during the LITE experiment.

In a test of NASA's Simplified Aid for EVA Rescue (SAFER) backpacks, STS-64 mission specialists Mark Lee and Carl Meade made a spacewalk of 6 hours and 51 minutes on September 16. The SAFER devices were designed as a backup device for spacewalking astronauts who might become untethered while conducting extravehicular activities.

The crew also deployed and retrieved the Shuttle Pointed Autonomous Research Tool for Astronomy (SPARTAN-201), which collected data about the solar wind and the Sun's corona during two days of free flight, and the Shuttle Plume Impingement Flight Experiment (SPIFEX), which was carried on the end of the shuttle's robotic arm while collecting data about the potential impact of the shuttle's Reaction Control System thrusters on space structures like the Mir space station or the future International Space Station (ISS). Discovery landed at Edwards Air Force Base on September 20, 1994.

Linenger next relocated to Star City, Russia to train at the Gagarin Cosmonaut Training Center (GCTC) for a long-duration mission aboard the Russian space station Mir. He was also trained to oversee more than 100 experiments in a variety of disciplines, as he would serve as the chief scientist responsible for American experiments aboard the station.

On January 12, 1997, he began his remarkable long-duration mission as a member of the STS-81 crew of the space shuttle Atlantis, lifting off from the Kennedy Space Center (KSC) in Florida on his way to Mir. STS-81 was the fifth shuttle mission to feature a docking of the U.S. shuttle and the Russian space station. Linenger's stay aboard Mir was part of an ongoing program of cooperation between the United States and Russia that maintained a continuous American presence aboard the station. In addition to the crew exchange, the shuttle also delivered more than three tons of equipment and supplies to Mir during five days of linked flight.

With his transfer from Atlantis to Mir, Linenger replaced fellow NASA astronaut John Blaha, who returned to Earth with the STS-81 crew.

Representing the best aspects of their nations' new cooperation in space, Linenger and Mir 23 cosmonaut Vasili Tsibliyev made an historic spacewalk on April 29, 1997. Primarily designed to test the new Russian Orlan-M spacesuit, their 4 hour and 59-minute EVA marked the first time an American had ever made an EVA from a foreign space station and the first time an American had ever made an EVA in a spacesuit that had been manufactured outside the United States. In addition to the successful test of the spacesuits, Linenger and Tsiblyev also installed and retrieved experiments on the outside of the station.

Other aspects of Linenger's stay aboard Mir were far less pleasant. On February 24, 1997, an oxygen-generating device aboard the station caused a fire to break out in the Mir Kvant-1 module. There were six crew members on board at the time: Linenger, Mir 23 cosmonauts Tsibliyev and Aleksandr Lazutkin, Mir 22 cosmonauts Valery Korzun and Alexander Kaleri, and visiting cosmonaut Reinhold Ewald of Germany.

Equipping themselves with goggles and masks and emptying the contents of three fire extinguishers on the flames, the crew was able to extinguish the fire quickly, although heavy smoke filled the station for five to seven minutes and heat from the fire damaged hardware and cables.

Linenger put his extensive medical expertise to good use, examining his fellow crew members and advising them on how best to protect themselves from the lingering effects of the fire. Fortunately, none of the crew was injured in the incident.

Additional difficulties arose throughout the flight. Several failures occurred in the station's environmental control systems; electrical power was lost for a time, and at one point the attitude control system failed, leaving the station tumbling through orbit.

Despite the frequently precarious conditions, the crew was able to achieve the goals lined out in the original mission profile, including the completion of the entire voluminous U.S. science program.

At the end of his remarkable stay, Linenger was replaced aboard Mir by Michael Foale and then returned to Earth on the space shuttle Atlantis during STS-84, landing at the KSC in Florida on May 24, 1997. He had lived and worked on the Russian space station for 123 days, and spent a launch-to-landing total of 132 days, 4 hours, and 1 minute in space during the mission.

Jerry Linenger spent more than 143 days in space during his remarkable career as an astronaut. He retired from NASA in 1998.

LINNEHAN, RICHARD M.

(1957–)

U.S. Astronaut

Richard Linnehan was born in Lowell, Massachusetts on September 19, 1957. In 1975, he graduated from Pelham High School in Pelham, New Hampshire. He attended Colby College in Waterville, Maine, and in 1980 he earned a Bachelor of Science degree in animal sciences from the University of New Hampshire. In 1985, he earned a Doctor of Veterinary Medicine degree from the Ohio State University.

Linnehan began his career in private veterinary practice in 1985, and then interned in zoo animal medicine and comparative pathology in a joint program of the Johns Hopkins University and the Baltimore Zoo.

Upon completion of his internship, he joined the Veterinary Corps of the U.S. Army and was commissioned as a captain. Assigned as chief clinical veterinarian for the U.S. Navy's Marine Mammal program at the Naval Ocean Systems Center in San Diego, California, he conducted research in anesthesia, orthopedics, drug treatment, and reproductive services for sea creatures such as whales, seals, and walruses.

He was selected as an astronaut candidate in 1992.

Linnehan first flew in space during STS-78 in June 1996. The mission was a milestone in international cooperation in spaceflight. The primary payload for the mission was the Life and Microgravity Spacelab, which represented the work of scientists from 10 countries and the personnel of five space agencies, including NASA, the European Space Agency (ESA), the French space agency Centre National d'Études Spatiales (CNES), the Canadian Space Agency (CSA), and the Italian space agency Agenzia Spaziale Italiana (ASI). The crew conducted more than 40 LMS experiments during the flight, which, at nearly 17 days, was the longest shuttle flight up to that time.

During his second space mission, STS-90 in 1998, Linnehan served as the payload commander for the Neurolab Spacelab experiments. The STS-90 crew conducted 26 experiments aimed at furthering understanding of the nervous system in tests designed by scientists at NASA, CSA, CNES, the German space agency Deutsches Zentrum für Luft- und Raumfahrt e.V. (DARA), the ESA, and the National Space Development Agency (NASDA) of Japan.

On March 1, 2002, Linnehan launched aboard the shuttle Columbia for STS-109, the fourth Hubble Space Telescope (HST) maintenance and repair mission. Among the most complex and ambitious spaceflights of the entire shuttle program, the HST servicing missions stretched the expertise and abilities of the astronauts and mission controllers to the utmost.

With fellow mission specialist John Grunsfeld, Linnehan performed the first, third, and fifth of the five spacewalks necessary to repair and service the HST, accumulating more than 21 hours in EVA. The HST received new solar arrays, a new camera, a new power control unit, and was outfitted with an experimental cooling system.

In March 2008, Linnehan served as a member of the STS-123 crew that delivered the pressurized component of the Japanese Kibo Laboratory to the International Space Station (ISS). During that mission, he made three long spacewalks totaling more than 22 hours.

Richard Linnehan has spent more than 48 days in space, including more than 43 hours in EVA, during four space missions.

LIU, BUOMING (LIU BUOMING)

(1966–)

Chinese Taikonaut

Buoming Liu was born in Heilongjiang province, in the People's Republic of China, on October 29, 1966. In 1985, he joined the People's Liberation Army (PLA), and he then attended military pilot school, from which he graduated in 1989. He has risen to the rank of lieutenant colonel in the PLA Air Force, and has accumulated more than 1,000 flight hours during his career as a military pilot.

He was chosen as one of the first group of Chinese taikonauts (cosmonauts) in 1998. He served as a member of the backup crew for the Shenzhou VI mission, which took place in October 2005.

On September 25, 2008, Liu lifted off from the Jiuquan Satellite Launch Center in Gansu Province, with crew mates Zhigang Zhai and Haipen Jing in Shenzhou VII during China's third manned spaceflight. Once in orbit, he conducted a stand-up EVA of about 14 minutes in support of Zhigang Zhai, who left the Shenzhou VII capsule completely in order to retrieve an experiment package from the spacecraft's exterior. The EVAs were the first spacewalks for the Chinese space program.

Shenzhou VII returned to Earth on September 28, 2008. During his historic first spaceflight, Buoming Liu spent 2 days, 20 hours, and 28 minutes in space.

LONCHAKOV, YURY V.

(1965–)

Russian Cosmonaut

Yury Lonchakov was born in Balkhash, in the Dzhezkazkansk region of Kazakhstan, on March 4, 1965. He graduated from high school in Aktyubinsk in 1982 and then attended the Orenburg Air Force Pilot School, where he earned a degree with honors as a pilot-engineer in 1986. He went on to attend Zhukovsky Air Force Academy, where he again graduated as an honors student, in 1998.

Earning positions of successively greater responsibility as a pilot and military officer, Lonchakov has risen to the rank of colonel in the Russian Air Force. A Class

1 Air Force pilot, he has flown a wide array of aircraft and has accumulated more than 1,400 flight hours. He has also made more than 500 parachute jumps during his military career.

In 1997, he was chosen for training as a cosmonaut.

Lonchakov first flew in space as a mission specialist during the STS-100 flight of the U.S. space shuttle Endeavour. Launched on April 19, 2001, the STS-100 crew visited the International Space Station (ISS), where they installed the station's Remote Manipulator System (RMS) robotic arm and delivered 6,000 pounds of equipment and supplies to the ISS Expedition 2 crew. The flight returned to Earth at Edwards Air Force Base in California on May 1, 2001, after 186 orbits and a flight of 11 days, 21 hours, and 30 minutes.

During his second space mission, Lonchakov served as flight engineer for the first flight of the redesigned TMA class of Soyuz spacecraft when he lifted off with Soyuz TMA-1 commander Sergei Zalyotin and European Space Agency (ESA) astronaut Frank De Winne on October 30, 2002 to travel to the ISS. Lonchakov was originally a member of the backup crew for the Soyuz TMA-1 flight, but he moved up to the prime crew after plans fell through for the American entertainer Lance Bass of the musical group 'N Sync to join the flight as a space tourist.

Lonchakov and his crew mates visited with the ISS Expedition 5 crew at the station and returned to Earth on November 20, 2002 aboard Soyuz TM-34, the last of the Soyuz TM-series. During his second flight, Lonchakov added another 10 days, 20 hours, and 53 minutes to his career total time in space.

On October 12, 2008, Yury Lonchakov lifted off from the Baikonur Cosmodrome in Soyuz TMA-13 on a remarkable third spaceflight, this time serving as flight engineer for the ISS Expedition 18 crew. Lonchakov docked Soyuz TMA-13 at the ISS two days later, and he and Expedition 18 commander Michael Fincke began their six-month stay aboard the station.

LOPEZ-ALEGRIA, MICHAEL E.

(1958–)

U.S. Astronaut

Michael Lopez-Alegria was born in Madrid, Spain on May 30, 1958 and was raised in Mission Viejo, California. In 1976, he graduated from Mission Viejo High School and then attended the United States Naval Academy, where he received a Bachelor of Science degree in systems engineering in 1980.

He served in Pensacola, Florida as a flight instructor and as a mission commander, and was next assigned, in 1986, to participate in a cooperative program of the Naval Postgraduate School in Monterey, California and the U.S. Naval Test Pilot School in Patuxent River, Maryland.

In 1988, he earned a Master of Science degree in aeronautical engineering from the U.S. Naval Postgraduate School. He is also a graduate of the Harvard University's Kennedy School of Government program for senior executives in national and international security.

He was serving as a program manager and engineering test pilot at the Naval Air Test Center when NASA selected him for astronaut training. During his outstanding military career, he has accumulated more than 5,000 hours of flying time. He is a captain in the U.S. Navy.

Lopez-Alegria first flew in space during STS-73 in 1995, the second flight of the U.S. Microgravity Laboratory (USML-2). The STS-73 crew worked around the clock in two shifts to conduct the USML experiments, using the pressurized Spacelab module in the shuttle's payload bay. Lopez-Alegria was responsible for overseeing the crew's activities during his shift.

After his first space mission, Lopez-Alegria served as NASA's director of operations at the Gagarin Cosmonaut Training Center (GCTC) in Star City, Russia.

Lopez-Alegria's second space mission, STS-92 in October 2000, was the 100th flight of the space shuttle program. Tasked with delivering and attaching the Z1 Truss and Pressurized Mating Adapter 3 (PMA-3) to the International Space Station (ISS), the STS-92 crew achieved their space construction work during seven days of docked operations. He participated in two of the four spacewalks necessary to installing the PMA-3 docking port and the Z1 Truss, accumulating a total of more than 14 hours in EVA.

After his second spaceflight, Lopez-Alegria led the ISS Crew Operations Branch of NASA's Astronaut Office.

On November 23, 2002, he lifted off aboard the shuttle Endeavour for STS-113, which was devoted to the installation of the 28,000-pound P1 Truss at the ISS. Operating from the ISS Quest Airlock rather than from the space shuttle, Lopez-Alegria and his EVA partner John Herrington made three spacewalks during STS-113, accumulating about 20 hours in EVA.

In their first spacewalk, on November 26, attached the P1 truss to the ISS. Two days later, they installed TV cameras to the outside of the station, and in their third spacewalk, on November 30, they added another seven hours to their EVA total while finishing their work.

On September 18, 2006, Lopez-Alegria launched aboard Soyuz TMA-9 to travel to the ISS, where he served a seven-month stay as commander of ISS Expedition 14. During his Expedition 14 mission, he participated in five spacewalks, accumulating more than 33 hours in EVA. He returned to Earth on April 21, 2007.

During his Expedition 14 mission at the ISS, Michael Lopez-Alegria established a new U.S. record for the longest stay in space, at 215 days, 8 hours, and 22 minutes. In his career as an astronaut, he has spent more than 257 days in space, including more than 67 hours in EVA.

LOUNGE, JOHN M.

(1946–)

U.S. Astronaut

John M. "Mike" Lounge was born in Denver, Colorado on June 28, 1946. In 1964, he graduated from Burlington High School and then attended the United States

Naval Academy, where he received a Bachelor of Science degree in 1969. He also received a Master of Science degree in astrogeophysics from the University of Colorado in 1970.

Lounge received his initial training as a flight officer at Pensacola, Florida and was then trained as a radar intercept officer in the F-4J aircraft. He served aboard the USS *Enterprise* during a nine-month deployment to Southeast Asia while attached to Fighter Squadron 142 (VF-142), and flew 99 combat missions during the Vietnam War. He also served aboard the USS *America* for seven months, in the Mediterranean Sea.

He returned to the Naval Academy in 1974 to serve as a physics instructor. In 1976, he was assigned to the Navy Space Project Office in Washington, D.C. as a staff project officer. He resigned his regular U.S. Navy commission in 1978, the same year he began his association with NASA.

In his initial assignment in the space agency's Payload Operations division, Lounge was lead engineer for satellites launched from the space shuttle. He also served as a member of the Skylab Reentry Flight Control Team, which was responsible for overseeing the safe de-orbiting of the Skylab space station in July 1979. For his work in that capacity, he shared in the Johnson Space Center (JSC) Superior Achievement Award.

NASA selected Lounge for training as an astronaut in 1980. His initial technical assignments in the Astronaut Office included service as a member of the launch support team for the first three space shuttle missions, and he was subsequently assigned to the planned STS-61F flight, which was to have integrated the liquid-fueled Centaur upper stage rocket into the shuttle, but that mission was subsequently canceled after the January 1986 loss of the space shuttle Challenger and the resulting changes in the shuttle program.

Lounge first flew in space as a mission specialist aboard the space shuttle Discovery during STS-51I, the 20th mission of the space shuttle program. Launched on August 27, 1985, the crew deployed three communications satellites during the flight and located, captured, and re-deployed the SYNCOM IV-3 satellite (also known as LEASAT-3) in a remarkable repair mission that allowed ground controllers to properly position the satellite in the correct orbit after the end of the STS-51I mission.

During the flight, Lounge deployed the Australian AUSSAT communications satellite and operated Discovery's Remote Manipulator System (RMS) robotic arm. The crew returned to Earth at Edwards Air Force Base in California on September 3, 1985, after a flight of 7 days, 2 hours, 17 minutes, and 42 seconds.

Lounge again flew aboard Discovery on STS-26, the first shuttle flight after the Challenger disaster. Launched on September 29, 1988, the STS-26 crew successfully deployed a Tracking and Data Relay Satellite, TDRS-C, during the mission and performed 11 mid-deck scientific experiments during the four-day flight. Discovery landed at Edwards Air Force Base on October 3, 1988.

After his second mission, Lounge was named chief of NASA's Space Station Support Office, a position he held until he left NASA in 1991.

On December 2, 1990, he lifted off on a remarkable third trip into space, this time as a flight engineer aboard the shuttle Columbia for the STS-35 ASTRO-1 mission. In the first shuttle flight devoted to astronomical observations, the STS-35 worked in alternating shifts to make around the clock ultraviolet and X-ray observations. An unexpected drama unfolded when the data display units on the observatory's ultraviolet telescopes failed, but the crew was able to fine-tune the instruments manually according to the aiming instructions of mission support teams on the ground at the Marshall Space Flight Center. The shuttle had to make over 200 maneuvers to accommodate the proper pointing of the telescopes to achieve the mission's objectives. Columbia made 142 orbits during the flight and landed on December 10, 1990.

Mike Lounge spent more than 482 hours in space during his career as an astronaut.

He resigned from NASA in June 1991 to pursue a private sector career, and subsequently became director of Space Shuttle and Space Station Program Development at the Boeing Corporation.

LOUSMA, JACK R.

(1936–)

U.S. Astronaut

Jack Lousma was born in Grand Rapids, Michigan on February 29, 1936. He graduated from Ann Arbor High School in Ann Arbor, Michigan and then attended the University of Michigan, where he received a Bachelor of Science degree in aeronautical engineering in 1959.

Lousma joined the U.S. Marine Corps in 1959 and completed his initial training as a pilot at the U.S. Naval Air Training Command, receiving his wings in 1960. He served as an attack pilot while attached to VMA-224, 2nd Marine Air Wing and was then assigned to VMA-224, 1st Marine Air Wing, at Iwakuni, Japan. He also served with VMCJ-2, 2nd Marine Air Wing at Cherry Point, North Carolina, as a reconnaissance pilot.

He attended the U.S. Naval Postgraduate School, where he earned a Master of Science degree in aeronautical engineering in 1965.

During his outstanding military career, Lousma accumulated more than 5,300 hours of flying time in a variety of military and civilian aircraft, including 4,500 hours in jet aircraft. He had risen to the rank of colonel in the U.S. Marine Corps by the time of his retirement from the service.

NASA selected Lousma for training as an astronaut in April 1966, as one of its Group Five selection of pilot astronauts. He was a member of the astronaut support crew for the Apollo 9, Apollo 10, and Apollo 13 missions, and he later served as a member of the backup crew for the Apollo-Soyuz Test Project (ASTP).

Lousma piloted an Apollo command module on his first trip into space when he lifted off on the Skylab 3 long-duration mission on July 28, 1973. His crew mates for the flight included Apollo veteran and Moonwalker Alan Bean, who commanded Skylab 3, and Group Four astronaut Owen Garriott, who served as the scientist astronaut for the flight. Skylab 3 was the second manned flight to the first American space station; the station itself had been designated Skylab 1 when it launched unmanned on May 14, 1973.

The crew endured an uncomfortable first few days when the Apollo craft that would bring them back to Earth at the end of their mission was found to be leaking fuel. Fortunately, the problem was resolved without posing serious danger to the astronauts or the mission, and Lousma, Bean, and Garriott subsequently settled into a remarkably productive routine, working 12-hour days to complete an intense program of scientific experiments that included studies of the Sun, life sciences investigations, astronomical observations, and photography. They far exceeded the original goals that NASA scientists had set for the Skylab 3 science program by the end of the flight, in both the number of experiments conducted and the amount of data collected.

Lousma, Bean, and Garriott also made regular television broadcasts throughout their two-month stay on Skylab 3 and participated in the regular daily regimen of exercise necessary to combat the losses in blood plasma volume and bone density that routinely occur during long-duration stays in space.

On August 6, 1973, Lousma and Garriott made a 6 hour and 31-minute spacewalk while installing a new sunshade over Skylab's work area, shielding it from the harshest exposure to the Sun and lowering the temperature inside the station to a comfortable level. They also changed the film in the station's Apollo Telescope Mount (ATM), Skylab's main instrument for making astronomical observations.

In his second EVA, on September 22, Lousma again worked with Garriott while they changed the ATM film cartridge and repaired one of the instrument's gyroscopes; Lousma added 4 hours and 31 minutes to his total spacewalking time for the mission.

The Skylab 3 flight came to an end on September 25, 1973, when Lousma and his crew mates left the station and returned to Earth, splashing down in the Pacific Ocean after a total flight of 59 days and 11 hours. During his two Skylab 3 spacewalks, Lousma accumulated 11 hours and 2 minutes in EVA.

During their long, productive stay in space, Lousma, Bean, and Garriott made extensive studies of the Sun and captured 16,000 images and 18 miles of magnetic tape containing data gleaned from their studies of Earth resources. They also performed 333 medical experiments.

For his participation in the superb Skylab 3 flight, Lousma was awarded the Robert J. Collier Trophy, the Dr. Robert H. Goddard Memorial Trophy, and the Marine Corps Aviation Association's Exceptional Achievement Award.

Lousma served as commander of his second spaceflight when he launched aboard the space shuttle Columbia during STS-3 on March 22, 1982, with pilot Gordon Fullerton. STS-3 was an eight-day orbital flight test of the space shuttle.

During the flight, Lousma and Fullerton tried out the shuttle's robotic arm, the Remote Manipulator System (RMS), placed the shuttle in various orbital positions to measure its interaction with the Sun, and deployed a group of experiments (OSS-1) for the Office of Space Science.

At the end of the eight-day flight, Fullerton and Lousma made the first diverted landing of the shuttle program, as wet weather forced NASA mission controllers to move the scheduled landing from Edwards Air Force Base in California to the Northrup Strip in White Sands, New Mexico.

Columbia returned to Earth at the New Mexico site on March 30, 1982 in the midst of a dust storm, the veteran crew enduring a difficult landing that resulted in damage to the shuttle's brake system that fortunately did not result in injuries.

Lousma was inducted into the International Space Hall of Fame in 1982. He left NASA in 1983.

Jack Lousma spent more than 67 days in space, including more than 11 hours in EVA, during his career as an astronaut.

LOVELL, JAMES A., JR.

(1928–)

U.S. Astronaut

James Lovell was born in Cleveland, Ohio on March 25, 1928. He attended the University of Wisconsin and the United States Naval Academy, where he earned a Bachelor of Science degree in 1952.

After earning his wings as a naval aviator and serving in a variety of assignments, Lovell attended the U.S. Naval Test Pilot School at Patuxent River, Maryland, from which he graduated in 1958. He remained at the Naval Air Test Center in Patuxent River after graduation, and served as a test pilot and as program manager for the F4H Phantom Fighter aircraft.

In 1961, he graduated from the University of Southern California's Aviation Safety School, and then served with Fighter Squadron 101 at Naval Air Station Oceana, Virginia as a safety engineer.

James Lovell (right) celebrates the successful conclusion of the Gemini 12 mission with crew mate Buzz Aldrin on the deck of the USS *Wasp*, November 15, 1966. [NASA/courtesy of nasaimages.org]

During his outstanding military career, Lovell accumulated over 7,000 hours of flying time, including 3,500 hours in jet aircraft. He retired from the U.S. Navy in 1973 with the rank of captain.

NASA selected Lovell as one of its second group of astronauts, in September 1962. His initial assignments at the space agency included service as a member of the backup crew for Gemini 4.

Lovell first flew in space during Gemini 7, lifting off on December 4, 1965 with Frank Borman. Borman served as commander of the flight, with Lovell as pilot. The Gemini 7 flight successfully extended NASA's longest single flight experience to nearly 14 days, significantly advancing the agency's understanding of the effects that extended stays in space might have on its early astronauts. Both Lovell and Borman emerged unharmed and in good spirits from their long odyssey in orbit.

They also played a role in the success of the Gemini 6 mission, in which Walter "Wally" Schirra and Thomas Stafford mastered rendezvous procedures by using Gemini 7 as a target. The Agena rocket that had originally been intended as the target for Schirra and Stafford had exploded shortly after it was launched on October 25, which led to Gemini 6 being postponed. Knowing that the Gemini 7 flight was scheduled for December, NASA mission planners decided to substitute Gemini 7 for the lost Agena, and their willingness to improvise worked well, as Schirra and Stafford were able to successfully rendezvous with their fellow astronauts' craft for more than five hours.

During their long Gemini 7 flight, Lovell and Borman participated in extensive medical investigations, including a study of the astronauts' sleeping patterns (the Inflight Sleep Analysis experiment) to monitor brain activity via an eletroencephalogram (EEG), careful measurements of the amount of calcium in the astronauts' systems before, during, and after the flight, and an investigation into the ways in which the stresses of spaceflight might be measured through analysis of urine samples, in an experiment known as the Bioassays of Body Fluids.

They also carried out technology and engineering experiments, Earth observations, and measurements. Their regularly scheduled work, rest, and eating periods gave mission managers insight into how astronauts could perform during longer stays in orbit.

Lovell and Borman completed 206 orbits during their historic flight and returned to Earth on December 18, 1965, splashing down in Gemini 7 in the Atlantic Ocean after a flight of 13 days, 18 hours, 35 minutes, and 1 second.

After his landmark first flight, Lovell served as backup commander during the Gemini 9 mission, which was flown by Thomas Stafford and Eugene Cernan in June 1966.

On November 11, 1966, Lovell set out on his second spaceflight as commander of Gemini 12, with pilot Edwin "Buzz" Aldrin.

In the final mission of the Gemini program, Lovell and Aldrin successfully demonstrated rendezvous and docking procedures, and Aldrin performed several important EVAs. Their successful flight greatly expanded the knowledge and expertise necessary for the rapidly approaching Apollo program.

After long months of innovative training and careful engineering of EVA equipment, including the hand and foot restraints and spacecraft railings that would later become standard equipment for spacewalking astronauts, Aldrin carried out three remarkable EVAs during Gemini 12, demonstrating the techniques necessary for doing useful work in the weightless environment.

Lovell and Aldrin also located an Agena target vehicle from a distance of 436 kilometers, despite the loss of their radar system, and achieved rendezvous and docking, with Aldrin calculating the proper rendezvous positions and Lovell expertly piloting the Gemini 12 capsule to rendezvous and dock with the Agena.

Lovell also piloted the Gemini craft during an innovative tethered flight experiment, with the manned vehicle and the unmanned target linked by a long cable. He first moved Gemini 12 into a position in which the tether extended vertically relative to the capsule, with the attached Agena target vehicle at the top of the vertical line. He found it difficult to keep the tether taut between the two vehicles, but was able to achieve both the primary purpose of the experiment, the generation of a period of artificial gravity, and the secondary goal, unattended stationkeeping (keeping the two spacecraft at a constant chosen distance).

In addition to their spectacular EVA and flying activities, Lovell and Aldrin also worked on a program of 14 scientific experiments, including the photographing of a solar eclipse. They splashed down on November 15, 1966 after making a controlled re-entry—another important goal of the Gemini 12 mission profile—after completing 59 orbits during a flight of 3 days, 22 hours, 34 minutes, and 31 seconds.

In 1967, Lovell began a long-time association with the promotion of physical fitness when he accepted a special assignment from President Lyndon Johnson to act as the president's consultant for physical fitness and sports. He took on the additional role of chairman of the President's Physical Fitness Council at the request of President Richard Nixon in 1970, and served in both capacities until 1978. He has since continued to serve as a consultant to the council.

Lovell's third space mission, the December 1968 Apollo 8 flight, ultimately became known as the "Christmas Apollo," as he and crew mates Frank Borman and William Anders made a remarkable broadcast from lunar orbit on Christmas Eve.

The December 21 1968 launch of Apollo 8 promised to end a particularly tumultuous year in American history on a positive note; the exploits of Lovell, Borman, and Anders were welcome news for television and radio audiences that had suffered through months of frightening developments, including a major setback in the Vietnam War, the assassinations of Dr. Martin Luther King and Senator Robert F. Kennedy, and rioting at the Democratic National Convention.

Interest in the Apollo program had grown substantially in the period following NASA's recovery from the tragic Apollo 1 fire in January 1967 and the first manned flight after the fire, Apollo 7. For the Apollo 8 crew, which would make the second manned flight of the program, NASA officials had set a remarkable goal: to make the first manned flight to the Moon and to become the first human beings to enter lunar orbit.

Although the lunar orbit objective had originally been planned for a later mission, the success of the unmanned Soviet Zond 6 flight, which had sent a vehicle to the Moon and back, caused NASA's top officials to reconfigure Apollo 8 as a manned lunar flight. Lovell, Borman, and Anders enthusiastically embraced the new plan despite its risks.

An audience of 500 million television viewers followed their exploits as they made regular broadcasts on their way to the Moon, and then, on Christmas Eve, the Apollo 8 crew members made one of the most enduring transmissions of the entire Apollo program when they read from the Bible's Book of Genesis, reciting the Judeo-Christian story of how the world began.

Anders began with the first four verses of Genesis, followed by Lovell and then Borman. In keeping with both the spirit and practice of the Christmas holiday and providing a moment of genuinely peaceful reflection for all those who were disheartened by the persistent problems of the previous year, the "Christmas Apollo" broadcast became a cherished symbol of the higher purposes of the space program for many observers.

The Apollo 8 mission also achieved its engineering and scientific goals, and Lovell and his crew mates returned stunning photographic images of the Moon and the Earth. They completed 10 orbits of the Moon and successfully tested their spacecraft and its systems while spending a total of 20 hours in lunar orbit, a particularly important achievement in light of NASA's plans for the rapidly approaching first lunar landing mission. Most importantly, Lovell, Borman, and Anders safely returned to Earth on December 27, 1968, after a flight of 6 days and 3 hours.

After making the first manned flight to the Moon, Lovell served as backup commander to Neil Armstrong for Apollo 11, the first lunar landing mission.

Lovell began his fourth spaceflight on April 11, 1970, as commander of Apollo 13.

The flight suffered unprecedented setbacks even before its launch, when primary crew member Thomas "Ken" Mattingly, who Lovell and Apollo 13 lunar module pilot Fred Haise had trained with throughout their long preparations for the flight, was suddenly removed from the mission by NASA medical personnel after he was exposed to the German Measles. John "Jack" Swigert moved up from the backup crew to replace Mattingly as the Apollo 13 command module pilot just days before the start of the flight.

Once they were safely in orbit, Lovell, Haise, and Swigert put the pre-launch stress of the last-minute crew substitution behind them and turned their attention to the Apollo 13 flight plan, which called for Lovell and Haise to become the fifth and sixth astronauts to land on the lunar surface. During their trip to the Moon, however, a fault in an oxygen tank aboard their spacecraft's service module caused the tank to explode, putting the crew in mortal danger. Any thoughts of completing the lunar landing portion of the flight were immediately abandoned as the crew rushed to move from the Odyssey command module to the Apollo 13 lunar module Aquarius.

The life support systems aboard Odyssey failed shortly after the explosion, and even as mission controllers monitoring the situation suggested that Lovell, Haise, and Swigert move to the lunar module, the astronauts themselves had come to the same conclusion and were already in the midst of making the transfer. Originally intended to support Lovell and Haise for about 40 hours while they traveled to the lunar surface and back to Odyssey, the Aquarius lunar module was pressed into service as a sort of lifeboat that would need to support all three crew members for several days while NASA support teams worked out procedures for their safe return to Earth.

NASA officials immediately activated all available personnel and resources within the space agency and at many of the industrial contractors and subcontractors that had designed and built various components of the Apollo spacecraft and its systems and equipment.

Although they faced unprecedented peril, Lovell and his crew mates remained calm and focused during their ordeal while NASA engineers and support teams carefully simulated on the ground each of the difficulties faced by the crew in orbit, in what all those involved quickly recognized as the most perilous mission of the entire U.S. space program up to that time. The crew members carefully conserved their supplies of food and water while waiting for communicators in Mission Control to transmit the details of the revised flight plan and procedures for bringing the Odyssey command module back to life for the return trip through the atmosphere, as the lunar module was not designed to survive re-entry.

Having absorbed the initial shock of the accident, Lovell and his crew mates faced their next major challenge when the environmental control system in the lunar module became overwhelmed and required immediate repair. Engineers simulated a solution for the crew to put into effect—a makeshift adaptation of filters that had originally been intended only for use in the command and service modules—and the quick fix proved adequate enough to keep the crew safe and well, despite the severe limitations of their circumstances.

Lovell, Haise, and Swigert also endured a series of nerve-wracking firings of the lunar module's engine to guide the linked command, service, and lunar modules into the proper trajectory for their return to Earth. They carried out the course corrections with great skill and precision, despite the difficulty of having to fly through debris left over from the explosion.

Four difficult days passed while the Apollo 13 crew traveled to the Moon and then made its way back to Earth using a "free return trajectory," which mission planners had, after careful consideration, chosen as the best way to return the astronauts safely to Earth. The decision to have the spacecraft continue on to the Moon and then return, rather than simply turning around immediately after the accident, was influenced by the amount of fuel a quick-turnaround would require.

Once they had successfully circled the Moon and made the trip back to the edge of the Earth's atmosphere, Lovell, Haise, and Swigert faced the most harrowing part of their entire ordeal. Swigert carried out a long sequence of tasks necessary to revive Odyssey while still preserving enough electrical power to enable the spacecraft to safely withstand re-entry. Lovell and Haise completed the transfer from the lunar module to the command module, and after separating Odyssey from the lunar and service modules, the three astronauts prepared for their return to Earth.

Despite the substantial damage it had sustained as a result of the accident, the Apollo 13 Odyssey command module remained intact during the fiery plunge through the Earth's atmosphere, and Lovell, Haise, and Swigert splashed down safely in the Pacific Ocean on April 17, 1970 after a harrowing 5 days, 22 hours, and 54 minutes in space. They emerged exhausted and ill from their ordeal, but without lasting damage to their health.

Apollo 13 became known in NASA lore as the agency's "successful failure," an audacious mission that, while not achieving the objectives of its mission profile, put the collective abilities of astronauts, engineers, scientists, and space agency officials to their most extreme test and found them capable of achieving the overriding goal of the space program: the safe return of the individuals who were sent to explore outer space.

James Lovell spent more than 29 days in space during his remarkable career as an astronaut.

Continuing his education concurrently with his space career, Lovell attended the Harvard Business School's advanced management program, from which he graduated in 1971. He retired from the U.S. Navy and from NASA on March 1, 1973 to pursue a private sector career.

Lovell began his post-NASA career with the Bay-Houston Towing Company in Houston, Texas, where he subsequently became president and chief executive officer in 1975. In 1977, he became president of Fisk Telephone Systems, and in 1981 became group vice president of Centel Corporation's Business Communications Systems. He retired from Centel Corporation on January 1, 1991 as executive vice president and a member of the company's board of directors.

Among the many honors he has received for his achievements as an astronaut, Lovell was awarded the Presidential Medal for Freedom, the Robert J. Collier Trophy, the Robert H. Goddard Memorial Trophy, the Haley Astronautics Award of the American Institute of Astronautics and Aeronautics (AIAA), the U.S. Air Force Thomas D. White Space Trophy, the H. H. Arnold Trophy and the Institute of Navigation Award. He was also a three-time recipient of the Harmon International Trophy and was twice a co-recipient of the American Astronautical Society Flight Achievement Award.

LOW, G. DAVID

(1956–2008)

U.S. Astronaut

David Low was born in Cleveland, Ohio on February 19, 1956. In 1974 he graduated from Langley High School in McLean, Virginia, and then attended Washington & Lee University, where he received a Bachelor of Science degree in physics and engineering in 1978.

He then attended Cornell University, where he earned a Bachelor of Science degree in mechanical engineering in 1980. He began his professional career at the Jet Propulsion Laboratory at the California Institute of Technology in 1980, in the Spacecraft Systems Engineering section. His work there involved the development of several unmanned space missions, including the Galileo spacecraft that was eventually launched from the space shuttle Atlantis on July 13, 1985 during STS-34.

In 1983, he earned a Master of Science degree in aeronautics and astronautics from Stanford University and then served as the principal spacecraft systems engineer for the Mars Geoscience/Climatology Observer Project. He became an astronaut in 1985.

Low first flew in space as a mission specialist aboard the space shuttle Columbia during STS-32, which launched on January 9, 1990. The 11-day flight set a new record for the longest shuttle mission up to that time (a record previously held by the STS-9 crew, who flew from November 28 to December 8, 1983).

The STS-32 crew deployed the SYNCOM IV-F5 defense communications satellite (also known as LEASAT 5) and successfully retrieved the Long Duration Exposure Facility (LDEF), an orbiting cylinder containing 57 experiments designed to test the effects of long exposure to the space environment on a variety of materials. The LDEF had been deployed during STS-41C in April 1984 and was originally intended to be retrieved on an earlier flight, but the January 1986 Challenger accident had delayed its retrieval until STS-32. Columbia returned to Earth during a nighttime landing at Edwards Air Force Base on January 20, 1990, after 173 orbits.

On his second space mission, STS-43, Low flew aboard the shuttle Atlantis as a mission specialist. At launch on August 2, 1991, the shuttle was heavier than any previously flown shuttle, thanks largely to its hefty payloads, which included the Tracking and Data Relay Satellite-5 (TDRS-E). The total weight, including the shuttle itself, was 235,000 pounds. The satellite was successfully deployed about six hours after launch, and during the rest of the flight the crew performed 32 scientific experiments in life sciences, materials science, and physical science. Atlantis touched down at the Kennedy Space Center (KSC) in Florida on August 11, 1991, after 142 orbits.

On June 21, 1993, Low lifted off on a remarkable third spaceflight, this time as a payload commander for the STS-57 mission aboard the shuttle Endeavour. He and crew mate Peter Wisoff were scheduled to perform a spacewalk during the flight to try out various EVA tasks for future missions, but as things turned out, they also carried out a remarkable repair job during the flight that salvaged a key component of the STS-57 mission profile.

The problem arose when the crew captured the European Retrievable Carrier (EURECA) on flight day three. Developed by the European Space Agency (ESA), EURECA had been deployed during the STS-46 flight of the shuttle Atlantis in 1992. When Endeavour retrieved the spacecraft on June 24, ground controllers found that they could not get EURECA's antennae to return to the proper position for the flight back to Earth. Improvising a solution, Low and Wisoff began their scheduled EVA the following day by making an unscheduled detour to manually fold the antennae, thus allowing for proper stowage of EURECA. Their total spacewalk, including the impromptu salvage duties, took 5 hours and 50 minutes.

STS-57 also marked the first flight of the pressurized SPACEHAB laboratory, which the crew used to perform 22 experiments.

The mission was extended twice because of poor weather at KSC in Florida before Endeavour finally landed there on July 1, 1993, after 155 orbits.

David Low accumulated more than 714 hours in space, including more than 5 hours in EVA, during his career as an astronaut.

Low continued to serve NASA after his third spaceflight, initially helping to transition the Space Station Freedom project to the International Space Station (ISS) program, as a member of the Russian Integration Team in 1993. The following year, he became manager of the EVA Integration and Operations Office, and in 1995 he worked as a liaison between NASA and the U.S. Congress, as an assistant in the agency's Legislative Affairs Office.

Low left NASA in 1996 to take a position with the Launch Systems Group of Orbital Sciences Corporation in Dulles, Virginia.

David Low died on March 15, 2008. He was survived by his wife, JoAnn, and their three children, his mother, and several brothers and sisters.

LU, EDWARD T.

(1963–)

U.S. Astronaut

A veteran of three space missions who visited the legendary Russian space station Mir and spent six months aboard the International Space Station (ISS) in 2003, Edward Lu was born in Springfield, Massachusetts on July 1, 1963. He graduated from R. L. Thomas High School in Webster, New York in 1980, and then attended Cornell University, where he earned a Bachelor of Science degree in electrical engineering in 1984 and was recognized as a Cornell University Presidential Scholar. He went on to attend Stanford University, where he was awarded a Masters fellowship sponsored by the Hughes Aircraft Company, and where he earned a Ph.D. in applied physics in 1989.

Lu began his career as a visiting scientist at the High Altitude Observatory in Boulder, Colorado. In 1992, he was also attached to the University of Colorado's Joint Institute for Laboratory Astrophysics, while continuing his work at the High Altitude Observatory.

He subsequently pursued his research interests as a postdoctoral fellow at the Institute for Astronomy in Honolulu, Hawaii. Lu's wide-ranging research interests are reflected in the many scientific articles that he has published, covering subjects such as solar flares, plasma physics, statistical mechanics, and asteroids.

In addition to his outstanding academic career, Lu also pursued training as a pilot and earned a commercial pilot's license with instrument and multi-engine ratings. He has accumulated more than 1,300 hours of flying time.

NASA selected Lu as a member of its Group 15 selection of pilots and mission specialist astronauts in 1995. He underwent a one-year period of intensive training at the Johnson Space Center (JSC) in Houston, Texas, and became qualified for assignment as a space shuttle mission specialist in 1996. His initial duties as an astronaut included service in the computer support branch of NASA's Astronaut Office, and he also served as lead astronaut for space shuttle training and as lead astronaut for space station training.

Lu first flew in space as a mission specialist aboard STS-84, the sixth docking of an American space shuttle and the Russian space station Mir, in May 1997.

Launched aboard the shuttle Atlantis on May 15, the STS-84 crew was commanded by Charles Precourt, and also included pilot Eileen Collins and Lu's fellow mission specialists Carlos Noriega, Jean-Francois Clervoy of the European Space Agency (ESA), Elena Kondakova of the Russian Space Agency (RSA), and NASA's Michael Foale, who replaced fellow NASA astronaut Jerry Linenger aboard Mir.

After docking with Mir on May 16, 1997, the STS-84 crew and Mir 23 crew members Vasili Tsibliyev and Aleksandr Lazutkin transferred 7,500 pounds of equipment and supplies from the shuttle to the station, and moved experiment samples and other materials onto the shuttle for their return to Earth. Linenger also returned with the STS-84 crew.

Atlantis also carried a pressurized SPACEHAB Double Module and Biorack facility during STS-84, which the crew used to conduct an extensive program of scientific experiments.

The shuttle undocked from Mir on May 21 and returned to Earth on May 24, 1997 after a flight of 9 days, 5 hours, 19 minutes, and 56 seconds.

Lu next flew in space aboard Atlantis during STS-106, for which he served as a mission specialist and as payload commander. Launched on September 8, 2000 under the command of Terence Wilcutt, along with pilot Scott Altman and Lu's fellow mission specialists Daniel Burbank, Richard Mastracchio, and Russian space agency cosmonauts Yuri Malenchenko and Boris Morukov, STS-106 was designed to prepare the ISS for its first long-duration crew.

Two days after launch, Atlantis docked with the space station and the crew began unloading supplies from the Progress-M1 3 supply vehicle, which had launched unmanned on August 6 and automatically docked with the ISS.

With his fellow mission specialist Yuri Malenchenko, Lu conducted a 6 hour and 14-minute spacewalk on September 11, 2000 to connect communications cables and electrical power lines between the station's Zvezda and Zarya modules and to install the ISS magnetometer. During the EVA, Lu and Malenchenko reached a point 100 feet above the space shuttle's cargo bay, the farthest distance any tethered spacewalker had achieved up to that time.

The crew spent an extra day in space to finish their chores at the station, preparing the ISS for its first resident crew by transferring nearly three tons of equipment and supplies from the shuttle and the Progress cargo vehicle and performing a variety of maintenance and repair tasks.

After more than five days of docked operations, Atlantis undocked from the ISS on September 17, and the STS-106 crew returned to Earth on September 20, 2000, after 185 orbits and a flight of 11 days, 19 hours, 12 minutes, and 15 seconds.

On April 25, 2003, Lu began a remarkable third space mission, during which he returned to the ISS to serve as flight engineer of the station's seventh long-duration crew. He lifted off from the Baikonur Cosmodrome on April 25 aboard Soyuz TMA-2 with ISS Expedition 7 commander Yuri Malenchenko, who had previously visited the ISS with Lu during STS-106. During their second visit to the station, they would live and work together in orbit for more than six months.

Their Expedition 7 scientific program included experiments in biotechnology, fluid dynamics, medical research, and protein crystal growth. Lu also witnessed the first wedding to be conducted with a groom in space, when Malenchenko was married via a long-distance ceremony on August 10.

Their replacements arrived on October 18, when the Expedition 8 crew arrived with ESA astronaut Pedro Duque of Spain, who visited the station to conduct the ESA Cervantes program of scientific research. Lu and Malenchenko returned to Earth with Duque on October 28, 2003. They accumulated more than 184 days in space during their ISS Expedition Seven mission.

Edward Lu spent more than 206 days in space, including 6 hours and 14 minutes in EVA, during his three space missions.

Lu has continued to serve as a NASA astronaut since the completion of his third spaceflight, working in the exploration branch of the Astronaut Office.

LUCID, SHANNON W.

(1943–)

U.S. Astronaut

Shannon Lucid was born in Shanghai, China on January 14, 1943. In 1960, she graduated from Bethany High School in Bethany, Oklahoma, and then attended the University of Oklahoma, where she received a Bachelor of Science degree in chemistry in 1963.

She served as a teaching assistant in the department of Chemistry at the University of Oklahoma for one year following her graduation, and then worked at the Oklahoma Medical Research Foundation as a senior laboratory technician from 1964 to 1966, when she joined the Kerr-McGee company in Oklahoma City as a chemist.

Shannon Lucid has flown in space five times. Here she works out during her long-duration Mir mission in 1996. [NASA/courtesy of nasaimages.org]

In 1969, she joined the department of Biochemistry and Molecular Biology at the University of Oklahoma's Health Science Center, where she worked as a graduate assistant while pursuing her graduate education. She received a Master of Science degree in biochemistry from the university in 1970 and a Ph.D. in biochemistry in 1973.

Lucid next worked as a research associate at the Oklahoma Medical Research Foundation beginning in 1974, until NASA selected, in 1978, for training as an astronaut. She is a commercial, instrument, and multi-engine rated pilot.

She first flew in space as a mission specialist during STS-51G, which launched aboard the shuttle Discovery on June 17, 1985. The STS-51G crew deployed three communications satellites during the busy flight, including the Mexican MORELOS-A satellite and the ARABSAT-A spacecraft of the Arab Satellite Communications Organization. The international crew included payload specialists Patrick Baudry of France and Sultan Salman Al-Saud, the first citizen of Saudi Arabia to fly in space. The mission also featured the High Precision Tracking Experiment (HPTE), a laser tracking test related to the proposed Strategic Defense Initiative (SDI). The seven-day flight ended with Discovery's landing at Edwards Air Force Base in California on June 24, 1985.

Lucid's second space mission was STS-34, aboard the shuttle Atlantis in October 1989. The main payload for STS-34 was the Galileo Jupiter orbiter and probe, which became the first spacecraft to orbit Jupiter and the first probe to enter the planet's atmosphere.

The crew deployed the Galileo craft six and a half hours into the flight, starting it on its six-year journey to Jupiter. At the end of that journey, long after the STS-34 mission, Galileo successfully entered orbit around Jupiter and released its probe into the planet's atmosphere on July 13, 1995. Both the orbiter and the probe returned a great deal of scientific data, including evidence of a previously unknown radiation belt around the planet, the possibility of water beneath the surface of Jupiter's moon Europa, and ongoing volcanic activity on the moon Io. Galileo also discovered magnetic fields on Io and Ganymede and measured winds of more than 600 kilometers on Jupiter.

After successfully deploying Galileo and conducting a variety of scientific experiments, the STS-34 crew returned to Earth on October 23, 1989.

On her third spaceflight, STS-43 in 1991, Lucid again served as a mission specialist aboard Atlantis. At lift-off on August 2, the shuttle was heavier than any previously flown shuttle, thanks largely to its hefty payloads, which included the Tracking and Data Relay Satellite-5 (TDRS-E). The total weight, including the shuttle itself, was 235,000 pounds. The satellite was successfully deployed about six hours after launch, and the rest of the flight was largely dedicated to scientific experiments. Atlantis touched down at KSC on August 11, 1991, after 142 orbits.

Lucid lifted off on a fourth spaceflight on October 18, 1993, during the second Spacelab flight dedicated to life sciences, STS-58, aboard the shuttle Columbia. At just over 14 days, STS-58 was also the longest flight of the shuttle program up to that time. The crew performed a variety of Spacelab experiments during the mission in the areas of regulatory physiology, the cardiovascular and cardiopulmonary systems, the musculoskeletal system, and neuroscience, and also conducted engineering tests and extended duration medical experiments. Columbia returned to Earth at Edwards Air Force Base in California on November 1, 1993.

With the completion of her STS-58 flight, Lucid set a new record for the most time in space by an American astronaut, and she had not yet begun the epic mission that would define her career and set a new benchmark for living and working in space.

After the STS-58 mission, she spent a year in Star City, Russia, training for a long-duration flight aboard the Russian space station Mir.

On March 22, 1996, Lucid lifted off aboard the STS-76 flight of the shuttle Atlantis. The third docking mission between a shuttle and Mir, STS-76 marked the beginning of a continuous U.S. presence aboard the Russian space station, as part of the two nation's initial arrangements for their further cooperation in the International Space Station (ISS) program. During five days of linked flight, the shuttle crew delivered two tons of equipment and supplies to Mir 21 cosmonauts Yuri Onufrienko and Yury Usachev, and Lucid transferred from Atlantis to Mir to begin an amazing four and a half month stay aboard the station.

During her epic stay on Mir, Lucid performed scientific experiments in the areas of life sciences and physical sciences, and, along with her Russian crew mates, participated in the daily routines of long-duration life aboard a space station, which typically include a significant emphasis on daily exercise aimed at lessening the losses in bone density and blood plasma volume that typically occur during long stays in space.

At the end of her long-endurance flight, after 188 days, 4 hours, and 14 seconds in space, Lucid returned to Earth aboard the shuttle of Atlantis during STS-79, the fourth shuttle-Mir docking. By the time her mission aboard Mir came to a close, she was serving with the Mir 22 crew of Valery Korzun and Alexander Kaleri. Korzon and Kaleri had arrived at the station on August 17, 1996, accompanied by French cosmonaut Claudie Andre-Deshays (who, during the mission, became the first French woman to fly in space and is also known by her married name, Claudie Haigneré), who made a brief visit before returning to Earth on September 2.

When the STS-79 Atlantis crew arrived at Mir later in September, they delivered NASA astronaut John Blaha, who replaced Lucid and became the next American to live and work on Mir as part of the continuous American presence aboard the station.

Lucid returned to Earth on Atlantis on September 26, 1996, having established a new U.S. record for long-duration spaceflight and a new world record for the longest stay in space by a woman. Her mark of more than 188 days was surpassed by Sunita Williams in 2007.

Shannon Lucid spent 223 days in space during her remarkable career

In February 2002 she became NASA's Chief Scientist, serving at the agency's headquarters in Washington, D.C. for a little over a year, before returning to her management role in the Astronaut Office at the Johnson Space Center (JSC) in Houston in September 2003.

LUNA PROGRAM

(1959–1976)

Soviet Lunar Exploration Program

Consisting of a series of increasingly sophisticated scientific satellites launched by the Soviet Union with the purpose of exploring the Moon, the Luna program was intended to build on the Soviet's successful early Sputnik launches. The program was carried out over the entire course of the first era of space exploration and coincided with the Soviet-American space race and each nation's attempts to be the first to send a manned spacecraft to the Moon. The unmanned Luna series was conducted independently of the Soviet Zond program, which was responsible for the development of the planned Soviet manned lunar missions, but the two projects were closely coordinated.

Over the course of the 24 named flights of the Lunar program (there were also a number of launch failures and flights that failed for a variety of reasons that were either concealed completely at the time or given other designations), the Soviets

First photographed by Luna 3, the Moon's Tsiolkovskii crater was named for pioneering Russian space theorist Konstantin Tsiolkovskii. [NASA/courtesy of nasaimages.org]

successfully placed seven spacecraft into orbit around the Moon, impacted the lunar surface four times, and completed eight soft landing missions that returned three samples of lunar surface materials to Earth and deployed two small robotic rovers, known as Lunokhods.

The rovers were among the Luna program's most innovative achievements, and the sample return missions also represented a major achievement, albeit one that was entirely overshadowed by the manned U.S. Apollo landings. During the early years of the Luna program, several launches achieved important milestones, including the placing of the first spacecraft into orbit around the Sun (Luna 1), the first successful lunar impact (Luna 2), the first return of photographs of the far side of the Moon (Luna 3), the first soft landing and first detailed photos of the lunar surface (Luna 9), and the first placement of a spacecraft in lunar orbit (Luna 10).

Luna vehicles also carried out a diverse array of observations and measurement of the lunar environment, returning data about the Moon's gravity, radiation and temperature, and the composition of the lunar crust.

Luna 1 (Launched January 2, 1959)

A 261-kilogram (575.4 pounds) spherical satellite equipped with radio and telemetry equipment and scientific instruments, Luna 1 was launched toward the Moon on January 2, 1959, apparently with the intent of its impacting the surface, but it passed the lunar surface at a distance of 5,995 kilometers two days later. It continued on to enter orbit around the Sun between the orbits of Earth and Mars.

Data returned by the craft's scientific equipment, which included a Geiger counter, magnetometer, micrometeorite detector, and scintillation counter led to the discovery of the solar wind and the realization that the Moon had no magnetic field.

Luna 2 (Launched September 12, 1959)

The first spacecraft to impact another planetary body, Luna 2 was launched on September 12, 1959. Similar in design and instrumentation to Luna 1, the Luna 2 satellite was designed to emit a cloud of sodium gas during the second day of its flight to help engineers track its progress and as an experiment intended to study how gases behave in space. The vehicle crashed as planned in the Palus Putredinus region of the Moon, near the Aristedes, Archimedes, and Autolycus craters, on September 14, 1959.

Luna 3 (Launched October 4, 1959)

Launched on October 4, 1959—two years to the day after the launch of Sputnik 1—the Luna 3 satellite was the first spacecraft to return photos of the far side of the Moon. Although the images were rudimentary by later standards, the successful acquisition of data about the far terrain of the lunar surface was a remarkable achievement for the time, and the images appeared to show mountains and several dark regions. Twenty-nine photographs were captured, 17 of which were successfully transmitted to Earth.

Luna 4 (Launched April 2, 1963)

The first of the Lunar series to feature a heavier, more sophisticated spacecraft, Luna 4 was launched on April 2, 1963 and was first placed in an Earth orbit before continuing its flight toward the Moon (the first three Luna satellites had been launched in a direct ascent trajectory, meaning that they were sent directly toward the Moon without first entering Earth orbit). Probably intended as a lunar impact mission, the 1,422-kilogram (3,134.97 pounds) Luna 4 was successfully placed on the proper course, but sailed past the Moon on April 5, 1963 at a distance of slightly more than 8,336 kilometers (5,179.75 miles).

Luna 5 (Launched May 9, 1965)

Likely intended to make a soft landing on the lunar surface, the Luna 5 satellite was launched on May 9, 1965. The vehicle suffered a malfunction as it neared the end of its flight and crash-landed on the lunar surface.

Luna 6 (Launched June 8, 1965)

Launched toward the Moon on June 8, 1965, Luna 6 flew past the lunar surface at a distance of greater than 159,612 kilometers (99,178.29 miles).

Luna 7 (Launched October 4, 1965)

The Soviets launched Luna 7 on October 4, 1965, apparently with the intent of attempting a soft landing on the lunar surface. A malfunction at the end of the flight caused the spacecraft to crash onto the surface in the lunar region known as the Sea of Storms.

Luna 8 (Launched December 3, 1965)

Apparently another attempt to make a soft landing on the lunar surface, Luna 8 launched on December 3, 1965 and crashed in the same area as Luna 7.

Luna 9 (Launched January 31, 1966)

On February 3, 1966, Luna 9 became the first spacecraft to land successfully and intact on the lunar surface. The spacecraft was built in two pieces, a bus that carried the Luna 9 landing probe to the Moon and the lander itself, which was designed with overlapping "petals" that unfolded after landing.

Launched atop a four-stage A-2-E launch vehicle on January 31, 1966, Luna 9 was equipped with a television system that successfully returned images of the lunar surface showing the area around the spacecraft to a distance of nearly 1.5 kilometers.

Luna 10 (Launched March 31, 1966)

Launched on March 31, 1966, Luna 10 was the first spacecraft to enter orbit around another planetary body. Equipped with a variety of instruments for scientific study, including a magnetometer, a meteorite detector, and a gamma ray spectrometer, Luna 10 entered lunar orbit on April 3, 1966. The spacecraft also carried radio transmitting equipment that relayed the Soviet anthem "The Internationale" to Earth. Luna 10 completed 460 orbits of the Moon before the end of its mission, on May 30, 1966.

Luna 11 (Launched August 24, 1966)

A scientific satellite designed to measure the chemical composition of the lunar surface, study the Moon's gravitational and radiation environment, and to study meteorites in lunar orbit, Luna 11 was launched on August 24, 1966. The spacecraft was successfully inserted into lunar orbit on August 28, 1966 and made 277 orbits before it failed on October 1, 1966.

Luna 12 (Launched October 22, 1966)

Launched on October 22, 1966, Luna 12 successfully entered into orbit around the Moon three days later, and its television system returned photos of the lunar surface. It continued to transmit data until January 19, 1967 and made 602 orbits of the Moon.

Luna 13 (Launched December 21, 1966)

The Soviets launched Luna 13 on December 21, 1966 and successfully landed it intact on the lunar surface three days later. The spacecraft returned panoramic photographic images of the area around its landing, and was equipped with instruments designed to make studies of the lunar surface, but apparently failed within several weeks of landing.

Luna 14 (Launched April 7, 1968)

Luna 14 was launched on April 7, 1968 and successfully entered into orbit around the Moon on April 10, 1968. Equipped with a variety of instruments, the spacecraft returned scientific data about long-distance radio communications, cosmic rays, and solar charged particles.

Luna 15 (Launched July 13, 1969)

The first of a third series of Luna vehicles, Luna 15 was launched on July 13, 1969 and traveled to the Moon at the same time that the American Apollo 11 crew was attempting the first manned lunar landing mission. Although Soviet space officials would not confirm the spacecraft's mission, they reassured NASA officials that the tiny craft would not interfere with the activities of the Apollo 11 crew. Varying accounts over the years have speculated that Luna 15 was perhaps intended as a sample return mission, which would have allowed the Soviets to salvage some of their pride in light of their failure to beat the U.S. to the surface with a manned landing.

What is known about the flight is that Luna 15 entered lunar orbit several days after its launch and completed 52 orbits of the Moon before crashing on July 21, 1969 into the area known as the Sea of Crises.

Luna 16 (Launched September 12, 1970)

A remarkable first in the field of robotic space probes, Luna 16 was launched on September 12, 1970. Soviet space engineers had designed the craft to make a soft landing on the lunar surface and then drill a small hole in the crusty landscape to collect sample material that would then be launched from the Moon via the spacecraft's ascent stage and returned to Earth in a tiny soil sample container.

Luna 16 landed in the region of the Moon known as the Sea of Fertility on September 20, 1970, achieving the first nighttime lunar landing. The craft is thought to have returned photographic images of the area around the landing site and to have collected the planned sample of surface material, which was deposited to the sample container for the return trip to Earth.

The vehicle's ascent stage left the surface on September 21 with 101 grams of sample material sealed in the vehicle's re-entry module, which landed in Dzhezkazgan, Kazakhstan on September 24, 1970 and was successfully recovered.

Luna 17 / Lunokhod 1 (Launched November 10, 1970)

Trumping the remarkable achievements of the Luna 16 mission, Luna 17 was the first spacecraft to deploy a remote-controlled rover, Lunokhod 1, to facilitate an in-depth examination of the lunar surface. Launched on November 10, 1970, Luna 17 reached lunar orbit five days later and then landed intact in the region known as the Sea of Rains on November 17.

Lunokhod 1 rolled out of the spacecraft's main body a short time later, traveling down a ramp to reach the lunar surface. Equipped with four television cameras, scientific instruments, and a probe for collecting and testing soil samples, Lunokhod 1 surpassed even the most optimistic aims of its designers when it returned some 20,000 images, including 200 panoramic images, and carried out 500 tests of the lunar soil. The rover continued to operate until October 4, 1971.

Luna 18 (Launched September 2, 1971)

With a mission profile similar to that of Luna 16, Luna 18 was launched on September 2, 1971 and entered lunar orbit five days later. After 54 orbits of the

Moon, when Soviet mission managers attempted to land Luna 16 on the surface, the spacecraft crash-landed, ending the mission before a surface sample could be collected or returned.

Luna 19 (Launched September 28, 1971)

Launched on September 28, 1971, Luna 19 successfully entered lunar orbit on October 3, 1971, where it conducted scientific studies and returned photographs of the lunar surface.

Luna 20 (Launched February 14, 1972)

The third attempt by the Soviet space program to obtain a sample of lunar surface materials, Luna 20 was launched on February 14, 1972 and entered lunar orbit four days later. The spacecraft landed in the Apollonius Highlands region of the Moon on February 21, 1972.

Like Luna 16, Luna 20 successfully returned television images of the area around the landing site and dug into the surface to collect a soil sample. The vehicle's ascent stage blasted off from the Moon on February 20 with a sample weighing 30 grams, and the return capsule landed on February 25, 1972 and was subsequently recovered.

Luna 21 / Lunokhod 2 (Launched January 8, 1973)

Launched on January 8, 1973, Luna 21 landed the second remote-controlled rover vehicle, Lunokhod 2, on the lunar surface.

Luna 21 entered lunar orbit on January 12, 1973, and on January 15 the spacecraft was directed onto the lunar surface, where it landed in the LeMonnier Crater. The spacecraft unfolded shortly afterward, exposing two ramps for the deployment of the Lunokhod 2 rover.

Operated by teams of mission managers on Earth, Lunokhod 2 rolled onto the Moon's surface on January 16, where it returned television images of the area around the landing site and of the lander itself. The rover was equipped with a heater to maintain its systems during the cold lunar nights and also featured rechargeable batteries that were restored to full operating capacity several times during the mission via the craft's solar panels.

The Lunokhod 2 rover also collected and tested soil samples, and Soviet scientists used its laser retro-reflector instrument to conduct laser locating experiments from Earth that were similar to those that NASA carried out with instruments deployed by the Apollo lunar explorers.

Lunokhod 2 returned more than 80,000 television images before its mission was completed in early June 1973.

Luna 22 (Launched May 29, 1974)

Launched on May 29, 1974, Luna 22 successfully entered orbit around the Moon on June 2, 1974. It conducted extensive scientific studies of the lunar surface, collecting data about the surface gamma ray emissions and the composition of surface materials during its long mission, which ended in November 1974.

Luna 23 (Launched October 28, 1974)

Luna 23, the Soviets' fourth attempt at a lunar sample return flight, was launched on October 28, 1974 and landed several days later in the region of the Moon known as the Sea of Crises. The spacecraft did not achieve the planned soft landing that mission managers had hoped for, and was unable to collect and return the desired samples. Communications with Luna 23 were lost within three days after the spacecraft's impact.

Luna 24 (Launched August 9, 1976)

The final flight of the Luna program, Luna 24 launched on August 9, 1976 and landed on the Moon on August 18, 1976, near the spot where Luna 23 had earlier impacted the surface in the Sea of Crises. The Luna 24 craft fared far better than its predecessor, however, and ultimately collected and returned a far greater sample than either of the previous successful sample return flights, Luna 16 and Luna 20.

Mission managers remotely deployed the drill at the end of the spacecraft's robotic arm to collect samples that, collectively, weighed just over 170 grams. Then, safely sealed within a return container, the samples were launched in the spacecraft's ascent stage on August 19 and returned to Earth on August 22, 1976, when they landed near Surgut, Siberia and were recovered.

LUNAR ORBITER PROGRAM

(1966–1968)

U.S. Lunar Exploration Program

A key component of U.S. unmanned lunar exploration in the mid-1960s, the Lunar Orbiter Program gathered photographic and mapping data about the lunar surface in anticipation of the later Apollo Moon landings and contributed to scientific understanding of the Moon's gravitational environment. With the Ranger and Surveyor programs, the Lunar Orbiter project made it possible for Apollo mission planners and NASA scientists to increase the safety of the lunar landing missions and to select landing sites that were most likely to yield the most productive explorations.

In contrast to the flights of the earlier Ranger program, which had begun primarily as a means of exploring and photographing the lunar surface for scientific purposes and was later transformed in its final missions to a support project for the planned Apollo program, the first of the Lunar Orbiters were devoted to close inspection of 20 sites considered for possible Apollo landings, and the final two missions were used for expanded mapping of the surface for scientific research purposes.

The Lunar Orbiter spacecraft. [NASA/courtesy of nasaimages.org]

The Lunar Orbiter Program was managed by the staff of NASA's Langley Research Center in Hampton, Virginia.

A little more than 1.5 meters tall, with a base that was exactly 1.5 meters in diameter, the Lunar Orbiter spacecraft featured four solar panels that extended to a length of about 3–3.25 meters from end to end. The four panels generated 375 watts of power, a portion of which was stored in nickel-cadmium batteries to provide power when the spacecraft was unable to draw solar power. Each of the five spacecraft in the series was launched atop an Atlas-Agena launch vehicle.

Scientific equipment carried aboard the Lunar Orbiters included instruments to measure micrometeoroid impacts and radiation, and the craft's communications system included a 10-watt transmitter and two S-band antennas, including one dedicated to the transmission of photographs.

Equipped with a high resolution camera that was able to gather images of the lunar surface with both an 80 millimeter wide angle lens and a 610 millimeter telephoto lens, the Lunar Orbiters recorded images on 70 millimeter film which was purposely shifted slightly while the camera was taking photos, to prevent the images from being blurred by the motion of the spacecraft. The film was then processed on board the spacecraft, and the resulting images were scanned and transmitted to Earth.

In its five launches, the Lunar Orbiter project gathered 2,180 high resolution and 882 medium resolution photographs of the lunar surface that accounted for approximately 99 percent of the terrain on both the near and far side of the Moon. Mission controllers were able to make detailed photographic studies of 20 potential landing sites for the later Apollo program, and used three of the spacecraft, Lunar Orbiter 2, Lunar Orbiter 3, and Lunar Orbiter 5, to test the Earth-based tracking stations that would later be used for Apollo, and to help determine orbital information for the Apollo flights.

The first flight in the series achieved an historic "first" for the U.S. space program when it became the first U.S. spacecraft to enter into orbit around another celestial body.

One of the most important, and least expected, results of the Lunar Orbiter program was the discovery that the Moon's gravitational field was not uniform. Apollo mission planners were able to take into account the substantial variations in gravity that exist over certain large craters on the lunar surface, and as a result were able to accurately estimate landing coordinates for the Apollo lunar modules.

The total cost of the Lunar Orbiter Program, including the design, launch, and control of the five spacecraft and the collection and return of the images, was estimated at $163 million.

Lunar Orbiter 1 (August 10, 1966–October 29, 1966)

Designed to photograph prospective landing sites for the subsequent unmanned Surveyor and manned Apollo missions, Lunar Orbiter 1 was launched on August 10, 1966. After successfully achieving its intended Earth parking orbit and then being placed on the proper trajectory, the spacecraft traveled to the Moon and became the first U.S. spacecraft to be successfully placed into orbit around a celestial body, with initial lunar orbital parameters of 189.1 kilometers by 1,866.8 kilometers

Lunar Orbiter 1 took the first photograph of the Earth from the Moon and also returned 42 high resolution and 187 medium resolution photos of the lunar surface, including clear photos of the lunar far side. A malfunction in the spacecraft's imaging system resulted in higher-than-expected blurring in the high resolution photos, but the spacecraft's scientific instruments worked as expected.

At the end of its mission, after it had completed 576 trips around the Moon, Lunar Orbiter 1 was purposely crashed on the lunar far side on October 29, 1966 to avoid any chance that it might interfere with communications during later flights.

Lunar Orbiter 2 (November 6, 1966–October 11, 1967)

The second Lunar Orbiter spacecraft was launched on November 6, 1966, and successfully entered into orbit around the Moon 92.5 hours after lifting off from

Cape Kennedy. The spacecraft's initial orbit of 196 kilometers by 1,850 kilometers was altered several times during its mission to accommodate the photographic and scientific goals of the flight.

During seven days of photo operations, Lunar Orbiter 2 captured and transmitted 609 close-up and 208 wide-angle photos of the lunar surface, including an image of the Copernicus Crater that was prominently featured in the U.S. news media. The spacecraft's instruments transmitted data about the first micrometeorite impacts recorded during the program, and the vehicle was also used to test the Earth-based tracking stations that would later be used during the Apollo program before it was purposely crash-landed on the lunar surface on October 11, 1967 to avoid the possibility that its communications system might interfere with other spacecraft during later flights.

Lunar Orbiter 3 (February 5, 1967–October 9, 1967)

Launched on February 5, 1967, Lunar Orbiter 3 was successfully placed in lunar orbit three days later, on February 8, 1967, and ultimately returned 477 high resolution and 149 medium resolution images of the lunar surface despite a malfunction in the frame advance mechanism of its camera system, which resulted in a quarter of the spacecraft's total picture-taking harvest being lost. The images gathered by Lunar Orbiter 3 were designed to assist Apollo mission planners in choosing sites for the later manned Apollo landings.

Among the images that were successfully returned by Lunar Orbiter 3, one in particular created great excitement because it showed the Surveyor 1 spacecraft on the lunar surface. Surveyor 1 had launched on May 30 of the previous year and landed on the Moon on June 2, 1966.

Lunar Orbiter 3 was also used, with Lunar Orbiter 2 and Lunar Orbiter 5, to test the Earth-based tracking stations that would later be used for Apollo, before the spacecraft was purposely crashed onto the lunar surface on October 9, 1967 to avoid the chance that it might interfere with the communications system of other vehicles during later flights.

Lunar Orbiter 4 (May 4, 1967–October 31, 1967)

The fourth spacecraft in the Lunar Orbiter series was launched on May 4, 1967. In a departure from the mission profile of the first three spacecraft in the series—which had successfully returned the data necessary for planning the later manned Apollo landing flights—Lunar Orbiter 4 and Lunar Orbiter 5 were used to conduct high resolution photographic surveys of the lunar surface to expand scientific knowledge of the Moon, to aid in mapping the surface, and to collect scientific data via their on-board instruments.

Lunar Orbiter 4 was placed in a near polar lunar orbit of 2,706 kilometers by 6,111 kilometers. Despite a series of malfunctions in its on-board photographic equipment, the spacecraft returned 419 high resolution and 127 medium resolution images that detailed 99 percent of the near side of the Moon. The fourth

Lunar Orbiter was then tracked for several months until it ran out of fuel and fell from lunar orbit. The exact date of its descent and final resting place on the Moon were not determined with absolute certainty, but NASA estimates place the impact of Lunar Orbiter 4 as having occurred no later than October 31, 1967.

Lunar Orbiter 5 (August 1, 1967–January 31, 1968)

The final mission of the successful lunar photography and mapping series, Lunar Orbiter 5 was designed to make a high resolution photographic survey of the lunar far side and to collect scientific data about the Moon via its on-board instruments.

Lunar Orbiter 5 entered orbit around the Moon on August 5, 1967, in a near polar orbit of 194.5 kilometers by 6,023 kilometers. It collected photographic data for 12 days and transmitted 633 high resolution and 211 medium resolution photos to Earth.

At the end of the photographic and scientific data collection phase of its mission, Lunar Orbiter 5 was used to provide orbital data for Project Apollo, as part of the Apollo Orbit Determination Program, and was also used, with Lunar Orbiter 2 and Lunar Orbiter 3, to test the Earth-based tracking stations that would later be used for Apollo

Lunar Orbiter 5 was purposely crashed onto the lunar surface on January 31, 1968 to avoid the possibility that its communications systems might interfere with the flights of other spacecraft during later missions.

LUNAR PROSPECTOR

(1998–1999)

U.S. Lunar Exploration Mission

The first U.S. spacecraft dedicated solely to the study of the Moon since the final Apollo lunar landing in 1972, the Lunar Prospector program was part of NASA's Discovery initiative of economic space exploration vehicles (the Lunar Prospector mission was estimated to cost about $63 million).

The Lunar Prospector spacecraft launched on January 7, 1998 with a primary goal of studying the Moon's polar regions in hopes of answering the long-debated scientific question of whether water ice exists on the lunar surface. The mission was also designed to study the types and amounts of natural resources located on the Moon.

The spacecraft carried a variety of instruments, including a gamma ray spectrometer and neutron spectrometer, to collect data about the presence of the water ice and to quantify the amounts of the elements that are found on the Moon, a magnetometer and electron reflectometer to study the Moon's magnetic field; an alpha particle spectrometer to study radon outgassing events on the lunar surface, and a Doppler gravity experiment to collect data about lunar topography.

Returning an impressive amount of data over the course of its 19-month mission, Lunar Prospector's most highly anticipated result was the conclusive finding of evidence that water ice does indeed exist on the lunar surface. Water ice was detected in the Aitken Basin of the lunar South Pole—as scientists expected—and in even greater abundance in the craters of the lunar North Pole. The data returned by Lunar Prospector indicates the presence of 10 to 300 million tons of water ice at the lunar poles.

On July 31, 1999, Lunar Prospector ended its mission with a controlled crash onto the lunar surface near the south pole in an effort to detect the water ice indicated by the data it had collected, but no water ice was found as a result of the maneuver.

LYAKHOV, VLADIMIR A.

(1941–)

Soviet Cosmonaut

A veteran of three spaceflights, including two long-duration missions during which he spent a total of more than 300 days in space, Vladimir Lyakhov was born in the town of Antratsit, Ukraine, on July 20, 1941.

He was educated in Antratsit and then attended Air Force military aviation school, where he received his initial training as a pilot. In 1964, he graduated from Kharkov Higher Air Force School and subsequently served as a pilot and senior pilot in the Soviet Air Force. From 1964 to 1966, he served on Sakhalin Island while attached to FAR 777, ADV division 24, and he then served in the village of Pereslavka, in the Khabarovsk region, as a senior pilot attached to FAR 300, ADF division 29.

Lyakhov was chosen for training as a cosmonaut as a member of the Air Force Group Four selection of cosmonaut candidates in May 1967. After a period of intensive training, he became a cosmonaut in August 1969.

He received the rating of second class military pilot in 1972 and became a first class military pilot in 1974. In 1975, he completed correspondence studies at the Gagarin Air Force Academy, where he specialized in command staff tactical aviation.

As a pilot, Lyakhov accumulated over 4,500 hours of flying time in a variety of aircraft.

His initial assignments as a cosmonaut included service as a member of the backup crew for the Soyuz 29/Salyut 6 mission in 1978.

Lyakhov first flew in space as commander of the Soyuz 32/Salyut 6 mission in 1979, when he and flight engineer Valeri Ryumin lived and worked aboard the Salyut 6 space station for a then-record six months.

Launched aboard Soyuz 32 on February 25, 1979, Lyakhov and Ryumin traveled to Salyut 6 to become the station's third long-duration crew. They quickly settled into a productive routine of scientific experiments, medical observations, daily exercise, space station maintenance, and engineering tests. About a month into their stay, however, they were confronted with the sudden failure of one of the station's three fuel tanks. The malfunction did not pose any danger to their safety, but did mean that the station's main engine could not be used to correct its orbit, in the event that such maneuvers might be necessary.

Momentarily distracted by the malfunction, Lyakov and Ryumin faced a more pleasant prospect as they anticipated their first visitors at the station. As part of the Soviet Intercosmos program, which provided opportunities for citizens of nations aligned with or friendly to the Soviet Union to fly in space, Soviet cosmonaut Nikolai Rukavishnikov and guest cosmonaut Georgi Ivanov of Bulgaria were scheduled to arrive at the station in April for a brief visit with the resident crew.

More bad news emerged when the visiting crew arrived, however; the main engine of the Soyuz 33 spacecraft containing Rukavishnikov and Ivanov experienced a major malfunction during their flight, and they were forced to abandon their visit to Salyut 6. Fortunate to survive their ordeal, the Soyuz 33 landed safely on April 12.

Although the Soyuz 33 failure caused no immediate difficulties for Lyakhov and Ryumin, they still needed a replacement vehicle for their own Soyuz 32 spacecraft, which by early June would reach its 90-day safe operations limit. To alleviate that problem, Soviet mission controllers launched Soyuz 34 unmanned on June 6 and automatically docked it with Salyut 6. Lyakhov and Ryumin prepared Soyuz 32 for an unmanned trip back to Earth, and the Soyuz 32 Descent Module landed and was recovered on June 13.

In the wake of safety concerns spurred by the Soyuz 33 failure, no further manned flights were made to Salyut 6 during the remaining months that Lyakhov and Ryumin lived and worked aboard the station. As a result, they continued their medical, scientific, and maintenance activities, but received no visitors during their long stay.

They did receive supplies and equipment, however, during visits from three unmanned Progress cargo spacecraft. One piece of equipment in particular, the KRT-10 radio telescope, would cause yet another difficulty for the crew before the end of their long mission.

The telescope worked well and enabled Lyakhov and Ryumin to carry out a series of observations that were part of their scientific program. When they were finished using it, however, they realized that wires extending from the instrument's large dish antenna had become entangled in Salyut 6's docking mechanism.

As a result, on August 15, 1979, Lyakhov and Ryumin ventured outside of the space station for a spacewalk of 1 hour and 23 minutes. Armed with wire cutters, Ryumin was able to cut the antenna loose and free the docking port.

Then, with their long and frequently precarious stay aboard Salyut 6 finally complete, Lyakhov and Ryumin returned to Earth, landing on August 19, 1979 aboard Soyuz 34, after a then-record stay of 175 days, 35 minutes, and 37 seconds in space.

After his remarkable first flight in space, Lyakhov served as backup commander for the Soyuz 39 Intercosmos flight, which in March 1981 made Jugderdemidiyn Gurragcha the first citizen of Mongolia to fly in space.

He also served as backup commander for Soyuz T-8, in April 1983.

On June 27, 1983, Lyakov lifted off as commander of his second space mission, aboard Soyuz T-9, with flight engineer Aleksandr Aleksandrov. The two cosmonauts traveled to Salyut 7, where they lived and worked for nearly five months as the station's second long-duration crew.

No doubt benefiting from Lyakhov's previous long-duration experience, Lyakhov and Aleksandrov quickly settled into a productive routine of scientific experiments, observations, photography, and exercise.

On July 25, the cosmonauts experienced a scary moment when a micrometeorite hit one of the space station's windows. They were not injured in the incident, but the impact served as a sobering reminder of the fragile nature of living and working in space for an extended period.

Lyakhov and Aleksandrov were the first crew to successfully occupy Salyut 7 after it had been expanded by the addition of the Cosmos 1443 module, which had been launched unmanned on March 2, 1983 and automatically docked with the station on March 10. Cosmos 1443 doubled the living space aboard the station, but also occupied a docking port that was needed for the docking of unmanned Progress resupply and refueling spacecraft. As a result, Cosmos 1443 was undocked on August 13, to accommodate the arrival of Progress 17.

Although the arrival of a Progress vehicle was normally a welcome break in the routine for a long-duration crew, Progress 17's refueling activities touched off a string of calamities for Lyakhov and Aleksandrov. A fault during the automated refueling process caused a leak in Salyut 7's propulsion system, which then failed as a result. The fault did not put the cosmonauts in long-term danger, but in the initial hours during which mission controllers and the crew tried to sort out the situation, Lyakhov and Aleksandrov were instructed to move out of the space station and to prepare their Soyuz T-9 craft for an emergency return to Earth, should the situation warrant an immediate evacuation.

The situation was oddly similar to the mishap that had occurred aboard Salyut 6 during Lyakhov's first space mission, and, as had been the case then, it was ultimately determined that the crew was not in imminent danger.

Lyakhov and Aleksandrov remained aboard Salyut 7 after the Progress 17 accident, continuing their assigned activities while awaiting the arrival of their replacements, Vladimir Titov and Gennady Strekalov, who were due to launch on September 26, 1983. In yet another strange parallel with Lyakhov's first experience, however, he and Aleksandrov did not receive their visitors as planned. Titov and Strekalov were nearly killed by a launch pad fire just before they were to have lifted off, which meant that the resident crew aboard Salyut 7 would have to remain well beyond the 90-day safe operations limit of their Soyuz T-9 vehicle.

Despite the difficulties involved, Lyakhov and Aleksandrov continued their long-duration stay with considerable courage, remaining focused on their activities, which, in November, included the task of installing new panels on the space station's solar arrays.

On November 1, 1983 they made their first spacewalk to begin the installation, spending 2 hours and 50 minutes in EVA. They added another 2 hours and 55 minutes to their EVA total when they finished the work on November 3.

They undocked Soyuz T-9 on November 23, 1983, and began a cautious return to Earth in the well-over-the-limit spacecraft. Fortunately, the return flight proceeded without any difficulty, and Lyakov and Aleksandrov landed safely after a total flight of 149 days, 10 hours, and 46 minutes.

Following his second spaceflight, Lyakhov served as backup commander for Soyuz TM-5, which flew in 1988. He also received the rating of first class cosmonaut in 1988.

During his third space mission, Lyakhov served as commander of the Soyuz TM-6 Intercosmos flight, which launched on August 29, 1988. In addition to being part of the Intercosmos program, the Soyuz TM-6 flight also called for the switching of Soyuz spacecraft at the Mir space station and for a partial exchange of crew members.

Lyakhov was accompanied on Soyuz TM-6 by Soviet research cosmonaut Valeri Polyakov and Abdul Ahad Mohmand, the first citizen of Afghanistan to fly in space. After a short visit with Mir long-duration crew members Vladimir Titov and Musa Manarov, Lyakhov and Mohmand left the station in Soyuz TM-5 on September 6, while Polyakov remained aboard Mir to begin his own long-duration stay.

A computer fault during their return to Earth created a tense situation for Lyakhov and Mohmand, when they missed their first opportunity for re-entry. A second attempt also failed, and they then passed a difficult 24 hours in orbit while waiting for mission controllers to diagnose and overcome the problem. The difficulty was eventually resolved, and Lyakhov and Mohmand landed safely on September 7, 1988, after a flight of 8 days, 20 hours, 26 minutes, and 27 seconds.

During his remarkable career as a cosmonaut, Vladimir Lyakhov spent more than 333 days in space, including more than 7 hours in EVA during three spacewalks.

Lyakhov became a department head at the Gagarin Cosmonaut Training Center (CGTC) in 1987, and was named deputy head of cosmonaut training division one in October 1988. In March 1993, he became deputy head of the cosmonaut corps, a position in which he served until September 1994, when, having reached the age limit for active duty service, he left the cosmonaut corps with the rank of colonel.

MAGELLAN VENUS RADAR MAPPER

(1989–1994)

U.S. Venus Exploration Spacecraft

Carrying sophisticated high-resolution radar mapping equipment to perform detailed mapping of the planet Venus, the Magellan spacecraft was deployed by the STS-30 crew of the space shuttle Atlantis on May 4, 1989.

David Walker commanded Atlantis during STS-30; he was accompanied on the flight by pilot Ronald Grabe and mission specialists Norman Thagard, Mary Cleave, and Mark Lee. With the launch of Magellan a little over six hours into their mission, the crew achieved the first deployment of a planetary probe from a manned spacecraft.

Partially assembled from hardware left over from other vehicles and ultimately operated in space for more than five years at an estimated cost of $680 million, the Magellan spacecraft was a little more than six meters long, with solar panels extending about nine meters from end to end, and a high-gain dish antenna measuring 3.7 meters (12 feet) in diameter, which served both as a science instrument for the radar mapping exercises and as a communication device.

The primary technological advance of the Magellan craft over the earlier U.S. Pioneer Venus mission and the Soviet Venera program, which had each mapped portions of the planet at lower resolutions, was the inclusion on Magellan of Synthetic Aperture Radar (SAR), which was capable of radar and topographic mapping of nearly the entire surface of the planet at much higher resolution than had previously been available.

After a journey of 15 months, Magellan entered into a highly elliptical orbit around Venus on August 10, 1990. Then, over the course of five eight-month cycles of collecting and transmitting data, the Magellan craft gathered and returned high-resolution radar mapping data for more than 99 percent of the surface of Venus. Magellan also returned extensive data about the planet's gravity field, first collecting data from its initial elliptical orbit and then gathering further information after an aerobraking maneuver adjusted the vehicle's orbital parameters so it entered a circular orbit.

At the end of its long, successful mission, Magellan entered Venus's atmosphere on October 11, 1994. NASA mission managers lost contact with the spacecraft the following day.

MAKAROV, OLEG G.

(1933–2003)

Russian Cosmonaut

A veteran of four space launches, including the first manned launch abort, Oleg Makarov was born in Udomlya, Russia on January 6, 1933.

He graduated from the Bauman Higher Technical School in Moscow and began his space career as an engineer at the Special Design Bureau Number One, where he helped to develop the Vostok spacecraft and the Salyut series of space stations.

Makarov was chosen to receive training as a cosmonaut as a member of the civilian specialist group two selection of cosmonaut candidates in 1966. His initial assignments as a cosmonaut included training for a flight to the Moon, where he would have entered lunar orbit as part of the Soviet Union's plans for a manned lunar landing. The flight was canceled, however, when Soviet space officials altered the focus of the Soviet program from a Moon landing to the development of space stations in Earth orbit.

Makarov made his first trip into space as flight engineer for Soyuz 12, the first manned test of the Soyuz spacecraft following the tragic Soyuz 11 accident, in September 1973. He was accompanied by Soyuz 12 commander Vasili Lazarev during the test flight, which followed a long period of reflection and redesign after the June 1971 Soyuz 11 accident that had resulted in the deaths of cosmonauts Georgy Dobrovolsky, Vladislav Volkov, and Viktor Patsayev. The Soyuz spacecraft and its systems, equipment, and procedures had been studied in minute detail and re-worked in the aftermath of the accident to better ensure the safety of future crews and the success of future missions.

The success of Soyuz 12 was doubly important to the Soviet space program, as the planned launch of the first of the Salyut series of space stations, which included the Almaz military reconnaissance stations, was fast approaching.

A series of unmanned tests of both the Soyuz modifications and the Salyut/Almaz stations had begun in mid-1972, and the Soyuz 12 flight was the first major step toward the future for both the Soyuz and the space station programs. The flight went well, and Makarov and Lazarev safely returned to Earth on September 29, 1973, after 1 day, 23 hours, and 16 minutes in space. Their flawless performance brought renewed confidence to the struggling Soviet space program and cleared the way for additional manned flights.

After his first spaceflight, Makarov served as a member of the backup crew for the Soyuz 17/Salyut 4 mission, which launched in January 1975.

Makarov's positive experiences as a cosmonaut took a tumultuous turn on April 5, 1975, when he and Vasili Lazarev endured the first manned launch abort in the history of spaceflight.

Intended to launch on that day to travel to the Salyut 4 space station, where they were to have lived and worked for about two months, Makarov and Lazarev were nearly killed when an electrical fault caused the stages of their launch vehicle to remain linked beyond the point at which they were supposed to have separated. Ground controllers engaged the launch escape system to separate the Soyuz capsule containing the crew from the rest of the launch vehicle, and Makarov and Lazarev were, as a result, shot away at an alarming speed. They reached an altitude of 180 kilometers during their brief trip and re-entered the atmosphere about 20 minutes after the aborted launch.

Mission controllers tracking their progress worried that they might land in China, but the spacecraft apparently landed on Soviet soil, about 900 miles from the Baikonur Cosmodrome, where they had begun their harrowing journey. They landed in a remote area on the side of a mountain, and were both injured in the ordeal. Their launch was subsequently re-named "Soyuz 18–1," to differentiate it from the next launch attempt, which was officially known as Soyuz 18.

In his first spaceflight after the abort, Makarov served as flight engineer with commander Vladimir Dzhanibekov aboard Soyuz 27, which lifted off on January 10 to fly to the Salyut 6 space station. The flight was the first Soyuz switching mission, in which one Soyuz spacecraft would be left at a space station for the future use of a long-duration crew while the Soyuz already at the station would be returned to Earth before the expiration of its 90-day safe operations limit, the period after which its onboard systems would begin to deteriorate.

Makarov and Dzhanibekov visited Salyut 6 long-duration crew members Yuri Romanenko and Georgi Grechko at the station, and conducted a variety of scientific experiments and other tests during their brief stay. They then left Soyuz 27 at Salyut 6 for Romanenko and Grechko to use when they returned to Earth at the end of their long-duration mission.

During his third spaceflight, Makarov added 5 days, 22 hours, and 59 minutes to his total career time spent in space.

Makarov next served as a member of the backup crew for the June 1980 flight of Soyuz T-2, the first manned test flight of the Soyuz T-series of spacecraft.

On November 27, 1980, he launched aboard Soyuz T-3 with commander Leonid Kizim and research engineer Gennady Strekalov. Constituting the first three-person crew of the Soviet space program since the Soyuz 11 accident more than nine years earlier, Makarov, Kizim, and Strekalov successfully refurbished the Salyut 6 space station during a stay of just under two weeks, ensuring that the station could continue to be operated and safely and would support additional crews. They returned to Earth aboard Soyuz T-3 on December 10, 1980, after a flight of 12 days, 19 hours, and 8 minutes.

Oleg Makarov spent more than 20 days in space during his career as a cosmonaut. He continued his career at NPO Energia after his final space mission, working on the Mir space station and Buran space shuttle programs.

Oleg Makarov died as a result of a heart attack in 2003.

MALAYSIA: FIRST CITIZEN IN SPACE.

See Shukor, Sheikh Muszaphar

MALENCHENKO, YURI I.

(1961–)

Russian Cosmonaut

Yuri Malenchenko was born in Svetlovodsk, Ukraine, on December 22, 1961. In 1983, he graduated from the S. I. Gritsevets Kharkov Higher Military Aviation School.

Following his graduation, he served as a pilot, senior pilot, and multi-ship flight leader, and he has risen to the rank of colonel in the Russian Air Force.

Malenchenko was selected for cosmonaut training in 1987, and served as backup commander for the Mir 15 long-duration mission.

He continued his education concurrently with his military service, and graduated from the Zhukovsky Air Force Engineering Academy in 1993.

Malenchenko began his first flight in space on July 1, 1994, aboard Soyuz TM-19. He traveled to the Mir space station, where he lived for four months as commander of the Mir 16 crew. With Mir 16 crew mate Talgat Musabayev, he made two spacewalks totaling more than 11 hours. During the first EVA, on September 9, they struggled for more than five hours to repair the thermal blanket on their Soyuz spacecraft, and during their second excursion four days later they performed maintenance on the space station.

Malenchenko also accomplished the first manual docking of a Progress supply spacecraft during his time as Mir 16 commander. The first manual docking of a Progress craft was a major advance, as the nearly two dozen unmanned Progress cargo vehicles that had been used prior to Mir 16 had been programmed to dock automatically with the station.

At the conclusion of Mir 16, Malenchenko returned to Earth aboard Soyuz TM-19 on November 4, 1994.

He next flew in space aboard the U.S. space shuttle Atlantis, during STS-106 in September 2000. The STS-106 crew prepared the International Space Station (ISS) for its first long-duration occupancy. Malenchenko and fellow STS-106 mission specialist Edward Lu of NASA made a spacewalk of more than six hours to attach the station's magnetometer and boom and to link power, data, and communications cables between the ISS and its Zvezda Service Module.

Malenchenko and Lu returned to the ISS in 2003 to serve a long-duration mission as the Expedition 7 crew. Malenchenko served as commander for the flight, which began with the launch of Soyuz TMA-2 on April 26, 2003.

The mission held a particularly poignant moment for Malenchenko when he became the first person ever to be married while in space. The long-distance union took place on August 10, 2003, as Malenchenko and his bride, Ekaterina Dmitriev, exchanged vows, she at the Johnson Space Center in Texas, he in the ISS in orbit some 240 miles above New Zealand. The unique wedding was testament to the long-held public view of the space program as the ultimate adventure of individuals whose imagination and joy are sufficient to overcome mere obstacles of distance and circumstances.

Malenchenko and Lu returned to Earth on October 28, 2003, after more than 184 days in space.

In October 2007 Malenchenko returned to the ISS for another long-duration mission, this time as a member of the ISS Expedition 16 crew. Launched on October 10 aboard Soyuz TMA-11, he served with NASA astronaut Peggy Whitson, who was the first female ISS commander. During his stay, he participated in a spacewalk of nearly seven hours. He returned to Earth in Soyuz TMA-11 on April 19, 2008, after more than 191 days in space.

Yuri Malenchenko has spent more than 523 days in space, including more than 24 hours in EVA, during his career as a cosmonaut

MALERBA, FRANCO

(1946–)

Italian Astronaut; First Citizen of Italy to Fly in Space

On July 31, 1992, Franco Malerba became the first citizen of Italy to fly in space when he lifted off aboard the space shuttle Atlantis as primary payload specialist during STS-46.

Malerba was born in Genova, Italy on October 10, 1946. He received his undergraduate education at the University of Genova, where he specialized in telecommunications and graduated cum laude in 1970 with a degree in electronics engineering.

He pursued research in biophysics as a research assistant at the Laboratorio de Biofisica e Cibernetica, a facility of the Italian National Research Council located in Genova, from 1970 to 1975. His work at the laboratory included experiments in membrane biophysics and biological membrane modeling with lipid bilayers.

In 1971, he served as a member of the University of Genova physics faculty as a lecturer teaching cybernetics and information theory, and in the summer of 1972 he studied computer-aided methods of signal detection for use in sonar data systems as a research assistant at the Nato Saclant Research Center in La Spezia, Italy.

Malerba also conducted research at the U.S. National Institutes of Health in Bethesda, Maryland, where, as a research fellow from 1972 to 1974, he designed a fast micro-spectrophotometer for research in photoreceptor biophysics.

While in the United States, he earned a private pilot's license, and later earned the equivalent license in Italy.

He earned a doctorate degree in physics at the University of Genova in 1974. Following graduation, he served as a reserve officer in the Italian Navy, initially as science lecturer at the Navy Academy while assigned to the destroyer San Giorgio. He was subsequently assigned to the Mariperman Technical Center in La Spezia.

Malerba also worked for the Digital Equipment Corporation in Italy and in other European countries, specializing in the development of microprocessor systems, computer networking, and telecommunications.

His association with the European Space Agency (ESA) began in 1977, when the agency chose him as one of its candidates for training as a payload specialist in anticipation of the first flight of the ESA Spacelab module aboard the American space shuttle. He worked in the Space Plasma Physics division of the ESA Space Science department at the European Space Technology Center (ESTEC) in Noordwijk, the Netherlands.

In 1989, he joined the Italian space agency, Agenzia Spaziale Italiana (ASI), when the agency selected him to receive training at the Johnson Space Center (JSC) in Houston as part of the joint ASI-NASA astronaut development program.

Malerba became the first Italian citizen to fly in space when he launched aboard the STS-46 flight of the space shuttle Atlantis on July 31, 1992 as the primary payload specialist responsible for the operation of the joint ASI-NASA Tethered Satellite System (TSS-1) experiment.

The STS-46 crew also included Claude Nicollier, who became the first citizen of Switzerland to fly in space. Nicollier deployed the European Retrievable Carrier (EURECA), which would remain in orbit until its retrieval during STS-57 in 1993.

Developed over the course of more than a decade through the careful collaboration of teams of engineers from NASA and ASI, the TSS experiment was the first major test of tethered spaceflight. The exercise called for linked flight between the shuttle and a satellite that would be deployed at a different orbital altitude from Atlantis for tethered flight at a distance of as much as 12.5 miles.

Despite all the careful preparations, however, a mechanical hitch foiled the test when the tether jammed after it had been unwound to a distance of just 840 feet. The crew attempted to release the stuck line for several days, but the experiment finally had to be written off as a failure. The TSS satellite was retrieved and returned to Earth. Atlantis landed at KSC on August 8, 1992.

Franco Malerba spent 7 days, 23 hours, 15 minutes, and 3 seconds in space during the STS-40 flight.

He has continued to serve ASI in the years since his historic space mission, helping to develop subsequent activities and flights involving Italian astronauts.

MANAKOV, GENNADI M.

(1950–)

Russian Cosmonaut

A veteran of two long-duration space missions during which he spent a total of more than 310 days in space, Gennadi Manakov was born in Yefimovka, in the Andreyev district of Russia, on June 1, 1950. He was educated in Yefimovka and then attended the Kuybyshev Aviation Institute and the Armavir Higher Military Aviation School.

From 1967 to 1969 Manakov received flight training at the school of the Public Organization of the Army, Aviation, and Fleet Support, and in 1973 he graduated from the Armavir Air Defense Higher Military Aviation School of Pilots, where he received a pilot-engineer diploma with a specialty in command of fighter aviation.

He served in the Kiev Air Defense district in Kramatorsk as a pilot while attached to Engineering Aviation Regiment 636 in Air Defense Division Nine. In 1974, he was transferred to Engineering Aviation Regiment 894 of Air Defense Division 19, and the following year he became a senior pilot with Engineering Aviation Regiment 636. He also served as a senior pilot and as deputy commander for political issues for Engineering Aviation Regiment 865 in the Far Eastern Air Defense District, in Yelizovo.

Manakov received the rating of first class military pilot in 1977.

In August 1978 he was named deputy commander of political issues for Engineering Aviation Regiment 153, in Air Defense Corps 16, and in 1980 he was assigned to the Air Force State Red Banner Research and Test Institute as a test pilot, where he participated in tests of bombardment aviation. He also served as leading engineer and senior test pilot in Air Fighter Squadron Four before his selection as a cosmonaut candidate.

Continuing his education concurrently with his military career, Manakov undertook studies in aircraft manufacturing at the Akhtubinsk branch of the Moscow Aviation Institute, from which he graduated in 1985.

During his military career, Manakov accumulated 1,620 hours of flying time, in 42 different types of aircraft. He also served as a parachute instructor for the Soviet Air Force and made 248 parachute jumps. He received the rating of first class test pilot in 1987.

Manakov was chosen for training as a cosmonaut as a member of the Air Force Buran Group Two selection of cosmonaut candidates in September 1985, with the original intent of his flying missions aboard the Soviet space shuttle Buran. When the Buran program was phased out, Manakov was also considered a member of the Air Force Group Nine selection of cosmonaut candidates. After a period of intensive training, he was qualified as a cosmonaut in January 1988.

His initial assignments as a cosmonaut included service as a member of the backup crew for the Soyuz TM-9/Mir long-duration mission in 1990. Manakov also received the rating of second class cosmonaut in 1990.

Manakov first flew in space as commander of the Soyuz TM-10/Mir 7 long-duration mission. Lifting off on August 1, 1990, Manakov and flight engineer Gennady Strekalov traveled to Mir, where they would live and work for more than four months. During their long stay, they conducted a program of scientific experiments that included astronomical observations, biological and biotechnology experiments, and materials processing investigations, and they also performed the space station maintenance and regular physical exercise necessary to maintain a healthy environment aboard the space station.

They ventured outside of Mir on October 29, 1990 to make repairs to the hatch on the station's Kvant 2 module. During the spacewalk, Manakov and Strekalov spent 2 hours and 45 minutes in EVA.

Toward the end of their long flight, Manakov and Strekalov welcomed their replacements, Viktor Afanasyev and Musa Manarov, who arrived at Mir aboard Soyuz TM-11 with Japanese broadcast journalist Toyohiro Akiyama, who made regular daily broadcasts from the station during his eight days aboard Mir.

Manakov and Strekalov returned to Earth in Soyuz TM-10 on December 10, 1990, along with Akiyama.

After his first spaceflight, Manakov served as a member of the backup crew for the Soyuz TM-15/Mir 12 mission, which launched in July 1992.

On January 24, 1993, Manakov began a second spaceflight when he launched with Alexander Poleschuk aboard Soyuz TM-16 to travel to Mir for another long-duration stay, this time as a member of the Mir 13 crew. During his second visit to the station, Manakov spent six months in space.

He and Poleschuk made the first of two Mir 13 spacewalks on April 19, 1993. Poleschuk began the exercise by maneuvering Manakov into position at Mir's Kvant module with the Mir Strela boom. Manakov successfully installed an electric drive for a solar array onto a framework that had previously been installed on Kvant, but in the midst of the work, one of the two handles on the Strela boom became unattached and drifted away. The incident posed no threat to Manakov, who was still securely tethered to the station, but the loss of mobility made the Strela boom far less useful, and mission managers subsequently decided to postpone the mission's second EVA until a new handle for Strela could be delivered to the crew aboard the next unmanned Progress resupply spacecraft.

During their first Mir 13 EVA, Manakov and Poleschuk spent more than five hours outside the station.

The cosmonauts returned to their EVA work after the arrival of Progress M-19 in May, which delivered the new handle for the control unit on the Strela boom.

On June 18, Manakov and Poleschuk finished the installation of the solar array electrical drives during a 4 hour and 33-minute spacewalk. During their two Mir 13 EVAs, they spent a total of about 10 hours in EVA.

After successfully completing their long-duration mission, Manakov and Pole-schuk returned to Earth aboard Soyuz TM-16 with guest cosmonaut Jean-Pierre Haigneré of France, who had arrived at Mir on July 1 with Mir 14 crew members Aleksandr Serebrov and Vasili Tsibliyev arrived aboard Soyuz TM-17.

Manakov was next assigned to command the 1996 Soyuz TM-24/Mir 22 long-duration flight, but he was forced to give up the mission when a routine physical examination indicated that he might have a heart condition. He was subsequently discharged from the cosmonaut corps as a result of his condition.

During his remarkable career as a cosmonaut, Gennadi Manakov spent more than 309 days in space, including more than 12 hours in EVA during three spacewalks.

In December 1996, he was assigned to lead a special unit at the Gagarin Cosmonaut Training Center (GCTC) that was responsible for developing life support systems for use in a variety of special situations. He became a department head at the GCTC in 1997 and served in that capacity until July 2000, when he left active military service as a colonel.

MANAROV, MUSA K.

(1951–)

Russian Cosmonaut

A veteran of two long-duration space missions during which he made seven spacewalks and accumulated more than 540 days in space, Musa Manarov was born in Baku, Azerbaijan on March 22, 1951.

After receiving his initial education, Manarov joined the Soviet Air Force and attended the Moscow Aviation Institute, where he graduated with a degree in engineering in 1974. During his outstanding military career, he rose to the rank of colonel in the Soviet Air Force.

Manarov was chosen for training as a cosmonaut in December 1978, as a member of the civilian specialist group six selection of cosmonaut candidates.

He began his first spaceflight on December 21, 1987, as flight engineer for the Soyuz TM-4/Mir 3 long-duration mission. Launched with commander Vladimir Titov and research engineer Anatoli Levchenko from the Baikonur Cosmodrome, Manarov began a year-long odyssey during which he and Titov would live and work aboard Mir until December 1988.

The Soyuz TM-4 crew met Mir 2 crew members Yuri Romanenko and Aleksandr Aleksandrov at the station, and after a visit of about eight days, Manarov and Titov took over as the third long-duration crew while Levchenko returned to Earth with Romanenko and Aleksandrov in Soyuz TM-3. Manarov and Titov immediately established a productive routine of scientific work and station engineering and maintenance tasks, and prepared for the first spacewalk of their long mission.

They ventured outside of Mir for the first time on February 26, 1988, when they replaced one of the station's solar arrays during a spacewalk of about 4.5 hours.

Manarov and Titov also made two more EVAs during their Mir 3 stay, to repair X-ray instruments on the station's Kvant 1 module. They spent over five hours working on the X-ray detector on June 30, and returned to the problem again on October 20 to fix the X-ray telescope. During the October EVA, they also tested the new Orlan-DMA spacesuit, while adding more than four hours to their spacewalking total for the flight.

In their three trips outside of Mir, Manarov and Titov accumulated a total of 13 hours and 47 minutes in EVA.

They also hosted several visiting crews during their long stay, beginning with the arrival of the Soyuz TM-5 crew in June, which included guest cosmonaut Aleksandr Aleksandrov of Bulgaria. The Soyuz TM-5 crew visited for about a week, and then returned to Earth in Soyuz TM-4.

In August, a second visiting crew arrived, aboard Soyuz TM-6. Commanded by Vladimir Lyakhov, the Soyuz TM-6 crew also included research engineer Valeri Polyakov and guest cosmonaut Abdul Ahad Mohmand, the first citizen of Afghanistan to fly in space. Lyakhov and Mohmand visited for about a week and then returned to Earth in Soyuz TM-5, leaving Polyakov aboard the station with Manarov and Titov to begin his own long-duration stay. Polyakov would remain aboard Mir until April 1989, when he concluded his long stay and returned to Earth in Soyuz TM-7 with the Mir 4 crew.

Before closing out their remarkable long-duration Mir 3 stay, Manarov and Titov had the opportunity to see space history in the making when their replacements, Alexander Volkov and Sergei Krikalyov, arrived in November with guest cosmonaut Jean-Loup J. M. Chrétien of France. On December 9, 1988, Chrétien became the first individual from a country other than the Soviet Union or the United States to make a spacewalk, when he and Volkov spent nearly six hours outside of Mir while assembling an experimental grid structure.

Manarov and Titov concluded their long journey when they returned to Earth with Chrétien on December 21, 1988, aboard Soyuz TM-6. During his first flight, Manarov accumulated a remarkable total of 365 days, 22 hours, and 39 minutes in space. He and Titov were the first individuals in history to spend an entire year in space.

After his outstanding first spaceflight, Manarov served as a member of the backup crew for the Soyuz TM-10/Mir 7 mission, which launched in August 1990.

On December 2, 1990, he began a remarkable second long-duration space mission, this time as a flight engineer for the Soyuz TM-11/Mir 8 crew. Manarov traveled to the space station with Mir 8 commander Viktor Afanasyev and guest cosmonaut Toyohiro Akiyama, a Japanese broadcast journalist who made regular broadcasts during a week-long stay at Mir, before returning to Earth with the Mir 7 crew.

During his second six-month mission aboard Mir, Manarov participated in four more spacewalks, beginning with an excursion of more than five hours on January 7, 1991 when he and Afanasyev made repairs to the hatch on the Mir Kvant 2 module. They installed the Strela boom on the station during a 5.5-hour

EVA on January 23, 1991, and three days later, on January 26, they spent 6 hours and 20 minutes in EVA while installing brackets on the Kvant module that would later support solar arrays moved from the station's Kristall module. They completed their spacewalking chores on April 25, when they spent about 2.5 hours outside the station while inspecting a damaged antenna on the outside of the Kvant module.

Manarov added a total of 19 hours and 36 minutes to his career EVA time during the four Mir 8 spacewalks he made with Afanasyev.

On May 18, Manarov and Afanasyev welcomed their replacements, Anatoli Artsebarsky and Sergei Krikalyov, who arrived at Mir aboard Soyuz TM-12 with Helen Sharman, the first citizen of the United Kingdom to fly in space. After turning over command of the station to Artsebarsky and Krikalyov, Manarov and Afanasyev returned to Earth with Sharman on May 26, 1991. The two Mir 8 cosmonauts had been in space for a total of 175 days, 1 hour, and 52 minutes.

Musa Manarov spent more than 541 days, including 33 hours and 23 minutes in EVA during seven spacewalks, in his two missions aboard the space station Mir.

MANNED ORBITING LABORATORY (MOL)

(1963–1969)

U.S. Military Space Station Program

Although the U.S. space program was characterized from the founding of the National Aeronautics and Space Administration (NASA) in 1958 as a purely civilian initiative, much of the expertise and technology required for NASA's early manned spaceflights were derived from earlier work begun by the U.S. military.

In addition, the various branches of the military continued to develop space-related projects simultaneously but separately from NASA's program. Among the most ambitious and forward-looking of the early military projects, the U.S. Air Force Manned Orbiting Laboratory (MOL) was first announced in December 1963.

Introduced at a time when the Soviet Union was clearly outpacing the achievements of the United States in space, the MOL program was designed to launch and maintain in orbit a series of small space stations that were to be manned by members of the U.S. military for the purpose of spying on the activities of nations unfriendly to the United States and its allies. As the program developed, the MOL spacecraft was adapted from the Gemini spacecraft. Plans for the military space station called for a two-piece vehicle that linked a modified Gemini (the heat shield was outfitted with a hatch so the crew could transfer into the laboratory module) to a pressurized module in which the two-man crew would carry out its reconnaissance duties. The Douglas Aircraft Company developed both components for the Air Force.

MOL flights were initially designed to be launched aboard Titan IIIC rockets, and ultimately were to have been lifted into space by the Titan IIIM, which was

2B24070 Figure 3

An early conceptual illustration of the United States Air Force Manned Orbiting Laboratory. [NASA/courtesy of nasaimages.org]

capable of launching heavier payloads than the IIIC. The atmosphere inside the spacecraft was designed to be a mix of helium and oxygen, and the program's designers envisioned missions lasting about 30 days, during which crew members would conducted their surveillance using cameras and radar.

In November 1965, the Air Force selected its first group of MOL astronauts for training. Air Force service members Michael Adams, Albert Crews, Jr., Richard Lawyer, Lachlan Macleay, Francis Neubeck, and James Taylor were joined by U.S. Navy pilots John Finley and Richard Truly in the first group of eight MOL trainees.

As the training of the program's first astronauts began, the Air Force made the first—and ultimately, only—MOL test flight. Launched unmanned atop a Titan IIIC from Cape Canaveral in Florida on November 3, 1966, a prototype MOL flew a 33 minute sub-orbital flight while attached to the Gemini 2 space capsule, which had been flown earlier by NASA, was recovered, and was being re-used during the MOL test. The Gemini 2 capsule was again recovered at the end of the flight. The MOL separated from the Gemini craft shortly after the two vehicles exited the Earth's atmosphere, and the MOL then entered orbit, where it was used in the automatic deployment of three satellites.

Following the successful test flight and the first selection of trainees, the Air Force selected two more groups of MOL astronauts and set an ambitious schedule of launches that extended into the mid-1970s (all of which were later abandoned when the program was canceled).

Included in the Group 2 selection of MOL astronauts, who were chosen in June 1966, were Karol Bobko, Charles Fullerton, and Henry Hartsfield, Jr. of the Air Force, Navy pilot Robert Crippen, and Robert Overmyer of the U.S. Marine Corps.

Selected in June 1967, the four Group 3 MOL astronauts—James Abrahamson, Robert Herres, Robert Lawrence, Jr., and Donald Peterson—were all Air Force pilots.

As the MOL program matured throughout the 1960s, its progress was hampered by the success of the civilian space program, which garnered more attention and greater financial support, and by the widening American involvement in the Vietnam War, which steadily absorbed more and more Air Force personnel and resources.

Subsequent events made the MOL concept seem ill-fitting with the achievements of the civilian NASA program, and subsequent satellite technology achieved many of the goals sought by the designers of the MOL program with less risk and at less cost.

On June 10, 1969, the MOL program was canceled. Despite years of development, there had been only the single, unmanned MOL test flight in 1966. The technology developed for the program was later adapted for use in spy satellites.

At the time of the program's cancellation, NASA extended an offer to transfer those MOL astronauts under the age of 35 to its program. As a result, in August 1969 NASA made its Group Seven selection of astronauts, which included Karol Bobko, Robert Crippen, Gordon Fullerton, Henry Hartsfield, Robert Overmyer, Donald Peterson, and Richard Truly.

MARINER VENUS AND MARS PROBES

(1962–1975)

U.S. Venus and Mars Exploration Program

Representing the United States' first efforts to explore Venus and Mars, the Mariner program was responsible for the first spacecraft to perform a successful flyby of another planet, when Mariner 2 passed by Venus in 1962; the first spacecraft to perform a successful flyby of Mars and return the first photographs of the planet's surface, which were both achieved by Mariner 4 in July 1965; the first spacecraft to enter orbit around another planet, when Mariner 9 orbited Mars in 1971; and the first spacecraft to successfully fly by two planets, when Mariner 10 flew past Venus and Mercury. In the process, Mariner 10 also became the first craft ever to visit Mercury, in 1974.

Mariner 10's first photograph of the planet Mercury, March 24, 1974. [NASA/courtesy of nasaimages.org]

The 10 Mariner missions cost approximately $554 million.

Mariner 1 (July 22, 1962) and Mariner 2 (August 27, 1962) Venus Flybys

Considering how early they took place in the history of space exploration, the Mariner 1 and Mariner 2 missions represented bold attempts to explore the planet Venus.

The twin spacecraft were each slightly more than 3.5 meters tall and were powered by two solar wings that provided power directly to the craft and also recharged a 1,000 watt silver-zinc battery cell. Scientific equipment carried aboard each vehicle included instruments to detect cosmic dust, cosmic rays and other types of particles, a magnetometer, an infrared radiometer, a microwave radiometer, and a solar plasma spectrometer.

When Mariner 1 lifted off at Cape Canaveral in Florida on July 22, 1962, a fault occurred in its Atlas launch vehicle almost immediately, and the rocket and spacecraft were quickly lost in the Atlantic Ocean. Later investigation of the failure revealed a cause that bore witness to the amazing complexity of space missions and the way in which the smallest detail might ultimately impact the success or failure of any given flight. In the case of the Mariner 1 launch, the program that was responsible for guiding the operation of the Atlas launch vehicle had caused the vehicle to veer off course because a single hyphen had been mistakenly left out of the software.

Launched on August 27, 1962, Mariner 2 successfully reached Earth orbit, and after a second firing of its Agena rocket stage, began its trip to Venus. Mission controllers directed the spacecraft through a midcourse correction maneuver on September 4, and then, four days later, the spacecraft's attitude control system inexplicably failed for a period of three minutes before the craft's onboard gyroscopes reactivated it.

One of the satellite's solar panels failed during the flight, but the second panel generated enough power to complete the primary goal of achieving a Venus flyby. On December 14, 1962, Mariner 2 became the first spacecraft to successfully encounter another planet when it passed by the dark side of Venus at a distance of 34,773 kilometers. The vehicle then continued its journey toward the

Sun, where it entered into heliocentric orbit after sending its last signal to Earth on January 3, 1963.

During its encounter with Venus, Mariner 2 returned a great deal of scientific data about the atmosphere and surface climate of the planet, generally revealing a far more hostile environment than scientists had previously assumed, as surface temperatures were determined to be much higher than expected and the planet's atmosphere was found to consist primarily of carbon dioxide. The spacecraft also returned a treasure of other scientific data from Venus and during its trip through space.

Mariner 3 (November 5, 1964) and Mariner 4 (November 28, 1964) Mars Flybys

As had been the case with the first two spacecraft in the Mariner series, the Mariner 3 and Mariner 4 launches took place in quick succession and pursued a common goal—to encounter the planet Mars at a close distance.

Given the task of obtaining photographs of the Martian surface as well as collecting a variety of scientific data during its flight and in the vicinity of the planet, Mariner 3 was launched on November 5, 1964. Just nine hours into the flight, when the time came for the spacecraft to shed a protective shield that covered its onboard instruments, a malfunction occurred and the shield remained in place. As a result, Mariner 3 was unable to return data and ultimately failed to enter the trajectory necessary for its trip to Mars. The spacecraft continued on its silent flight, ultimately entering orbit around the Sun.

Mariner 4 took its turn at the attempt to encounter the Red Planet on November 28, 1964, when it was launched atop a multi-stage Atlas-Agena launch vehicle from Cape Canaveral in Florida.

On July 14, 1965, Mariner 4 achieved the first flyby of Mars. The spacecraft made its closest pass over the planet the following day, July 15, when it captured a series of 21 photographs as it passed within 9,846 kilometers of the surface. The photos were recorded onto the craft's onboard tape recorder and then sent to Earth over the course of several weeks.

Mariner 4 continued to transmit scientific data until October 1, 1965, when its position in space prevented it from communicating with Earth. It began sending signals again in 1967, registering a running account of its encounter with a micro-meteoroid shower before the end of its last transmission, on December 21, 1967.

Among the wealth of data returned by the spacecraft during its long flight, Mariner 4 gave humanity its first glimpse of Mars's landscape, measured the atmospheric pressure of the surface, and discovered daytime temperatures as low as -100 degrees centigrade.

Mariner 5 (June 14, 1967) Venus Flyby

Originally fashioned as a backup for the Mariner 4 Mars flyby mission, Mariner 5 was launched on June 14, 1967 with the intent of its instead traveling to Venus.

Mariner 5 successfully encountered Venus on October 19, 1967, making its closest approach to the planet at a distance of 4,000 kilometers. It returned useful data about the planet's atmosphere, including measurements of its charged particles and plasmas, and its ultraviolet emissions.

Mariner 6 (February 24, 1969) and Mariner 7 (March 27, 1969) Mars Flybys

Featuring identical spacecraft, the Mariner 6 and Mariner 7 spacecraft were intended to travel to Mars, where they were designed to collect data about the planet while flying past at a close distance.

Equipped with four solar panels and a high-gain parabolic antenna measuring one meter in diameter, Mariner 6 and Mariner 7 each stood 3.35 meters tall on an octagonal base made of magnesium. They were equipped with television cameras for photographing the Martian surface, with the images being stored onboard the vehicle on an analog tape recorder before being sent to Earth. Each spacecraft also carried a digital recorder for storing scientific data generated by the other onboard instruments, which included an infrared radiometer, an infrared spectroscope, and an ultraviolet spectroscope.

An innovative central computer directed the flight of each vehicle via a pre-programmed sequence that was loaded into the system's memory, but also had the capacity to be re-programmed as the mission progressed. The computer was capable of performing 53 direct commands, 5 control commands, and 4 quantitative commands.

An accident shortly before the scheduled lift-off of Mariner 6 nearly ruined the flight before it got under way. With the spacecraft sitting at the top of an Atlas-Centaur launch vehicle, the main valves of the Atlas rocket suffered a malfunction that threatened to topple the entire rocket stack. Two alert technicians at the launch site recognized the grave nature of the situation immediately and turned on pressurizing pumps that kept the launch vehicle standing. The Mariner 6 satellite was saved as a result of their fast response, and, after mating to a second Atlas-Centaur launcher, the spacecraft was successfully launched on February 24, 1969. The two ground crew technicians were honored with NASA's Exceptional Bravery Medal for acting without concern for their own safety to prevent the collapse of the rocket stack and save the Mariner 6 spacecraft.

Once it was successfully under way, Mariner 6 performed exceptionally well. The spacecraft passed by Mars at a closest distance of 3,431 kilometers on July 31, 1969. During its approach to the planet, Mariner 6 captured 49 photographs of Mars and then collected an additional 26 images during its close encounter.

Launched into a direct-ascent trajectory to Mars on March 27, 1969, Mariner 7 made its closest approach to the planet on August 5, at a distance of 3,430 kilometers from the surface. Mariner 7 captured 93 photographs of Mars during the initial phase of its encounter, and then, thanks to the flexibility of the spacecraft's onboard computer system, its flight was re-directed so that it could record images and scientific data from a different area of the planet than the region previously covered by Mariner 6.

During the period of its closest encounter, Mariner 7 captured an additional 33 photographs of the Martian surface.

Together, Mariner 6 and Mariner 7 were able to image about 20 percent of the surface of Mars. The two spacecraft returned valuable data about the planet's atmosphere, including measurements of its infrared and ultraviolet emissions and radio refractivity. They also returned data indicating that the polar cap in the planet's southern region consists primarily of carbon dioxide.

Mariner 8 (May 8, 1971) and Mariner 9 (May 30, 1971) Mars Orbiters

The Mariner 8 and Mariner 9 vehicles were intended to travel to Mars and enter into orbit around the planet. Standing 2.28 meters tall, each craft was equipped with four solar panels that were designed to initially provide 800 watts of power, which would degrade to 500 watts by the time the spacecraft reached Mars. Electricity provided by the solar panels was stored in a 20-ampere nickel-cadmium battery.

Electronic systems aboard the spacecraft included a central computer and sequencer that featured a memory capacity of 512 words and a digital reel-to-reel tape recorder with a 168-meter eight track tape capable of recording at 132 kilobits per second and playing back two tracks at a time at a maximum speed of 16 kilobits per second.

Mariner 8 was launched on May 8, 1971 and was promptly lost in the Atlantic Ocean after a failure in the Centaur stage of its multi-stage Atlas-Centaur launch vehicle.

Launched on May 30, 1971, Mariner 9 performed as expected during its 5.5-month journey, and flawlessly executed the 15 minute, 23 second firing maneuver that placed it into orbit around Mars on November 14, 1971, making Mariner 9 the first spacecraft ever to orbit another planet.

A massive dust storm in the Noachis region of Mars prevented Mariner 9 from beginning its planned mapping of the surface for several weeks, until December 1971. Once it got underway, the Mariner 9 mapping operation produced the first global mapping data for Mars, revealing surface features such as the Olympus Mons and the three Tharsis volcanoes in great detail.

From its perch in orbit, Mariner 9 achieved photographic coverage of the entire planet for the first time, returning 7,329 images, and also transmitted a large amount of quality data about Mars's atmosphere and surface. At the end of its remarkable mission, the spacecraft was shut down on October 27, 1972 and left in an orbit that will likely not decay for at least 50 years.

Mariner 10 (November 3, 1973) Venus and Mercury Flyby

Mariner 10, the first spacecraft to fly by two planets successfully and the first craft to explore the planet Mercury, was launched on November 3, 1973.

To achieve the double flyby, the spacecraft took a circuitous route to its rendezvous with Venus, where it would benefit from a gravity-assist from the ringed planet to continue its journey on to Mercury. Mariner 10 suffered a variety of difficulties along the way during its long trip, including a failure of several instruments whose protective shield failed to open as planned, the failure of heaters designed to protect the onboard television cameras, and periodic—but temporary—faults in the vehicle's star-tracker navigation system and on-board computer.

Despite its equipment problems, Mariner 10 achieved remarkable results. It made ultraviolet observations of the comet Kohoutek in January 1974, and then achieved its planned flyby of Venus at a distance of 5,768 kilometers on February 5, 1974. During the nearest pass, it recorded the first close-up ultraviolet images of the planet.

Marking the first time a spacecraft had ever received a gravity assist from one planet to aid in its journey to another, Mariner 10 got a boost from its Venus encounter and then continued on to Mercury. During the trip, the spacecraft suffered more mechanical difficulties when its attitude control system began to malfunction, but mission managers were able to work around the problem.

On March 29, 1974, Mariner 10 achieved the first flyby of Mercury, passing the planet at a distance of 704 kilometers. Taking photographs and gathering scientific data during the encounter and during two subsequent passes, at a distance of 48,069 kilometers on September 21 and at 327 kilometers on March 16, 1975, Mariner 10 was able to return images and data that revealed Mercury's barren, cratered surface in close detail for the first time. The data also found the planet to have no atmosphere, a small magnetic field, and a large core composed largely of iron. Mariner 10's mission at Mercury came to a close on March 24, 1975.

MARS EXPRESS

(2003)

ESA Mars Exploration Mission

A project of the European Space Agency (ESA), the Mars Express mission was primarily designed to study the geology and mineral composition of the Martian landscape via the Mars Express Orbiter and the Beagle 2 lander.

The spacecraft launched from Baikonur Cosmodrome on June 2, 2003 with seven experiments, including a high resolution stereoscopic camera, a visible and near-infrared spectrometer, infrared spectrometer, ultraviolet spectrometer, neutral and charged particle sensors, subsurface radar and altimeter, and a radio science experiment.

Mars Express arrived at Mars in December 2003, and released the Beagle 2 lander on December 19. The lander was scheduled to touch down on the surface after entering the Martian atmosphere on December 25, but it failed to send any signal after that time and was declared lost.

The Mars Express Orbiter successfully attained its orbit around Mars and achieved its data gathering and scientific mapping objectives.

MARS: SOVIET & RUSSIAN MISSIONS

(October 1960–)

The Soviet Union began a vigorous program of satellite launches designed to explore Mars as early as 1960. A bold initiative, given how early they took place in the history of space exploration, the flights met with frustrating malfunctions and, ultimately, failed to reach their destination until the flight of the Mars 2 satellite in 1971, which was the first Soviet spacecraft to enter into orbit around the Red Planet.

Soviet Mars launches occurred during three distinct periods: the early 1960s attempts (October 1960–November 1964), the forward-looking Mars program of the early 1970s (probably begun in 1969, but generally recognized as beginning with Cosmos 419 in May 1971 and ending with the Mars 7 mission, which launched in August 1973), and the Phobos Project of 1988.

In the years after the fall of the Soviet Union, Russia made one significant attempt at exploring Mars—the failed Mars 96 mission.

During the entire first era of space exploration—from the late 1950s to the mid-1970s—known Soviet successes in the exploration of Mars included four spacecraft placed in orbit around the planet, one lander which operated briefly on the surface, and several flybys.

Marsnik 1 & 2 (October 1960)

The Marsnik probes (also known as Korabl 4 and 5 and Mars 1960A and Mars 1960B) apparently represent the Soviet Union's first attempts to launch a planetary probe. There is some confusion over whether Marsnik 1, which was rumored to have been launched on October 10, 1960, was actually intended to travel to the planet, but the mission profile for Marsnik 2 clearly identifies that second attempt, on October 14, 1960, as a mission to Mars.

Confusion about early Soviet launches, and failed launches in particular, is common. The Soviet government frequently utilized the considerable achievements of its space program for propaganda purposes, and in that context often tried to bury the program's failures. As a result, officials in the Soviet space program sometimes renamed failed missions to disguise their true purpose, or even tried to disavow their existence altogether. Evidence arguing for the Marsnik 1 launch attempt can be found in the Soviet custom of launching planetary probes in twos, as they often did during their Venera program, which explored Venus. The Marsnik probes were nearly identical to the Venera 1 probe.

The primary goals of the Marsnik probes were to return images of Mars while at the same time testing the effect of long spaceflight on the spacecraft's instruments and testing radio communications from deep space.

Mars 1962A, Mars 1, and Mars 1962B (October–November, 1962)

Also known as Sputnik 22, Sputnik 23, and Sputnik 24, the Soviet Mars missions of late 1962 were aimed in the first two instances at flying by Mars, and in the case of the third launch, the intent was to deploy a small landing craft on the planet.

The first attempt, Mars1962A, was launched on October 24, 1962 at the height of the Cuban missile crisis, a time during which the United States and the Soviet Union seemed poised on the edge of nuclear confrontation over the Soviets' decision to move nuclear weapons to Cuba.

The satellite reached an Earth parking orbit as planned, but when it was given the command to maneuver into the proper trajectory for it to travel to Mars, a malfunction in the upper stage of the launch vehicle apparently caused the spacecraft to explode. Pieces of the debris were tracked by the radar equipment of the U.S. Ballistic Missile Early Warning System and were initially thought to be the beginning of a Soviet attack on the U.S. Military officials quickly determined the nonmalignant nature of the spacecraft fragments, however, before the situation escalated any further.

On November 1, 1962, the Soviets launched Mars 1, a small (3.3-meter) cylindrical satellite that was supposed to fly past Mars at a distance of 11,000 kilometers.

The Mars 1 spacecraft was equipped with solar panels, which stored power in a 42-ampere nickel-cadmium battery. It carried a television camera, magnetometer, a micrometeoroid detector, radiation sensors, a spectrograph, and a spectroreflexometer.

Mars 1 escaped Earth orbit as planned and was about half way through its eight-month journey to Mars when a failure in its onboard systems abruptly ended its communication with Soviet mission controllers. It continued its flight in silence, passing by Mars at a closest distance of 193,000 kilometers on June 19, 1963 and then continuing on into orbit around the Sun.

Although it did not achieve its intended results, Mars 1 did return valuable scientific data during the early portion of its flight, including measurements of the rate of micrometeoroid impacts during the Taurids meteor shower.

With the Mars 1962B mission, the Soviets intended to land a small spacecraft on the surface of Mars. The spacecraft was successfully launched on November 4, 1962, but as had been the case with the Mars 1962A flight several weeks earlier, the attempt to place the craft on the proper trajectory for the flight to Mars resulted in a malfunction that caused the vehicle to explode. Contemporaneous evidence indicated that the launch vehicle, or some portion of the multi-stage launcher, fell from orbit on December 25 and that the satellite re-entered the Earth's atmosphere and was destroyed on January 19, 1963.

Zond 2 (November 30, 1964)

Not to be confused with other spacecraft given the "Zond" prefix, the Zond 2 launch of November 30, 1964 was another in the series of Soviet attempts to reach Mars. Zond 2 was similar to the earlier vehicles aimed at Mars; it was equipped with a small landing craft, and, like Mars 1, it carried a television camera, magnetometer, a micrometeoroid detector, radiation sensors, a spectrograph, and a spectroreflexometer.

In addition to its standard gas engines, Zond 2 was also outfitted with six electrojet plasma ion engines, which were successfully tested during the flight. The spacecraft suffered a loss of electrical power when one of its two solar panels failed, and in a sequence of events similar to those that the Soviets had experienced with Mars 1, the communications system aboard Zond 2 failed in May 1965. As a result, no data could be returned when the craft achieved a flyby of Mars at a distance of 1,500 kilometers on August 6, 1965.

The Zond 3 spacecraft, which was launched on July 18, 1965, has long been thought to have been originally intended to follow Zond 2 to Mars. Its launch was delayed, however, for reasons that have not been made clear, and as a result its mission was redesigned as a lunar flyby.

Mars 1969A and Mars 1969B (March–April, 1969)

Because the Mars 1969A and Mars 1969B launches were not publicly acknowledged by Soviet space officials, it is difficult to determine if they were related to the early 1970s launches that gave the Soviets their first successes in their long attempts to reach the planet.

The vehicles involved in the two 1969 missions were thought to be identical in composition and instrumentation. Building on the designs of earlier probes, each of the 1969 spacecraft was equipped with larger solar panels and more and larger compartments to hold an expanded array of electronic systems and communications and camera equipment. Each probe carried three television cameras, a radiometer, a radiation detector, several spectrometers, and a device designed to detect the presence of water.

Although the mission profile for the two launches called for a six-month trip to Mars, the timing of the two 1969 Mars missions is interesting in light of the progress of the American Apollo program at that time. It is not at all unlikely that the Soviets had hoped to diminish the impact of the obviously impending manned lunar landing that the United States was about to attempt and which was ultimately achieved during Apollo 11 in July 1969. The Soviets were also planning for eventual manned lunar landings at the time, and were likely assuming that they would achieve the feat at some point after the first U.S. landing. In that context, it is not difficult to imagine that they might have attached some value to an attempt to obtain the first television images of the surface of Mars before the end of the year.

Whatever their intent, the two 1969 Soviet Mars launches did not go well. The Mars 1969A craft suffered a launch vehicle malfunction that caused it to explode about 7.5 minutes into its flight, and Mars 1969B fared even worse when one of its six first-stage rockets malfunctioned at launch, resulting in a spectacular crash three kilometers from the launch pad and the destruction of the vehicle less than a minute after lift-off.

Cosmos 419 & Mars 2–7 (May 1971–August 1973)

The most significant and sustained period of Soviet Mars launches began with the May 10, 1971 launch of Cosmos 419, which was apparently intended to travel to Mars and enter into orbit around the planet.

Cosmos 419 entered Earth orbit, but was stranded by an incorrectly set timer, which failed to direct the engine firing that would have placed the spacecraft into the proper trajectory for its journey to Mars. In a seemingly small error of mathematics at some point during the lengthy design of the complex launch vehicle and its systems, the timer had been instructed to direct the engine to fire one and a half years after the craft reached orbit, rather than the intended one and a half hours. The fault left the vehicle stranded in a rapidly decaying orbit, and the spacecraft was pulled back to Earth by the force of the planet's gravity on May 12, 1971 and destroyed during re-entry.

To conceal the mission's intent, and in keeping with the standard Soviet practice, the May 10, 1971 launch was given the name Cosmos 419—the "Cosmos" prefix being a catch-all designation for failed missions since the 1960s.

Shortly after the Cosmos 419 failure, the Soviets launched two identical spacecraft, Mars 2 and Mars 3, in quick succession. Both were intended to enter orbit around the planet and to send a small descent module down to the Martian surface. Each spacecraft was equipped with a television camera, a Lyman-alpha sensor, several types of photometers, a radiometer and radiotelescope, and an infrared spectrometer.

Mars 2 was launched on May 19, 1971 and reached Mars on November 27, 1971. It released its Descent Module as planned, but a fault in the module's onboard systems caused it to enter the atmosphere at the wrong angle and the small craft crashed onto the surface—the first human-made object to land on the Martian surface.

Although the Mars 2 Descent Module failed to return any data, the orbiter portion of the craft did successfully enter into a 1,380 kilometer by 24,940 kilometer orbit around the planet, achieving another first for the Soviets: it was the first Soviet Mars probe to achieve its mission profile. Mars 2 would make a total of 362 orbits of Mars before the announced completion of its mission in August 1972.

Launched on May 28, 1971, Mars 3 achieved a level of success similar to that of Mars 2. The spacecraft released its Descent Module on December 2, 1971, and the module achieved the first soft landing on the surface of Mars. Unfortunately, the lander arrived on the planet during an intense dust storm, and its instruments failed shortly after it touched down. As a result, it was able to transmit just 20 seconds of video before it fell silent.

The orbiter portion of the Mars 3 craft suffered a malfunction of a different sort, as it lost fuel and was as a result forced to enter a different orbit than originally planned. Despite the change in plan, Mars 3 was still able to return a large amount of valuable data about the planet's atmosphere and surface characteristics, and it made a total of 20 orbits before the announced end of its mission in August 1972.

After the successful Mars 2 and Mars 3 missions, the next group of Soviet Mars launches began in the summer of 1973. Mars 4, which lifted off on July 21 1973, and Mars 5, which launched on July 25, 1973, were both designed to travel to and enter into orbit around the planet. Each spacecraft was equipped with two television cameras and a variety of scientific instruments similar to those carried during the Mars 2 and Mars 3 flights.

Mars 4 reached its destination on February 10, 1974, but an electronic fault caused it to miss its orbit insertion maneuver, and as a result it passed Mars at a distance of 2,200 kilometers. Mars 5 reached the planet on February 12, 1974 and successfully achieved an orbit of 1,755 kilometers by 32,555 kilometers. After 22 orbits, during which the spacecraft returned 60 photographs and a wealth of scientific data, Mars 5 experienced an onboard fault and its communication system failed.

The two final launches of the successful early 1970s Soviet Mars missions occurred in August 1973, with the August 5 launch of Mars 6 and the launch of Mars 7 on August 9.

Mars 6 consisted of a small platform that was designed to fly past the planet after releasing a Descent Module lander that was designed to travel through the atmosphere and achieve a soft landing on the surface. When the spacecraft arrived at Mars on March 12, 1974, the platform performed as expected, separating from the Descent Module and then achieving a flyby of the planet at a distance of 1,600 kilometers.

Equipped with an array of scientific instruments and next-generation imaging equipment to return a panoramic record of the surface, the Mars 6 Descent Module successfully entered the Martian atmosphere and transmitted data for 3 minutes and 44 seconds—marking the first instance in which data was transmitted directly from the planet's atmosphere. Contact was then lost as the craft impacted the planet, however, and a fault in an onboard semiconductor was subsequently found to have fouled some of the information that had been retrieved.

Virtually identical in its systems, equipment, and instrumentation to Mars 6, the Mars 7 satellite reached Mars on March 9, 1974. A malfunction caused Mars 7 to release its Descent Module too soon, and the lander traveled past Mars at a distance of 1,300 kilometers.

After the frustrations of their early attempts to explore the planet, the Soviets did manage to achieve a number of impressive results with their early 1970s Mars missions. Their progress matched that of their superpower space rival, the United States, which achieved similar results at about the same time with its Mariner program.

Phobos Project (July 1988)

The Phobos 1 and Phobos 2 spacecraft were designed to orbit Mars and deploy two landing craft on the Martian moon Phobos. The mission plan for the Phobos project also called for observations of the interplanetary environment and the Sun.

Phobos 1 was launched on July 7, 1988, and Phobos 2 lifted off five days later, on July 12, 1988. Contact with Phobos 1 was lost on September 2, when the spacecraft failed to communicate with Phobos 2. The communication failure was later determined to be the result of a software error in a program uploaded to the spacecraft a few days before it failed. The software fault caused Phobos 1 to orient its solar array panels away from the Sun, which in turn caused its solar-powered batteries to run out of power and led to its loss of useful function.

Phobos 2 successfully completed its planned observations of the Sun, the interplanetary environment, and Mars and Phobos, and the exciting prospect of landing exploratory vehicles on Phobos drew near as Phobos 2 approached the Martian moon. Sadly, the spacecraft's on-board computer suffered a malfunction that caused it to lose contact with mission controllers on March 27, 1989, bringing the mission to a premature end.

Mars 96 (November 16, 1996)

An ambitious project of the Russian Space Agency, the Mars 96 program was designed to gather information about the evolution and composition of the Martian landscape and atmosphere.

The Mars 96 Orbiter, which carried two small stations to be deployed on the Martian surface and two penetrators designed to dig in the Martian soil, was launched on November 16, 1996. The spacecraft was scheduled to arrive at Mars on September 12, 1997, but it suffered a major malfunction and re-entered Earth's atmosphere just one day after launch. Pieces of the craft survived re-entry and crashed somewhere within an area thought to include parts of Chile, Bolivia, and a portion of the Pacific Ocean.

MARS: U.S. MISSIONS

(November 1964–)

American attempts to explore Mars began with the Mariner program of satellites, beginning with the Mariner 3 Mars flyby attempt in November 1964. The first phase of the United States' exploration of Mars culminated with Mariner 9 successfully entering into orbit around the Red Planet on November 24, 1971.

In a happy amalgam of forward-looking exploration and commemoration of the nation's past, the next leap forward in U.S. exploration of Mars occurred during the summer of 1976, when the Viking 1 and Viking 2 orbiters and landers arrived at Mars just as Americans were in the midst of Bicentennial celebrations marking the founding of the United States. The Viking spacecraft transmitted stunning, high quality images of Mars, and the valuable data they returned helped to advance scientific knowledge of the planet.

A panoramic view of the Martian landscape from the Mars Pathfinder spacecraft, July 1997. [NASA/ courtesy of nasaimages.org]

After the Viking mission, a decade and a half passed without a single U.S. Mars flight, until American interest in the planet was rekindled with the Mars Exploration program of the early 1990s. A series of launches throughout the decade and into the 21st century sought to expand knowledge of the planet's past—and, particularly, to determine if Mars ever featured an environment capable of harboring any form of life—and to catalog current conditions on Mars to aid in the planning of potential manned missions in the future.

Mariner Program Mars Probes (November 1964–October 1972)

The United States made its first attempt to explore Mars with the Mariner 3 and Mariner 4 missions, in 1964. Subsequent Mariner launches focusing on Mars included the Mariner 6 and Mariner 7 flyby missions in 1969 and the Mariner 8 and Mariner 9 orbiter launches of May 1971 (other spacecraft in the series were aimed at Venus).

Alternating amazing successes with frustrating failures, the Mariner Mars program eventually provided NASA scientists with their first on-site knowledge of Mars. Mariner 4 was the first spacecraft to successfully fly by the planet and return photos of the Martian surface, and Mariner 9 became the first craft to enter into orbit around another planet.

Incorporating both the Mars flights and other flights in the 10-spacecraft series, the Mariner program cost a total of approximately $554 million.

Designed to fly past Mars at a close distance, the Mariner 3 and Mariner 4 launches took place in quick succession in November 1964.

Mariner 3 launched on November 5, 1964, with the tasks of obtaining photographs of the Martian surface and collecting a variety of scientific data. Nine hours into its flight, a malfunction caused the spacecraft's communications system to fail, and Mariner 3 was unable to enter the trajectory necessary for its planned trip. The spacecraft continued on its silent flight, ultimately entering orbit around the Sun.

Mariner 4 was launched from Cape Canaveral, Florida on November 28, 1964, and achieved the first flyby of Mars on July 14, 1965. The spacecraft made its closest pass over the planet on July 15, and captured 21 photographs as it passed within 9,846 kilometers of the surface.

In October 1965, Mariner 4 stopped transmitting data when its position in space prevented it from communicating with Earth. It began sending signals again in 1967, and recorded the impacts of a micrometeoroid shower before making its last transmission on December 21, 1967.

Mariner 6 and Mariner 7 were twin spacecraft that were each intended to collect data about Mars during close passes near the planet.

The two spacecraft each carried television cameras for photographing the Martian surface, and a digital recorder for storing scientific data generated by their onboard instruments, which included an infrared radiometer, an infrared spectroscope, and an ultraviolet spectroscope.

Mariner 6 was nearly lost in an accident on the launch pad shortly before the vehicle's scheduled launch. With the satellite already mounted atop an Atlas-Centaur launch vehicle, the main valves of the Atlas rocket suffered a malfunction that threatened to topple the entire rocket stack. Two alert technicians at the launch site recognized the grave nature of the situation immediately and turned on pressurizing pumps that kept the launch vehicle standing. The Mariner 6 satellite was saved as a result of their fast response, and after it was mated to a second Atlas-Centaur launcher, the spacecraft was successfully launched on February 24, 1969.

Mariner 6 successfully passed by Mars at a distance of 3,431 kilometers on July 31, 1969. The spacecraft captured 49 photographs of the planet during its approach phase and then collected an additional 26 images during its close encounter.

Mariner 7 lifted off on March 27, 1969, and made its closest approach to Mars on August 5, at a distance of 3,430 kilometers. Mariner 7 captured 93 photographs of Mars during its approach and an additional 33 photographs during its closest pass.

During their separate flybys, Mariner 6 and Mariner 7 photographed about 20 percent of the Martian surface and returned valuable data about the planet's atmosphere and polar cap, including data indicating that the planet's southern region features high levels of carbon dioxide.

Launched on May 8, 1971 with the intent of traveling to Mars and entering into orbit around the planet, Mariner 8 was lost soon after launch because of a malfunction in the Centaur stage of its multi-stage Atlas-Centaur launch vehicle.

Mariner 9 was launched on May 30, 1971 and performed flawlessly throughout its long trip to Mars. On November 14, 1971, when it was successfully inserted into orbit around Mars, it became the first spacecraft to orbit another planet.

The scientific instruments on Mariner 9 completed the first global mapping of the planet, revealing previously unknown details about surface features such as the Olympus Mons and the three Tharsis volcanoes.

Mariner 9 also returned the first comprehensive photographic record of the planet, in the form of 7,329 images that were later instrumental in choosing landing sites for the mid-1970s Viking landers. The Mariner 9 spacecraft was shut down on October 27, 1972 and was left in an orbit designed not to decay for at least 50 years.

Viking 1 and Viking 2 Mars Orbiters and Landers (August 1975–November 1982)

NASA's Viking mission to Mars featured two identical spacecraft designed to travel to the planet, enter into orbit, and release landers equipped to photograph and collect data about the Martian surface. The successful flights rekindled public interest in NASA after a period of malaise in the years following the end of the Apollo lunar landing program.

The most expensive planetary exploration of its time, at a cost of approximately $1 billion, the Viking Mars vehicles and mission profile were based in part on data from the Mariner 9 mission, which, in November 1971, had been the first spacecraft to orbit another planet.

Each of the twin Viking spacecraft was powered in part by nuclear-powered Radioisotope Thermal Generators (RTGs), which included small amounts of plutonium-238. Each RTG unit could produce 30 watts of power.

A powerful high-resolution camera capable of capturing panoramic views in any direction from the landing site was mounted on each lander, and the vehicles also each had a robotic arm designed to collect samples of surface materials for testing in an on-board laboratory as part of the program's search for traces of Martian life forms.

The first Viking spacecraft was launched on August 20, 1975, atop a Titan-Centaur launcher, and entered into orbit around Mars on June 19, 1976.

On July 20, 1976, the Viking 1 lander made the first entirely successful landing on Mars, at Chryse Planitia. The spacecraft transmitted its first images shortly after landing, and after a pause of several days because of a temporary fault in its robotic arm, the lander collected its first sample of surface soil and tested it for signs of life.

Initial tests of the sample returned astounding results, as they appeared to indicate the presence of microscopic organisms. NASA scientists studying the result later concluded that the tests were probably faulty, and reflected unexpected chemical interactions in the testing facility rather than a true sign of extraterrestrial life.

On Earth, the confusing test results created a minor uproar for a short time among the American press and public, but the spectacular images being returned by the lander were received with genuine enthusiasm.

Shortly following the launch of its twin, Viking 2 was launched on September 9, 1975. It entered into orbit around Mars on August 7, 1976, and its lander touched down in Utopia Planitia. The Viking 2 lander settled onto the uneven surface at a slight tilt, but the awkward angle had no appreciable impact on its research activities, as it collected and transmitted high quality photographs and repeated the sampling and biological testing of its predecessor, yielding similar results.

The two Viking orbiters captured approximately 50,000 images of the planet and returned data that appeared to indicate that Mars harbors significant amounts of water ice beneath its surface. The images also revealed that Mars's thin atmosphere features clouds formed by water ice. The orbiters made close passes by the Martian moons Phobos and Deimos during their long flights in orbit around the planet. NASA mission managers shut down the Viking 2 orbiter on July 25, 1978 after a malfunction had caused the craft's attitude control system to fail. The Viking 2 orbiter made 706 orbits of Mars. The Viking 1 orbiter ended its mission on August 17, 1980, after 1,485 orbits.

After nearly four years of operations, the Viking 2 lander was shut down on April 11, 1980 after its batteries failed. NASA officials intended to end the mission of the Viking 1 lander at about the same time, but a strong protest from scientists led them to reconsider, and the agency subsequently announced that the surface probe would continue to return weekly Martian weather reports and photographs for several more years. Viking 1 continued to operate until November 13, 1982, when mission managers accidentally transmitted an improper command to the spacecraft, causing it to shut down.

Mars Observer
(September 25, 1992–August 21, 1993)

A joint effort of NASA, the Russian Space Agency, and the Centre National d'Etudes Spatiales (CNES) of France, the Mars Observer mission was designed to study the geology and climate of Mars. The spacecraft was also intended to acquire data from the Russian Mars 1994 mission.

Mars Observer was launched on September 25, 1992 and reached Mars in August 1993. Contact with the spacecraft was lost on August 21, 1993, three days before it was scheduled to enter orbit around the planet. The loss of communication was later thought to most likely have been caused by a rupture in the spacecraft's fuel line, which is suspected to have forced the craft into an uncontrollable spin, resulting in its destruction.

The total cost of the Mars Observer mission has been estimated at $813 million.

Mars Global Surveyor (Launched November 7, 1996)

The first spacecraft to be launched in NASA's Mars Surveyor Program, the Mars Global Surveyor (MGS) was designed to capture high resolution images of the Martian surface, to make studies of the planet's topography, gravity, weather, climate, and atmosphere, and to determine whether the planet has a magnetic field. The spacecraft was also designed to act as a data relay device for later U.S. and international missions.

Launched on November 7, 1996, the Mars Global Surveyor successfully began its orbit of Mars on September 12, 1997.

Mars Pathfinder and Sojourner Rover
(December 4, 1996–September 27, 1997)

The second project in NASA's Discovery series of low-cost planetary exploration missions, Mars Pathfinder proved that economically designed vehicles could land on Mars and gather useful scientific data about the planet.

With the Sojourner Rover aboard, Mars Pathfinder lifted off on December 4, 1996 and landed in the Ares Vallis region of Mars on July 4, 1997. NASA officials named the landing site the Sagan Memorial Station, in honor of American astronomer Carl Sagan (1934–1996).

Taking atmospheric measurements as it descended, Mars Pathfinder successfully demonstrated innovative landing techniques when it directly entered the Martian atmosphere (without first going into orbit around the planet), made a parachute landing, and bounced around on the surface while protected by air bags.

The spacecraft's three solar panels were successfully deployed, and the Sojourner Rover was released onto the surface on July 6.

Operated with a 10-minute time delay by controllers on Earth, Sojourner collected stunning images of the Martian landscape and conducted analyses of rock and soil samples.

The Mars Pathfinder mission successfully achieved its science objectives and attracted positive interest from the American public for both the Discovery initiative and future exploration of Mars. The mission ended on September 27, 1997, when communication with the Mars Pathfinder lander and the Sojourner rover were lost and could not be reestablished.

Mars Surveyor '98 Polar Lander and Climate Orbiter (January 3, 1999–December 1999)

Encompassing the Mars Polar Lander and the Mars Climate Orbiter, NASA's Mars Surveyor '98 program was designed to study the weather and climate on Mars.

The Mars Polar Lander was launched on January 3, 1999. Its payload included the Deep Space 2 (DS2) probes, which were designed to penetrate and analyze the Martian soil, and the Mars Microphone, which was expected to return sound samples. The spacecraft reached Mars on December 3, 1999 and was last known to be heading into the planet's atmosphere when it abruptly ceased to communicate with mission controllers on Earth.

The Mars Climate Orbiter was launched on December 11, 1998 and reached Mars on September 23, 1999. A navigation error during a brief communications blackout as the spacecraft passed behind the planet caused mission controllers on Earth to lose contact with the Orbiter, and communication was not reestablished. As a result, the spacecraft failed to establish its intended orbit around Mars and instead apparently entered the atmosphere at a low altitude that caused it to be destroyed by atmospheric stresses and friction. It was later determined that the navigation error was the result of ground personnel sending some commands to the Orbiter in English units, rather than converting them to the metric system.

Development of the spacecraft used in the Mars Surveyor '98 program cost $191.3 million.

Mars Odyssey Orbiter (Launched April 7, 2001)

Launched on April 7, 2001, the Mars Odyssey spacecraft was the surviving portion of the Mars Surveyor 2001 project, which originally called for both an orbiter and lander. The Mars Odyssey vehicle was the orbiter spacecraft envisioned in the original mission profile.

Consisting of an equipment module, which contained the craft's electronics and science instruments, and a platform equipped with cameras and additional instruments, the Mars Odyssey arrived at its destination on October 24, 2001 and initially entered an aerobraking orbit around Mars. Gradual adjustments then brought the spacecraft into an orbit of 400 kilometers by 400 kilometers on January 30, 2002.

The Mars Odyssey orbiter served as a communications relay for the Mars Exploration Rovers Spirit and Opportunity, which landed on the surface in January 2004. Although the original mission profile called for the Mars Odyssey mission to conclude in July 2004, NASA mission managers extended the spacecraft's mission through 2005 and on into 2006.

Among its other duties in Martian orbit, the Mars Odyssey spacecraft also captured data about radiation near the planet in order to help in the planning of potential future manned missions. Radiation measurements were gathered using the Mars Radiation Environment Experiment (MARIE).

Other instruments carried aboard the Mars Odyssey included the Thermal Emission Imaging System (THEMIS), which collected data about the planet's mineral composition, and a gamma-ray spectrometer to measure the amount of various elements.

Mars Spirit and Opportunity Rovers (Launched June 10, 2003 and July 6, 2003)

NASA launched the Mars Spirit and Opportunity Rovers in the summer of 2003.

Identical in design, systems, and equipment, the Spirit and Opportunity Rovers were given the task of studying the Martian surface to determine if the climate was ever likely to have supported any form of life. To that end, each rover was equipped with instruments to collect data about the planet's geology and mineralogy in order characterize the past Martian environment, when liquid water was likely present in abundance.

The rovers had a dune buggy-like appearance, each with six wide wheels for traction aligned alongside a rectangular body known as the vehicle's "warm electronics box." Each of the six wheels was equipped with a separate motor, and the front and rear wheels could be steered independently. Each vehicle was capable of traveling at a maximum of five centimeters per second and could withstand a tilt as severe as 45 degrees. Solar arrays on top of each rover were designed to provide as much as 140 watts of electricity, which was stored in rechargeable batteries.

Collectively known as the Athena suite of scientific equipment, instrumentation on each rover included a panoramic camera, an alpha particle X-ray spectrometer, a microscopic imager, a Mossbauer spectrometer, a thermal emission spectrometer, and a device known as the Rock Abrasion Tool.

The Spirit Rover was launched first, lifting off atop a Delta II rocket on June 10, 2003. Spirit arrived at Mars six months later, on November 20, and after floating through the planet's atmosphere beneath a parachute, bounced onto the surface of Mars at the center of a cushion of air bags on January 4, 2004.

Shortly after the landing, the airbags were deflated and stored within the outer shell that housed the rover, and the shell itself then unfolded, allowing Spirit to deploy and activate its solar arrays. At each stage of the descent, the spacecraft emitted a distinctive tone, giving anxious mission managers a succession of positive reports as the successful landing progressed.

Spirit landed in the Gusev Crater, about 15 degrees south of the planet's equator, and after nearly two weeks of engineering tests drove onto the surface of Mars on January 15, 2004 to begin collecting data.

The Opportunity Rover spacecraft was launched on July 8, 2003 and arrived at Mars in December. Following a descent phase virtually identical to that of its twin, Opportunity landed on Mars at a speed of approximately 50 kilometers per hour on January 25, 2004, at Terra Meridiani. Benefitting from their experience during the earlier landing of the Spirit Rover, mission managers were able to prepare Opportunity for its first drive onto the surface in just four days.

The remarkable high-resolution images returned by Spirit and Opportunity, along with NASA's regularly-issued accounts of the rovers' operations, led to intense public interest in the mission.

Although the initial plan for Spirit and Opportunity called for the rovers to transmit data for about three months, the exceptional results obtained during that period led NASA mission managers to extend the rovers' activities. Remarkably, both rovers were still collecting and transmitting useful data after five years on the surface of Mars. Opportunity concluded a study of Victoria Crater in October 2008, while Spirit struggled to draw sustaining sunlight through its dusty solar panels after weathering a dust storm in November 2008.

Mars Reconnaissance Orbiter
(Launched August 12, 2005)

Designed to gather information about the climate of Mars and to search for water in the atmosphere and beneath the planet's surface throughout the course of a Martian year (from November 2006 to November 2008), the Mars Reconnaissance Orbiter (MRO) was launched atop an Atlas launcher on August 12, 2005 and entered into orbit around the planet on March 10, 2006.

The MRO carried six instruments, including a high resolution camera, the High Resolution Imaging Science Experiment (HIRISE), which is capable of capturing stereo images of the planet, and the Shallow Subsurface Sounding Radar (SHARAD), which was designed by the Italian space agency Agenzia Spaziale Italiana (ASI) to help in the search for water beneath the surface.

The total cost of the MRO mission was estimated at approximately $720 million.

Mars Phoenix Lander
(August 4, 2007–November 2, 2008)

Launched August 4, 2007, the Mars Phoenix Lander landed in the northern arctic area of Mars on May 25, 2008.

Part of NASA's Mars Scout Program, which solicits plans for competitively developed, cost-efficient Mars missions, the Phoenix Lander was developed by a team at the University of Arizona and was adapted from the lander portion of NASA's 2001 Mars Surveyor project at a cost of $420 million.

In contrast to NASA's rover vehicles, the Phoenix Lander was a stationery craft that remained at its landing site, where it photographed the surrounding environment and used a robotic arm to retrieve soil samples for various on-board tests.

While the spacecraft's initial samplings seemed to indicate a soil climate similar to Earth's, subsequent tests identified perchlorate, a salt derived from perchloric acid that is generally inhospitable to the formation of most forms of life on Earth, among the Martian soil's chemical components. The inconsistency of the findings led to speculation that the spacecraft may have been contaminated before or during launch.

In early November 2008, the change of Martian seasons dimmed the amount of sunlight reaching the Phoenix Lander's solar arrays and the spacecraft lost the power necessary to operate. During its five months in Mars's arctic northern region, Phoenix confirmed the presence of water ice in the planet's soil and transmitted over 25,000 photographs of its otherworldly surroundings.

MATTINGLY, THOMAS K., II

(1936–)

U.S. Astronaut

A veteran of three spaceflights, including the Apollo 16 lunar landing mission, Thomas "Ken" Mattingly was born in Chicago, Illinois on March 17, 1936. He received his initial education in Florida, where he graduated from Miami Edison High School, and then attended Auburn University, where he graduated with a Bachelor of Science degree in aeronautical engineering in 1958.

He joined the U.S. Navy that same year, and after receiving training as a pilot, he was assigned to VA-35 aboard the aircraft carrier USS *Saratoga*, where he flew A1H aircraft. In July 1963 he was assigned to the USS *Franklin D. Roosevelt,* where he flew A3B aircraft while attached to VAH-11. He served in that capacity until 1965, and was then selected to attend the U.S. Air Force Aerospace Research Pilot School at Edwards Air Force Base in California.

During his outstanding military career, Mattingly accumulated 7,200 hours of flying time, including 5,000 hours in jet aircraft. He had risen to the rank of rear admiral in the U.S. Navy by the time of his retirement from the service.

NASA selected him for training as an astronaut as a member of its Group Five selection of pilot astronauts in April 1966. His initial assignments at the space agency included service as a member of the astronaut support crew for Apollo 8 and Apollo 11. He also helped to develop the spacesuit used during the Apollo program.

Mattingly was originally assigned to pilot the command module for the Apollo 13 flight, which flew in April 1970, but he was forced to give up the assignment because he was exposed to German measles when fellow astronaut Charles Duke was diagnosed with the illness three days before launch. Mattingly was replaced by his backup, John Swigert, who lifted off with commander James Lovell and lunar module pilot Fred Haise on April 11, 1970. The Apollo 13 spacecraft suffered an explosion in an on-board oxygen tank, and the crew was forced to forego the lunar landing portion of its mission.

As the original command module pilot for the flight, Mattingly played a key role in the Earth-based simulations of the conditions aboard Apollo 13 that allowed engineers to devise solutions to the problems Lovell, Haise, and Swigert faced during their harrowing six days in space. The exceptional courage of the crew and the carefully applied expertise of the support teams on Earth brought the difficult flight to a successful conclusion, and the Apollo 13 crew splashed down safely without lasting ill effects to their health.

On April 16, 1972, Mattingly lifted off on his first space mission, this time as command module pilot of Apollo 16, along with commander John Young and lunar module pilot Charles Duke.

A perilous success, the Apollo 16 flight overcame difficulties with its launch vehicle prior to the start of the mission, and after a relatively uneventful trip to the Moon, the backup control system in the three-compartment vehicle's Casper command module failed. Mattingly was alone in the command/service module complex at the time of the failure, detailing the situation within the spacecraft from lunar orbit to mission controllers on Earth while Young and Duke passed nearby in their lunar module Orion, awaiting permission to continue their descent to the lunar surface.

When NASA's engineering staff was satisfied that the loss of Casper's backup control system did not pose an immediate threat to Mattingly in the command/service vehicle or to the prospect of Young and Duke successfully landing on the Moon and then rejoining him, they allowed the mission to proceed.

Young and Duke landed in the Cayley Plains area of the Moon's Descartes Highlands on April 20. They made three long trips outside of the Orion lunar module, accumulating more than 21 hours in EVA while they explored the rough terrain, deployed scientific equipment, and drove the second Lunar Roving Vehicle (LRV) to a spot about four miles from their landing site to explore a plateau 700 feet above their Orion craft.

Alone in lunar orbit while his crew mates explored the surface, Mattingly carried out a program of more than 20 scientific experiments.

Despite the earlier worries, Young and Duke were able to rejoin Mattingly without difficulty, and the three astronauts were able to return to Earth safely in the Casper command module, despite the loss of the craft's backup control system.

Mission controllers were pleased enough the spacecraft's performance after the lunar landing to allow Mattingly to attempt the planned spacewalk that he had trained for as part of the original mission profile for the flight. On April 25,

1972, he ventured outside of the homeward-bound Apollo 16 to retrieve film from the mapping and panoramic camera mounted in the spacecraft's scientific instrument module (SIM). During a spacewalk of 1 hour, 23 minutes, and 42 seconds, he maneuvered himself effortlessly around the vehicle's service module, carefully inspecting the outside of the craft and performing the Microbial Ecological Evaluation Device experiment, which involved the exposure of the device to the space environment for a period of about ten minutes.

At the end of their remarkable lunar odyssey, Mattingly, Young, and Duke splashed down on April 27, 1972.

After his first spaceflight, Mattingly became the lead astronaut for support of NASA's shuttle program in the agency's Astronaut Office. He served in that capacity until 1978, and then, after a brief period as a technical assistant to the manager of the shuttle orbital flight test program, he was assigned to the Astronaut Office shuttle ascent and entry group in December 1979. He also served as the backup commander for the STS-2 and STS-3 shuttle missions.

Mattingly made his second flight in space as commander of STS-4, the space shuttle Columbia's final orbital test flight. Launched on June 27, 1982, Mattingly and pilot Henry Hartsfield, Jr. carried out a number of scientific experiments and deployed a classified payload for the U.S. Department of Defense during the week-long STS-4 mission. They also used the shuttle's Remote Manipulator System (RMS) robotic arm to deploy the Induced Environment Contamination Monitor (IECM), which was designed to measure the amount of contamination the shuttle produced in the near-Earth space environment, and they captured photos of dramatic lightning activity in the Earth's atmosphere.

Mattingly and Hartsfield returned to Earth on July 4, 1982 at Edwards Air Force Base in California, marking the shuttle's first landing on the base's 15,000-foot-long concrete runway. With the conclusion of STS-4, Mattingly added another 7 days, 1 hour, 9 minutes, and 31 seconds to his career total time in space.

In June 1983, Mattingly was given the responsibility of overseeing NASA's Astronaut Office support activities for the U.S. Department of Defense.

During his third space mission, as commander of STS-51C in 1985, the first shuttle mission devoted to the classified activities of the U.S. Department of Defense, Mattingly was accompanied by pilot Loren Shriver, mission specialists James Buchli and Ellison Onizuka, and payload specialist Gary Payton. Except for the veteran Mattingly, the entire crew was making its first flight in space. Launched aboard the space shuttle Discovery on January 24, the flight concluded with a landing at the Kennedy Space Center (KSC) in Florida on January 27, 1985, after a flight of 3 days, 1 hour, 33 minutes, and 23 seconds.

Ken Mattingly spent 504 hours in space, including 1 hour, 23 minutes, and 42 seconds in EVA, during his remarkable career as an astronaut. He left NASA in 1985.

Among the many honors he has received for his space achievements, Mattingly has been awarded the Ivan C. Kincheloe Award of the Society of Experimental Test Pilots, the Haley Astronautics Award of the American Institute of Astronautics and Aeronautics (AIAA), and the Vladimir M. Komarov Diploma of the Federation Aeronautique Internationale.

McARTHUR, WILLIAM S., JR.

(1951–)

U.S. Astronaut

A veteran of four space missions, including a long-duration stay aboard the International Space Station (ISS), William "Bill" McArthur was born in Laurinburg, North Carolina, on July 26, 1951. He graduated from Red Springs High School in Red Springs, North Carolina in 1969 and then attended the United States Military Academy (USMA) at West Point, New York, where he earned a Bachelor of Science degree in applied science and engineering in 1973.

He was commissioned as a second lieutenant in the U.S. Army and initially served with the 82nd Airborne Division. In 1975, he attended the U.S. Army Aviation School, and, upon graduation the following year, he was recognized as the leading student of his class and as a distinguished graduate, and was designated an Army aviator.

While assigned to the Army's 2nd Infantry Division in South Korea, McArthur served as leader of an aeroscout team and as commander of a brigade aviation section. He returned to the United States in 1978 to serve with the 24th Combat Aviation Battalion in Savannah, Georgia.

Continuing his education concurrently with his military career, McArthur attended the Georgia Institute of Technology, where he earned a Master of Science degree in aerospace engineering in 1983.

McArthur next served as an assistant professor at the USMA before attending the U.S. Naval Test Pilot School at Patuxent River, Maryland, from which he graduated in 1987. He subsequently served as an experimental test pilot.

McArthur had risen to the rank of colonel in the U.S. Army by the time of his retirement from the service in 2001. During his outstanding military career, he achieved the designation of Master Army Aviator and graduated from the Army Parachutist Course, the Jumpmaster Course, and the Command and General Staff Officers Course. In his career as a pilot and astronaut, he accumulated more than 4,500 hours of flying time in 39 different aircraft and spacecraft.

McArthur began his association with NASA in August 1987 when he was assigned to the Johnson Space Center (JSC) to work as a vehicle integration test engineer in the agency's space shuttle program.

NASA selected McArthur as a member of its Group 13 selection of pilots and mission specialist astronauts in 1990. His technical assignments as an astronaut have included service as chief of the flight support branch of the Astronaut Office and as NASA director of operations in Russia.

McArthur made his first flight in space as a mission specialist aboard the space shuttle Columbia during STS-58, which launched on October 18, 1993. The second shuttle flight devoted to life sciences study using the Spacelab module, STS-58 featured an extensive program of scientific study. The crew, which was commanded by John Blaha, included pilot Richard Searfoss and McArthur's fellow mission specialists Rhea Seddon, David Wolf, Shannon Lucid, and payload specialist Martin Fettman. During the mission they collected more than 650 samples while conducting cardiovascular research, regulatory physiology investigations, musculoskeletal experiments, and studies in neuroscience.

The STS-58 crew returned to Earth on November 1, 1993, after a flight of 14 days, 12 minutes, and 32 seconds—which set a record for the longest shuttle flight up to that time.

McArthur began his second spaceflight on November 20, 1995 when he lifted off as a mission specialist aboard the shuttle Atlantis during STS-74, the second docking of a U.S. space shuttle and the Russian space station Mir. McArthur and his STS-74 crew mates—commander Kenneth D. Cameron, pilot James Halsell, and mission specialists Jerry Ross and Canadian Space Agency (CSA) astronaut Chris Hadfield—traveled to Mir to deliver the Russian Mir Docking Module, which was designed to make future shuttle-Mir dockings easier.

On November 15, the members of the STS-74 and Mir 20 crews opened the hatches of their vehicles and began transferring equipment and supplies from the shuttle to the space station. They also removed medical samples from the station for return to Earth on the shuttle. Atlantis undocked from Mir in the early morning hours of November 18. Then, after a flight of 8 days, 4 hours, 30 minutes, and 44 seconds, the STS-74 crew landed at the Kennedy Space Center (KSC) in Florida on November 20, 1995.

On October 11, 2000, McArthur lifted off on an historic third space mission, STS-92, which marked the 100th flight of the space shuttle program. Launched aboard the shuttle Discovery, the STS-92 crew delivered and installed the ISS Z1 Truss segment and the station's Pressurized Mating Adapter-3 (PMA-3).

McArthur made two spacewalks during STS-92 with fellow mission specialist Leroy Chiao. Alternating EVAs with Michael Lopez-Alegria and Peter Wisoff, McArthur and Chiao made two of the four spacewalks necessary to install the Z1 Truss and the PMA-3.

McArthur and Chiao made the first spacewalk of the flight on October 13, 2000, starting the installation of the Z1 Truss by connecting electrical cables and moving two antennae to clear the way for their next EVA and accumulating 6 hours and 28 minutes of extravehicular activity. Lopez-Alegria and Wisoff began the PMA-3 docking port installation on October 16, and McArthur and Chiao then ventured outside the shuttle again on October 17, when they spent another 6 hours and 48 minutes in EVA while finishing the Z1 Truss installation. Their crew mates made the final EVA of the mission the following day.

The STS-92 crew next transferred equipment and supplies onto the ISS in preparation for the arrival of the station's first long-duration crew. Discovery undocked from the ISS on October 20 and returned to Earth on October 24, 2000 after a flight of 11 days, 19 hours, 12 minutes, and 15 seconds.

During his two STS-92 spacewalks, McArthur accumulated 13 hours and 16 minutes in EVA.

Following his third spaceflight, McArthur served as a member of the ISS Expedition 10 backup crew. He retired from the U.S. Army in 2001 with the rank of colonel, and then continued to serve his country as a NASA astronaut.

McArthur began his fourth flight in space on October 1, 2005, this time as commander of the ISS Expedition 12 crew, with flight engineer Valery Tokarev of the Russian Federal space agency Roscosmos. They launched from the Baikonur Cosmodrome in Soyuz TMA-7, along with American space tourist Gregory Olsen. Olsen spent about 10 days at the ISS and then returned to Earth with the Expedition 11 crew, while McArthur and Tokarev began their six-month stay aboard the station.

The Expedition 12 crew performed a variety of scientific work during their mission, and made two spacewalks devoted to space station maintenance and repairs and experiments.

McArthur and Tokarev first ventured outside the ISS on November 7, 2005, when they installed a camera and replaced a number of parts in various fixtures on the outside of the station during a spacewalk of 5 hours and 22 minutes. They made a second EVA on February 3, 2006, during which they deployed a spacesuit fitted with a radio transmitter (dubbed "SuitSat-1") and worked on several experiments. During their two Expedition 12 spacewalks, they spent 11 hours and 5 minutes in EVA.

They returned to Earth on April 8, 2006, aboard Soyuz TMA-7 with Marcos Pontes, the first citizen of Brazil to fly in space. Pontes had arrived at the ISS with the Expedition 13 crew.

Bill McArthur spent more than 224 days in space, including more than 24 hours in EVA, during four space missions.

McAULIFFE, S. CHRISTA CORRIGAN

(1948–1986)

U.S. Civilian Astronaut

Christa McAuliffe was born in Boston, Massachusetts on September 2, 1948. In 1966, she graduated from Marian High School in Framingham, Massachusetts and then attended Framingham State College, where she received a Bachelor of Arts degree in 1970.

She began her career as an American history teacher at Benjamine Foulois Junior High School in Morningside, Maryland, where she taught students in the eighth grade. From 1971 to 1978 she taught eighth grade English and eighth grade American history and ninth grade civics at Thomas Johnson Junior High School in Lanham, Maryland.

Continuing her education concurrently with her teaching career, she attended Bowie State College in Bowie, Maryland, where she received a Masters degree in education in 1978.

McAuliffe then relocated to Concord, New Hampshire, where she taught at Bundlett Junior High School for one year. In 1980, she taught ninth grade English at Bow Memorial High School in Concord, and in 1982 she transferred to Concord High School, where she taught American history, economics, and law and developed and taught a course on "The American Woman" for sophomores, juniors, and seniors.

An active and engaging teacher, Christa McAuliffe was a member of the National Council of Social Studies, a board member of the New Hampshire Council of Social Studies, a member of the National Education Association, the New Hampshire Education Association, and the Concord Teachers Association.

McAuliffe also taught Christian Doctrine classes at her church, was a member of the Junior Service League, and helped to raise funds for Concord Hospital and the Concord YMCA. She and her husband Steven served as a host family for inner-city youth as part of the A Better Chance (ABC) program.

In 1985, NASA chose her from a group of 11,000 applicants to become the first "Teacher in Space."

She received training to fly aboard the space shuttle as a payload specialist and was to have conducted lessons for broadcast in schools across the United States during STS-51L., which launched from the Kennedy Space Center (KSC) in Florida aboard the space shuttle Challenger on January 28, 1986.

Tragically, McAuliffe and her STS-51L crew mates were killed when a fault in one of the shuttle's huge solid rocket boosters caused a fuel leak that ignited and caused a massive explosion just 73 seconds after liftoff. She was 37 at the time of her death.

Christa McAuliffe was survived by her husband Steven and two children. She was posthumously awarded the Congressional Space Medal of Honor.

McCANDLESS, BRUCE, II

(1937–)

U.S. Astronaut

A member of NASA's Group Five selection of pilot astronauts and veteran of two spaceflights, Bruce McCandless was born in Boston, Massachusetts on June 8, 1937. He received his initial education in California, where he graduated from Woodrow Wilson Senior High School in Long Beach, and he then attended the United States Naval Academy, where he graduated with a Bachelor of Science degree in 1958 and ranked second in his class of 899 students.

He received his initial training as a pilot at the Naval Air Station in Pensacola, Florida and in Kingsville, Texas, and was designated a naval aviator in 1960. After completing additional training in weapons systems and aircraft carrier landing procedures, he was assigned to Fighter Squadron 102 in December 1960.

McCandless flew the F-6A Skyray and F-4B Phantom II aircraft during deployments aboard the USS *Forrestal* and the USS *Enterprise,* and was aboard the Enterprise during the tense Cuban Missile Crisis standoff between the United States and the Soviet Union in October 1962. As one of the American ships used in the blockade of Cuba, which was designed to prevent the Soviet Union from completing the installation of nuclear weapons on the island, the Enterprise played a key role in the United States' military response during the crisis.

In 1964, McCandless was briefly assigned to Attack Squadron 43 at Apollo Soucek Field in Oceana, Virginia as an instrument flight instructor.

Continuing his education concurrently with his military career, McCandless attended Stanford University, where he earned a Master of Science degree in electrical engineering in 1965. He is a member of the Institute of Electrical & Electronic Engineers.

In addition to his technical interests, McCandless is also a member of the National Audubon Society and former president of the Houston Audubon Society.

During his outstanding military career, McCandless accumulated over 5,200 hours of flying time, including 5,000 hours in a wide variety of jet aircraft including the F-4B Phantom II, F-6A Skyray, F-11 Tiger, T-1 Seastar, T-33B Shootingstar, T-34B Mentor, and TF-9J Cougar. He had risen to the rank of captain in the U.S. Navy by the time of his retirement from the service.

NASA selected him for training as an astronaut as a member of its Group Five selection of pilot astronauts in April 1966. His initial assignments at the space agency included service as a member of the astronaut support crew for Apollo 14, and he also served as a backup for Skylab 2 in 1973.

Putting his exceptional engineering skills to good use, McCandless helped to develop the M-509 astronaut maneuvering unit, which was tested during the Skylab program, and also played a key role in the creation of the Manned Maneuvering Unit (MMU), which he would later use in space. He received a patent for his work on the design of the EVA tool tethering system used during shuttle spacewalks.

McCandless first flew in space during the 10th space shuttle mission, STS-41B, aboard the space shuttle Challenger in 1984. Launched on February 3, 1984, McCandless and fellow mission specialist Robert Stewart achieved the first untethered spacewalks of the space shuttle era and successfully tested the Manned Maneuvering Unit (MMU) for the first time in space.

Veteran astronaut Vance Brand commanded the STS-41B mission, and the crew also included pilot Robert Gibson and mission specialists McCandless, Stewart, and Ronald McNair.

Invoking the glory days of the early U.S. space program, McCandless and Stewart were effortlessly able to direct themselves around the outside of the shuttle using the MMU backpacks, which were manufactured by the Martin-Marietta Company. The astronauts made their first spacewalk with the devices on February 7, 1984, when they spent 5 hours and 55 minutes in EVA while testing the MMUs and running through the procedures that would later be used in the retrieval of the Solar Maximum satellite, which was scheduled to be carried out during the next space shuttle mission.

On February 9, McCandless and Stewart donned the MMU backpacks for a second EVA. They made more tests related to the scheduled Solar Max satellite retrieval, practicing their ability to secure target items in the fashion that later spacewalkers would capture the satellite. McCandless and Stewart added another 6 hours and 17 minutes to their spacewalking total during the second EVA.

Media coverage of the MMU spacewalks was extensive, and the fascination of the American public rose in intensity with the second spacewalk by the astronauts that journalists dubbed "Flash and Buck," in a tip of the hat to science fiction heroes Flash Gordon and Buck Rogers. Looking for all the world like the heroes of comic books and movie serials, McCandless and Stewart quickly became icons of the new era of space travel and amply demonstrated the promise that the designers of the space shuttle program had originally envisioned when they first proposed the reusable "space plane."

Challenger landed on February 11, 1984, with the STS-41B crew making the first shuttle landing at the Kennedy Space Center (KSC) in Florida after a flight of 7 days, 23 hours, 15 minutes, and 55 seconds.

Continuing his education concurrently with his career as an astronaut, McCandless attended the University of Houston at Clear Lake, where he earned a Masters degree in business administration in 1987.

On April 24, 1990, he began his second spaceflight, this time as a mission specialist aboard the space shuttle Discovery during STS-31, the mission that included the deployment of the Hubble Space Telescope (HST). His made his second space mission with STS-31 commander Loren Shriver, pilot Charles Bolden, Jr., and fellow mission specialists Steven Hawley and Kathryn Sullivan.

The STS-31 crew traveled to a much higher orbit than usual for a shuttle flight to facilitate the proper deployment of the HST; Discovery achieved an altitude of 380 miles during the successful launch of the instrument.

In addition to the HST, Discovery also carried a number of experiments, including protein crystal growth investigations, a polymer membrane processing test, and an experiment that studied the behavior of electrical arcs in the space environment.

At the end of the flight, McCandless and his crew mates returned to Earth at Edwards Air Force Base in California on April 29, 1990, after a flight of 5 days, 1 hour, 16 minutes, and 6 seconds.

Bruce McCandless spent more than 13 days in space, including over 12 hours in EVA, during his two spaceflights.

Among many the many honors he has received for his achievements as an astronaut, McCandless was twice awarded the Victor A. Prather Award from the American Astronautical Society, the NASA Exceptional Engineering Achievement Medal, the Collier Trophy of the National Aeronautic Association, and the Smithsonian Institution National Air and Space Museum trophy.

McCOOL, WILLIAM C.

(1961–2003)

U.S. Astronaut

William McCool was born in San Diego, California on September 23, 1961. In 1979, he graduated from Coronado High School in Lubbock, Texas, and then attended the United States Naval Academy, where he received a Bachelor of Science degree in applied science in 1983, graduating second in his class of 1,083.

After completing his initial training as a pilot, he received training in EA-6B Prowler aircraft while serving at Whidbey Island, Washington, attached to Tactical Electronic Warfare Squadron 129. He served aboard the USS *Coral Sea* (CV-43) while assigned to Tactical Electronic Warfare Squadron 133, making two deployments to the Mediterranean Sea. During that time he was also designated a wing qualified landing signal officer.

Continuing his education concurrently with his military career, he attended the University of Maryland, where he received a Master of Science degree in computer science in 1985.

In 1989, McCool was chosen to participate in the cooperative program of the U.S. Naval Postgraduate School in Monterey, California and the U.S. Naval Test Pilot School in Patuxent River, Maryland. He graduated in 1992, receiving a Master of Science degree in aeronautical engineering from the Naval Postgraduate School and the Outstanding Student and Best DT-II Thesis awards from the Naval Test Pilot School.

He was then assigned to the Strike Aircraft Test Directorate at Patuxent River, where he worked as a test pilot in TA-4J and EA-6B aircraft, and flight tested the Advanced Capability EA-6B. He later returned to Whidbey Island, where he was assigned to Technical Electronic Warfare Squadron 32 and served as the squadron's Administrative and Operations Officer aboard the USS *Enterprise* (CVN-65).

During his outstanding military career, McCool accumulated more than 2,800 hours of flying time in 24 different types of aircraft, and made more than 400 landings on aircraft carriers.

NASA selected him for training as an astronaut in 1996. His initial assignments at the agency included service in the Computer Support Branch of the Astronaut Office, and he also served as technical assistant to the Director of Flight Crew Operations.

McCool made his first flight in space as pilot of the dedicated science and research STS-107 mission aboard Columbia in 2003. During 16 days in space, the STS-107 crew worked around the clock in two shifts to complete 80 experiments.

Tragically, the shuttle Columbia had been seriously damaged at launch (more seriously than NASA officials realized during the mission) and was destroyed by the stresses of re-entry into the Earth's atmosphere. McCool and his fellow crew members were killed when Columbia broke apart over the southern United States, just 16 minutes before the shuttle was scheduled to land, on February 1, 2003.

In his superb dual careers as a commander in the U.S. Navy and as an astronaut, William McCool devoted his life to the service of his country. He accumulated a total of 15 days, 22 hours, and 20 minutes in space during STS-107, and was posthumously awarded the Congressional Space Medal of Honor, the NASA Space Flight Medal, the NASA Distinguished Service Medal, and the Defense Distinguished Service Medal. He was survived by his wife and children.

McDIVITT, JAMES A.

(1929–)

U.S. Astronaut

A pioneering astronaut and veteran of two key space missions during the Gemini and Apollo programs, James McDivitt was born in Chicago, Illinois on June 10, 1929. He received his initial education in Michigan, where he graduated from Kalamazoo Central High School, and he also attended Jackson Junior College in Jackson, Michigan.

In 1951, he joined the U.S. Air Force, and after receiving his initial training as a pilot, he flew 145 combat missions in F-80 and F-86 aircraft during the Korean War.

He subsequently attended the Air Force Experimental Test Pilot School at Edwards Air Force Base in California and completed the Air Force Aerospace Research Pilot course.

Continuing his education concurrently with his military service, he attended the University of Michigan, where he graduated first in his class in 1959 and earned a Bachelor of Science degree in aeronautical engineering.

During his outstanding military career as a pilot, combat pilot, and test pilot, McDivitt accumulated more than 5,000 flying hours. He had risen to the rank of brigadier general in the U.S. Air Force by the time of his retirement from the service in June 1972.

NASA chose him for training as an astronaut as a member of its Group Two selection of pilot astronauts in September 1962.

McDivitt first flew in space as commander of Gemini 4, with pilot Edward White, whose duties during the flight included performing the first American spacewalk. The Gemini 4 mission and its EVA achievement put the United States on an equal footing with the Soviet Union in the Cold War space race between the two nations.

Gemini 4 was the eighth manned U.S. spaceflight, and the longest manned flight of the U.S. space program up to that time. In addition to the landmark EVA, the mission profile for the flight also included station keeping and rendezvous activities with the second stage of the Titan II launch vehicle, and tests of the Gemini capsule's ability to maneuver into different orbits. McDivitt and White were also assigned 11 scientific experiments to perform during their four days in space.

The Gemini 4 crew mates spent long hours training for their complicated flight in the Gemini

James McDivitt in May, 1965, during training for the Gemini 4 mission. [NASA/courtesy of nasaimages.org]

spacecraft simulator at NASA's Manned Spacecraft Center in Houston, and they studied reports about the world's first spacewalk, which had been made by Soviet cosmonaut Alexei Leonov during the Voskhod 2 flight in March 1965.

Although they were not able to achieve rendezvous with their spent Titan II rocket stage, their attempt provided valuable data that helped NASA engineers figure out how to accomplish the procedure on later flights.

White's spacewalk fared much better. McDivitt helped his crew mate to prepare for the EVA, and seeing how exhausted White was after the preparations, he convinced mission controllers to delay the excursion a bit. The extra 15 minutes of rest allowed White to emerge from the Gemini 4 capsule better focused and ready to take on the challenge of working in the space environment.

After White had spent about 16 minutes in EVA, McDivitt relayed word from Mission Control that it was time for him to return to the interior of the Gemini 4 spacecraft. Another five minutes passed while White prepared to return to his place within the craft, but getting back into the capsule proved more difficult than getting out of it, and McDivitt had to work hard to help White back into the vehicle. A struggle of some 15 minutes took place before the two astronauts were able to close the hatch.

The rest of the flight proceeded with far less drama, and McDivitt and White were ultimately able to complete their four-day mission successfully. They carried out a program of scientific experiments, gathered a large number of excellent photographs, and performed the navigational exercises they had been assigned.

An unexpected hazard arose at the end of the 48th orbit, when a malfunction in the spacecraft's computer system forced mission managers and the astronauts to forego their plans for an automatic re-entry. The astronauts returned to Earth safely despite the computer failure, however, splashing down on June 7, 1965 after 62 orbits in 4 days, 1 hour, 56 minutes, and 12 seconds.

During the remarkable Gemini 4 flight, McDivitt and White achieved the first U.S. spacewalk, made the first tentative tests of rendezvous procedures, and provided medical data that supported the planning of longer future flights.

McDivitt's second space mission, Apollo 9, was a key flight in the development of the Apollo lunar module. McDivitt lifted off as commander of Apollo 9 on March 3, 1969, along with David Scott and Russell "Rusty" Schweickart. Five days later, he entered the Apollo 9 lunar module Spider with Schweickart and began the first flight in space for the strange looking vehicle that would, in later versions, ferry astronauts to the surface of the Moon.

During the test flight, McDivitt and Schweickart fired the lunar module's engine to propel the craft into a higher orbit than the command module Gumdrop, and then, after cutting the vehicle's descent stage loose, they returned to the command module, docked, and rejoined Scott for the return trip to Earth. Apollo 9 returned to Earth on March 13 after a flight of 10 days and 1 hour.

James McDivitt spent more than 14 days in space during his career as an astronaut.

In May 1969, he was named manager of NASA's lunar landing operations, and in that capacity he was responsible for shaping the subsequent Apollo explorations of the lunar surface. After the first lunar landing, in July of that year, he became manager of the Apollo spacecraft program, and served in that capacity for the Apollo 12, 13, 14, 15, and 16 flights.

McDivitt left NASA in June 1972 to pursue a career in the private sector, initially as executive vice president of corporate affairs for Consumers Power Company. He subsequently served as executive vice president at Pullman Inc., and in October 1975 became president of the Pullman Standard division and Railcar division. He joined the Rockwell International Corporation in 1981, where he has served as senior vice president of government operations.

McMONAGLE, DONALD R.

(1952–)

U.S. Astronaut

Donald McMonagle was born in Flint, Michigan on May 14, 1952. In 1970, he graduated from Hamady High School in Flint, and then attended the United States Air Force Academy, where he received a Bachelor of Science degree in astronautical engineering in 1974.

He received his initial training as a pilot at Columbus Air Force Base in Mississippi, and was then assigned to Homestead Air Force Base in Florida, where he trained as a pilot in F-4 aircraft. He flew the F-4 at Kunsan Air Base in South Korea and was then assigned to Holloman Air Force Base in New Mexico; he subsequently served as an F-15 instructor pilot at Luke Air Force Base in Arizona.

McMonagle was selected to attend the U.S. Air Force Test Pilot School at Edwards Air Force Base in California in 1981, and received the Liethen-Tittle Award as the top graduate in his class the following year. In 1982, he became operations officer and project test pilot for the Advanced Fighter Technology Integration F-16 aircraft.

Continuing his education concurrently with his military career, McMonagle attended California State University at Fresno, where he received a Master of Science degree in mechanical engineering in 1985.

He then attended the Air Command and Staff College at Maxwell Air Force Base in Alabama, and was assigned to the 6513th Test Squadron at Edwards Air Force Base as the squadron's operations officer.

During his outstanding military career, McMonagle accumulated more than 5,000 hours of flying time in a variety of aircraft, including the F-4, F-15, F-16, and T-38. He had risen to the rank of colonel in the U.S. Air Force by the time of his retirement from the service.

NASA selected him for training as an astronaut in 1987, and he first flew in space as a mission specialist during the STS-39 flight of the space shuttle Discovery. Launched on April 28, 1991, STS-39 was the 40th flight of the space shuttle program and the first to carry both classified payloads devoted to the activities of the U.S. Department of Defense and unclassified payloads. The unclassified portion of the flight included the Air Force Program-675, the Infrared Background

Signature Survey (IBSS), and the Space Test Payload-1 (STP-1). Discovery returned to Earth on May 6, 1991, landing at the Kennedy Space Center (KSC) in Florida.

On his second spaceflight, McMonagle piloted the space shuttle Endeavour, during STS-54 in January 1993. Launched on January 13, the crew deployed the fifth Tracking and Data Relay Satellite (TDRS-F) on the first day of the flight, and mission specialists Gregory Harbaugh and Mario Runco made a spacewalk of 4 hours and 28 minutes in the shuttle's open payload bay to try out a variety of EVA techniques that would later be used in the construction of the International Space Station (ISS).

In another test tied to the future development of the ISS, the STS-54 flight also marked the first time that a fuel cell was shut down and restarted while a shuttle was in orbit. At the close of the successful mission, Endeavour landed at the KSC in Florida on January 19, 1993, after nearly six days in space.

On November 3, 1994, McMonagle lifted off on a remarkable third space mission, as commander of the STS-66 ATLAS-3 flight of the space shuttle Atlantis.

The STS-66 crew worked around the clock to operate the Atmospheric Laboratory for Applications and Science (ATLAS-3), which was making its third shuttle flight, and the Cryogenic Infrared Spectrometers and Telescopes for the Atmosphere-Shuttle Pallet Satellite (CRISTA-SPAS), which had been jointly developed by NASA and the German space agency Deutsches Zentrum für Luft- und Raumfahrt e.V. (DARA).

Using the ATLAS-3 and CRISTA-SPAS instruments, the crew gathered extensive data about the Earth's atmosphere and the Sun's energy output, and how they affect the Earth's ozone layer. The overall program of study, including the earlier flights of the instruments, was designed to study the Earth's energy balance and atmospheric change during an 11-year solar cycle.

At the end of STS-66, Atlantis returned to Earth at Edwards Air Force Base in California on November 14, 1994 after 175 orbits.

Donald McMonagle spent more than 605 hours in space during three space shuttle flights.

In 1996, he established NASA's Extra-Vehicular Activity Project Office, which is responsible for developing and managing spacesuits and EVA tools for use in future spacewalking activities aboard the shuttle and at the ISS.

In 1997, he became manager of launch integration activities at the KSC in Florida and was made chair of the Mission Management Team, a position which carries the ultimate authority for deciding whether or not to proceed with any given launch.

McNAIR, RONALD E.

(1950–1986)

U.S. Astronaut

Ronald McNair was born in Lake City, South Carolina on October 21, 1950. In 1967, he graduated from Carver High School in Lake City and then attended North

Carolina A&T State University, where he was named a Presidential Scholar from 1967 to 1971. He graduated magna cum laude from North Carolina A&T in 1971, earning a Bachelor of Science degree in physics.

He next attended the Massachusetts Institute of Technology (MIT), where, while pursuing his Doctorate, he conducted groundbreaking experiments involving the interaction of high-pressure CO_2 laser radiation and molecular gases. He was a Ford Foundation Fellow from 1971 to 1974, and also received fellowships from the National Fellowship Fund and from NATO during his time at MIT. In 1975, he studied laser physics at E'cole D'ete Theorique de Physique in Les Houches, France, and was named the Omega Psi Phi Scholar of the Year.

McNair earned a Ph.D. in physics from MIT in 1976.

Widely regarded by experts in his field, McNair published the results of his research on lasers and molecular spectroscopy in technical papers, and gave presentations about his work throughout the United States and in other nations.

McNair began his professional career at Hughes Research Laboratories in Malibu, California, where he worked as a staff physicist. He developed methods for using lasers to separate isotopes and for use in photochemistry, and he also worked on the integration of electro-optic laser modulation in communications between satellites in space. His other research at Hughes involved the development of ultra-fast infrared detectors and ultraviolet remote sensing in the atmosphere.

In addition to his high-tech research, McNair put his outstanding academic expertise to use in the pursuit of a vibrant interest in martial arts. A fifth-degree black belt Karate instructor, he won an Amateur Athletic Union (AAU) Karate Gold Medal in 1976, finished first in five regional black belt Karate championships, and conducted research on the scientific foundations of the martial arts. He was also a performing jazz saxophonist and a member of the American Association for the Advancement of Science, the American Optical Society, and the American Physical Society. He served as a member of the board of trustees of the North Carolina School of Science and Mathematics and the MIT Corporation Visiting Committee, and was a visiting physics lecturer at Texas Southern University.

Among many awards and honors he received during his career as a scientist, McNair was honored by his alma mater North Carolina A&T State University in 1978, when it awarded him an honorary Doctorate of Laws degree. The National Society of Black Professional Engineers honored him with its Distinguished National Scientist Award in 1979, and he also received an honorary Doctorate of Science degree from Morris College in 1980 and from the University of South Carolina in 1984.

NASA selected him for astronaut training in 1978, and he first flew in space as a mission specialist aboard the space shuttle Challenger during STS-41B, in February 1984. Launched on February 3, the flight featured a complex mission profile that saw mission specialists Bruce McCandless and Robert Stewart make the first untethered space walks of the shuttle program, using Manned Maneuvering Units (MMUs) for the first time in orbit. A number of technical problems hindered the deployment of two satellites, but the crew was able to conduct a number of experiments during the eight-day flight.

McNair served as the primary operator of the shuttle's Remote Manipulator System (RMS) during its first use, and he also oversaw the operation of several of the experiments flown aboard the mission.

After a flight of 128 orbits in 7 days, 23 hours, 15 minutes, and 55 seconds, the crew made the first shuttle landing at the Kennedy Space Center in Florida, when Challenger touched down on Runway 15 on February 11, 1984.

McNair was again assigned to fly as a mission specialist aboard Challenger during STS-51L, which launched from the KSC on January 28, 1986.

Tragically, McNair and his crew mates were killed when a fault in one of the shuttle's huge solid rocket boosters caused a fuel leak that ignited and caused a massive explosion just 73 seconds after liftoff. He was 35 years old at the time of his death.

Ronald McNair is survived by his wife, Cheryl, and their two children. He was posthumously awarded the Congressional Space Medal of Honor.

MEADE, CARL J.

(1950–)

U.S. Astronaut

Carl Meade was born at Chanute Air Force Base in Illinois on November 16, 1950. In 1968, he graduated from Randolph High School at Randolph Air Force Base in Texas, and he then attended the University of Texas, where he graduated with honors in 1973, receiving a Bachelor of Science degree in electronics engineering.

He next attended the California Institute of Technology, where he was a Hughes Fellow. He graduated in 1975 with a Master of Science degree in electronics engineering and then worked as an electronics design engineer at Hughes Aircraft Company in Culver City, California.

Meade joined the U.S. Air Force and received his undergraduate pilot training at Laughlin Air Force Base in Texas, where he was recognized as a distinguished graduate in 1977. He flew the RF-4C aircraft while serving at Shaw Air Force Base in South Carolina while attached to the 363rd Tactical Reconnaissance Wing, and was later selected to attend the U.S. Air Force Test Pilot School, where he was again recognized as a distinguished graduate and received the Liethen-Tittle Award as the outstanding test pilot of his class.

As a test pilot assigned to the 6,510th Test Wing at Edwards Air Force Base in California, he flew in F-5E, RF-5E, and F-20 aircraft, and took part in extensive tests of the F-4E aircraft. He also served as a test pilot in the F-16 Combined Test Force, flying in F-16A and F-16C aircraft. In 1985, he returned to the Test Pilot School as a test pilot instructor and also served as a program manager for avionics systems test training.

During his outstanding career as a pilot, test pilot, and flight instructor, Meade has accumulated more than 4,800 hours of flying time in jet aircraft and has flown 27 different types of aircraft. He retired from the U.S. Air Force in 1996 with the rank of colonel.

A registered professional engineer, Meade also pursues his passion for flying as a member of the Experimental Aircraft Association, joining other enthusiasts in building his own aircraft.

He became an astronaut in 1985. Meade's initial technical assignments at NASA included work in the Shuttle Avionics Integration Laboratory, and he also participated in the flight testing of crew escape systems. He has served as the Astronaut Office representative for projects involving the Space Shuttle Main Engines (SSMEs) and the shuttle's solid rocket boosters (SRBs) at the Marshall Space Flight Center (MSFC) in Huntsville, Alabama.

Meade first flew in space as a mission specialist during STS-38, which launched aboard the space shuttle Atlantis on November 15, 1990. The seventh shuttle mission devoted to the classified activities of U.S. Department of Defense, STS-38 was also the first shuttle mission in more than five years to conclude with a landing at the Kennedy Space Center (KSC) in Florida, touching down on November 20, 1990 after 80 orbits and nearly five days in space.

His second space mission was STS-50 in 1992, which set a new record for the longest shuttle flight up to that time. Launched on June 25, 1992 aboard the shuttle Columbia, the STS-50 crew made 221 orbits in 13 days, 19 hours, 30 minutes, and 4 seconds. Meade was a mission specialist during the flight, which advanced the study of microgravity through the first flight of the U.S. Microgravity Laboratory (USML-1). The crew conducted experiments in crystal growth, fluid physics, fluid dynamics, biological science, and combustion science.

STS-50 was extended by one day because of poor weather at the originally scheduled landing site, at Edwards Air Force Base in California, and the landing was subsequently diverted to the KSC in Florida, where Columbia touched down on July 9, 1992.

After his second spaceflight, Meade served as NASA's lead astronaut for Rendezvous and Docking Operations.

On September 9, 1994, he launched on a remarkable third spaceflight, aboard the shuttle Discovery during STS-64. With fellow mission specialist Mark Lee, Meade made the first untethered American spacewalk in a decade.

Designed to test NASA's new Simplified Aid for EVA Rescue (SAFER) backpacks, the historic EVA took place on September 16, 1994. Meade and Lee spent 6 hours and 51 minutes outside the shuttle during the successful test.

The STS-64 crew also successfully tested an innovative optical radar system, the Lidar in Space Technology Experiment (LITE), and deployed and retrieved a SPARTAN free-flyer satellite.

Part of NASA's "Mission to Planet Earth" initiative, the radar exercise tested an experimental system that utilized laser pulses instead of radio waves. The LITE instrument was trained on a variety of targets, including cloud structures, dust clouds, and storm systems, among others. Groups in 20 countries around the globe collected data using ground- and aircraft-based radar instruments to help verify the data collected during the LITE experiment.

The Shuttle Pointed Autonomous Research Tool for Astronomy (SPARTAN-201) collected data about the solar wind and the Sun's corona during two days of free flight. Discovery landed at Edwards Air Force Base in California on September 20, 1994, after a flight of 10 days, 22 hours, and 51 minutes.

Carl Meade spent more than 29 days in space, including more than 6 hours in EVA, during his career as an astronaut.

He left NASA in 1996 for a position as deputy project manager at the Lockheed Skunk Works, working on the company's experimental X-33 vehicle.

MELROY, PAMELA A.

(1961–)

U.S. Astronaut

Pamela Melroy was born in Palo Alto, California on September 17, 1961. In 1979, she graduated from Bishop Kearney High School in Rochester, New York, and then attended Wellesley College, where she graduated with a Bachelor of Science degree in physics and astronomy in 1983. In 1984, she graduated from the Massachusetts Institute of Technology with a Master of Science degree in Earth and planetary sciences.

Melroy was commissioned in the U.S. Air Force through the Air Force ROTC program, and received her initial training as a military pilot at Reese Air Force Base in Lubbock, Texas. She then served as a pilot, aircraft commander, and instructor pilot at Barksdale Air Force Base in Bossier City, Louisiana until 1991, and participated in the Just Cause, Desert Shield, and Desert Storm deployments, during which she accumulated more than 200 hours of flying time in combat and combat support roles.

Following her graduation from the U.S. Air Force Test Pilot School at Edwards Air Force Base in California, she served as a test pilot for several years prior to her selection as an astronaut candidate in December 1994.

During her outstanding military career, Melroy accumulated more than 5,000 hours of flight time. She retired from the Air Force in February 2007, with the rank of colonel.

Melroy first flew in space as pilot of the space shuttle Discovery during STS-92, which lifted off from the Kennedy Space Center (KSC) on October 11, 2000. The 100th flight of the space shuttle program, STS-92 had as its primary objective the delivery and installation of the International Space Station (ISS) Z1 Truss and Pressurized Mating Adapter 3, a task that the crew performed flawlessly over the course of seven days of docked operations at the station. The STS-92 crew touched down at Edwards Air Force Base in California on October 24, after a total flight of just under 13 days in space.

She again served as the pilot for her second spaceflight, STS-112, which launched aboard the space shuttle Atlantis on October 7, 2002. The STS-112 crew added the S1 Truss to the ISS, during a flight of slightly fewer than 11 days.

With the launch of the STS-120 mission of the space shuttle Discovery on November 23, 2007, Melroy became the second woman to command a space shuttle flight (Eileen Collins was the first, as commander of STS-93 in July 1999).

The STS-120 crew delivered and installed the ISS Harmony Node 2 module and carried out improvised repairs to one of the space station's solar arrays. The flight also achieved a transfer of ISS crew members, delivering Daniel Tani to the station to serve as part of the long-duration Expedition 16 crew and returning Clayton Anderson to Earth at the end of his long-duration flight of more than 151 days.

At the end of STS-120, Melroy and her crew returned to Earth at KSC on November 7, 2007, after a flight of 15 days, 2 hours, and 23 minutes.

In her three space missions, Pamela Melroy has accumulated more than 38 days in space.

MERBOLD, ULF D.

(1941–)

ESA Astronaut; First Citizen of West Germany to Fly in Space

The first citizen of West Germany and first European Space Agency (ESA) astronaut to fly in space and a veteran of spaceflights with both the U.S. and Russian space programs, Ulf Merbold was born in Greiz, Germany on June 20, 1941. In 1960, he graduated from high school and relocated to West Germany (at the time, the country was divided into East and West, prior to the country's reunification after the fall of the Soviet Union).

He attended the University of Stuttgart, where he earned a diploma in physics in 1968 and a Doctorate degree in physics in 1976.

Merbold began his career as a scientist at the Max Planck Institute for Metals Research in Stuttgart. The ESA chose him for training as a payload specialist astronaut in 1978, as one of three candidates for the first flight of the ESA-designed Spacelab aboard the U.S. space shuttle.

On November 28, 1983, Merbold became the first citizen of West Germany and the first ESA astronaut to fly in space when he lifted off as a payload specialist on the first Spacelab mission, STS-9, aboard the space shuttle Columbia.

Gemini and Apollo veteran—and Apollo 16 moon walker—John Young served as commander of STS-9. The crew also included pilot Brewster Shaw, mission specialists Owen Garriott and Robert Parker, and Merbold's fellow payload specialist Byron Lichtenberg.

After years of cooperative development by NASA and the ESA, the eagerly anticipated first Spacelab flight achieved excellent results. Crew members conducted 73 scientific experiments in a variety of disciplines, including astronomy, physics, atmospheric physics, Earth observation, life sciences, materials sciences, space plasma physics, and technology.

Toward the end of the flight, a series of disturbing difficulties arose, beginning with the failure of two of Columbia's on-board computers and culminating in two of the shuttle's three auxiliary power units catching fire during the landing at Edwards Air Force Base in California. The crew emerged from the equipment failures without suffering any harm, however, and the flight was celebrated for the successful Spacelab work. From launch to landing, the STS-9 crew was in space for 10 days, 7 hours, 47 minutes, and 24 seconds.

After his historic first space mission, Merbold served as the backup payload specialist for the STS-61A Spacelab D-1 mission, which was devoted to research designed by German scientists.

In 1986, he was appointed to lead the astronaut office of the German national space agency Deutsches Zentrum für Luft- und Raumfahrt e.V. (DARA), and he joined the ESA European Space Technology Center (ESTEC) in Noordwijk, The Netherlands, to help develop the agency's Columbus module for the International Space Station (ISS).

Merbold's second flight in space began on January 22, 1992, when he launched aboard the space shuttle Discovery during STS-42, the first flight of the International Microgravity Laboratory (IML-1).

Ronald Grabe commanded STS-42, which was a milestone mission in international scientific cooperation in space. The crew for the flight included pilot Stephen Oswald, mission specialists Norman Thagard, David Hilmers, and William Readdy, and Merbold's fellow payload specialist Roberta Bondar, of the Canadian Space Agency (CSA).

Working around the clock in alternating shifts, the STS-42 crew conducted an intensive program of scientific research, which included observations of the effect of the microgravity environment on the human nervous system and on other forms of life, including shrimp eggs, lentil seedlings, fruit fly eggs, and bacteria. The crew also conducted crystal protein growth experiments, growing crystals from enzymes, mercury iodide, and a virus.

Mission managers extended the flight by one day to allow time for the crew to complete their research work. On January 30, 1992, Discovery returned to Earth at Edwards Air Force Base in California after a flight of 8 days, 1 hour, 14 minutes, and 44 seconds.

After his second space mission, Merbold served as science coordinator for the experiments flown aboard the STS-55 flight of the space shuttle Columbia in 1993, which was the second German Spacelab mission.

On October 3, 1994, Merbold became the first ESA astronaut to fly aboard a Russian Soyuz spacecraft when he launched with Aleksander Viktorenko and Yelena Kondakova aboard Soyuz TM-20 to travel to the Mir space station for EuroMir '94, a program of scientific research designed by European scientists.

When they arrived at the station, Merbold, Viktorenko, and Kondakova met Mir 16 crew members Yuri Malenchenko, Talgat Musabayev, and Valeri Polyakov, who was in the midst of a record-setting long-duration mission of more than 437 days.

Merbold lived and worked aboard Mir for a month while conducting his assigned scientific work, and then returned to Earth aboard Soyuz TM-19 with Malenchenko and Musabayev on November 4, 1994.

During the EuroMir '94 mission, Merbold added 31 days, 12 hours, and 36 minutes to his career total time in space.

Ulf Merbold spent more than 49 days in space during his remarkable career as an astronaut.

After his third space mission, Merbold worked at the ESA's microgravity promotion division at ESTEC.

MERCURY (PLANET): MAJOR FLIGHTS.

See Mariner 10; Messenger

MERCURY PROGRAM

(1961–1963)

First Manned U.S. Space Program

With the founding of NASA on October 1, 1958, the United States committed itself to the exploration of space by a large corps of civilian engineers, administrative personnel, and astronauts supported by a massive complex of experts from private industry, academia, and the military. President Dwight D. Eisenhower made known his preference that the U.S. Congress should charter NASA as a civilian agency, and the members of the U.S. House and Senate acted in accordance with his wishes.

The new organization's Space Task Group, led by Robert Gilruth, was given the goal of placing America's first astronauts into space, and that goal became the central purpose of what soon became known as Project Mercury—the United States' first manned space program.

Faced with the unprecedented challenge of creating every aspect of hardware, software, and procedure necessary to launch human beings into space, the engineers of the Mercury program responded with speed and accuracy. The program fascinated the American public, which was increasingly concerned about the space-related progress of the nation's superpower rival, the Soviet Union. The Soviets had beaten the United States into space with the launch of Sputnik 1, the first unmanned satellite to reach orbit, on October 4, 1957. Even though the first unmanned U.S. satellite, Explorer 1, followed fewer than four months later, on January 31, 1958, the Soviets' Sputnik achievement sorely tested the confidence and patience of the American public and the U.S. news media.

As a result of this climate of competition, NASA personnel faced intense pressure to mount a manned spaceflight as quickly as possible. To their enduring credit, they managed to achieve the first Mercury flights in a remarkably short period of time while maintaining the high priority on safety that was championed by the agency's leadership—including its first administrator, T. Keith Glennan, and deputy administrator, Hugh Dryden—and adopted as a core value by engineers and managers throughout the organization.

Careful development marked each stage of the huge effort to create the spacecraft, its launch vehicle, and all their related systems, software, equipment and procedures. The same was true regarding the process of selecting, evaluating, and training the first astronauts.

Popularly known as a "space capsule," the Mercury spacecraft was designed and built by the McDonnell Aircraft Corporation, which manufactured 20 of the

America's first astronauts, the Mercury Seven. Front, left to right: Walter "Wally" Schirra, Donald "Deke" Slayton, John Glenn, Scott Carpenter; back, left to right: Alan Shepard, Virgil "Gus" Grissom, Gordon Cooper. [NASA/courtesy of nasa images.org]

unique vehicles at its plant in St. Louis, Missouri for use by NASA during the Mercury program. Barely large enough to accommodate a single occupant of relatively slight build and no more than medium height, the Mercury craft was cone-shaped and measured 6 feet, 10 inches long and 6 feet, 2.5 inches at its widest point. It featured a round cylinder at its top, where the spacecraft's parachutes were stowed, and the craft then sloped gradually from the bottom of the cylinder to its wider, round base at the bottom. Within the crew cabin, the astronaut had very little room to move around, which led some of the military pilots who chose not to apply for a spot in the program to dismiss the whole idea of flying in space as being the work of "a man in a can."

Whatever the drawbacks of the initial U.S. spacecraft design, however, few could argue against the impressive engineering achievement of the Mercury capsule's ablative heat shield, which covered the entire bottom of the vehicle. Crucial to the survival of both the astronaut and the spacecraft, the smooth, delicately curved shield made it possible for the Mercury capsule to endure the intense heat of re-entry into the Earth's atmosphere.

Capable of generating a thrust of 75,000 pounds, the U.S. Army's Redstone rocket, which had been developed by a team led by Wernher von Braun, was selected for the first manned Mercury flights. The first two flights in the program would be sub-orbital—designed to send an astronaut just beyond the edge of the Earth's atmosphere, to an altitude of 100 or more miles, but not into orbit.

For the orbital Mercury flights, which required a more powerful rocket, NASA mission planners chose the U.S. Air Force's Atlas Intercontinental Ballistic Missile (ICBM), which was capable of a thrust of 360,000 pounds. NASA procured eight Redstone rockets and nine Atlas launchers for the Mercury program, at a total cost of $25.5 million.

At the same time that work was progressing on the Mercury spacecraft and launch vehicles, NASA engineers also designed an extensive network of communications tracking stations that was global in scope and involved the most advanced technology of the time. A total of 18 tracking stations were built for the program, and a large team of communications professionals was specially trained to staff each location.

The individuals chosen to fly the Mercury space missions were introduced to the public on April 9, 1959. Collectively known as the "Mercury Seven," the first U.S. astronauts were all drawn from the U.S. military, in accordance with the suggestions of President Eisenhower, who wished to avoid the unseemly crush of potential candidates that would likely have resulted if NASA were to have announced an open audition for astronaut candidates.

In addition to active military service, the president also suggested that NASA limit its first selection of astronaut candidates to those with experience as test pilots with 1,500 or more flight hours, who had graduated from college, were certified to fly in jet aircraft, were under the age of 40, in excellent health, and were less than five feet, eleven inches tall, so they could fit in the cramped Mercury capsule.

Having met all those criteria, the Mercury Seven underwent an intense program of training and psychological evaluation. Then, after passing all the tests that the infant space agency could devise to determine their suitability for the manned space-flight program, the astronauts settled into mission-specific training that involved elaborate simulations of the activities that they would carry out during their actual flights in space. Their military experience obviously served them well during the grueling astronaut selection process, but it seems equally apparent that their good humor, long experience as pilots, patience, endurance, and competitive nature were also of great value as they began their new careers as America's first space heroes.

The Mercury Seven astronauts were Malcolm Scott Carpenter, Leroy Gordon Cooper, Jr., John Herschel Glenn, Jr., Virgil I. "Gus" Grissom, Walter M. "Wally" Schirra, Jr., Alan B. Shepard, Jr., and Donald K. "Deke" Slayton.

Carpenter, Schirra, and Shepard were members of the U.S. Navy at the time of their selection by NASA; Cooper, Grissom, and Slayton represented the U.S. Air Force; Glenn was a member of the U.S. Marine Corps.

Together, they each worked toward the goal of flying in space, a goal that all would accomplish during the program with the exception of Slayton, who was diagnosed with a heart condition by NASA medical personnel and removed from active duty as an astronaut. Slayton would go on to play an instrumental role in the space agency's history as the single individual most responsible for selecting crew members for later missions, and in 1975 he finally realized his initial goal of flying in space, as a member of the U.S. crew of the Apollo-Soyuz Test Project (ASTP).

A series of unmanned test launches, some of which carried chimpanzees as test subjects, preceded the manned Mercury launches. The chimps were successfully recovered at the end of each test, providing mission planners with not only the confidence but also the data and experience necessary to attempt a manned flight.

Alan Shepard became the first U.S. astronaut to fly in space when he was launched from NASA's Cape Canaveral launch facility in Florida on May 5, 1961. He missed being the first human being to fly in space by just 23 days; the Soviet Union had launched cosmonaut Yuri Gagarin into orbit on April 12.

Shepard gave his Mercury spacecraft the name "Freedom 7," establishing a custom that each of his fellow Mercury astronauts would follow by providing a unique symbolic moniker for each particular mission. Grissom followed with "Liberty Bell 7;" Glenn named the spacecraft in which he made the first U.S. orbital flight "Friendship 7;" Carpenter's Mercury craft was "Aurora 7;" Schirra circled the Earth in "Sigma 7;" and Cooper spent America's first full day in space aboard "Faith 7."

Shepard's Freedom 7 success ignited a euphoric wave of adoration for the astronauts and an intense public interest in the space program.

In the climate of public fascination and increasing Cold War tension between the United States and the Soviet Union, which quickly became epitomized by the nascent space race between the two countries, U.S. President John F. Kennedy made a remarkable proposal to Congress on May 25, 1961. Shepard's brief flight of just over 15 minutes was the nation's only spaceflight experience at the time, and yet Kennedy put forth the broad outline of a plan to send a U.S. astronaut to the Moon, to land on the lunar surface, and then to return safely to Earth.

Fully understanding the enormous complexity involved in such a venture, the president set an end of the decade goal for the proposed Moon landing, in effect giving NASA a little more than eight years to accomplish the feat. He asked Congress for the first portion of the enormous amount of money he knew the new venture would require, which the best estimates of the time foresaw as growing to as much as $9 billion by 1967.

In the larger context of the lunar landing program envisioned by President Kennedy, the six manned Mercury missions became a strong foundation for later flights in the pioneering years of the U.S. space program. During the Mercury program, NASA established virtually all of the organizational support systems, the extensive network of primary contractors and sub-contractors, and the intricate, detailed procedures that were successfully utilized during the subsequent Gemini, Apollo, and Skylab programs and the ASTP mission, and which became the model for the expanded operations of the space shuttle era.

The space agency employed the services of 1,300 employees during the Mercury program, and called upon the services of 18,000 military personnel for each mission. Twelve primary contractors, 75 major sub-contractors, and 7,200 tertiary sub-contractors were involved in the program, representing a total of some two million individuals who made a contribution of one sort or another to the effort—a remarkable demonstration of the capabilities of a great democracy in action, the likes of which had not been seen in the United States since the accelerated industrial production of wartime material by U.S. factories during World War II.

Throughout the length of the Mercury program, from just before Shepard's flight in April 1961 to Cooper's mission, the last of the series, in May 1963, the United States and the Soviet Union were deeply engaged in active competition in space and increasingly perilous sparring on Earth. During that time period, the United States backed the abortive Bay of Pigs invasion of Cuba and the Soviet Union instigated the building of the Berlin Wall and tested a 100-megaton nuclear weapon in the Earth's atmosphere, breaking a three-year moratorium on such tests. Further, the two nations maneuvered toward—and ultimately away from—the brink of a potential nuclear confrontation over the Soviet Union's attempt to deploy nuclear weapons in Cuba during the Cuban Missile Crisis.

Thus in one of its farthest-reaching consequences, the Mercury program helped to establish a new, peaceful field of endeavor on which the superpower rivals could compete, as each sought to demonstrate the superiority of its technology and, by extension, its ideology and way of life.

MERCURY 1: FREEDOM 7

(May 5, 1961)

First Manned U.S. Spaceflight

Alan Shepard became the first American to be launched into space when he lifted off in the Freedom 7 Mercury capsule atop a Redstone rocket from launch pad LC-5 at Cape Canaveral, Florida on Friday, May 5, 1961. His 15-minute sub-orbital trip into space reached an altitude of 116.5 miles, and during his journey he traveled a total distance of 303 miles.

Alan Shepard prepares for his Freedom 7 mission. On May 5, 1961, Shepard became the first American to fly in space. [NASA/courtesy of nasaimages.org]

Originally scheduled for May 2, the launch of Freedom 7 was postponed for two days by poor weather on that date, and then delayed again by unfavorable weather on May 4. The initial launch scrub on May 2 took place 2 hours and 20 minutes before lift off, before Shepard had entered the space capsule.

Perhaps only slightly less anxious than Shepard himself, a television audience of millions carefully followed the first manned U.S. space launch, marveling at the novelty of the technical details of the flight, which began with a countdown—a methodical ticking off of the time between the start of final prepa-rations for a launch and the moment when the rocket's engine ignited and the launch vehicle and crew capsule lifted off. The countdown that ultimately resulted in the launch of Freedom 7 began at 8:30 A.M. on May 4.

As the public soon learned, the long countdown that at first glance seemed the very model of high tech efficiency was actually subject to frequent interruption, either in the form of a planned pause, known as a "built-in hold," or for emergency inspections or repairs. In either case, the progress of the countdown would be stopped until the scheduled duties were carried out or the unexpected difficulties were resolved.

In the case of the May 4 to May 5 countdown for Freedom 7, a built-in hold began when the countdown reached 6 hours and 30 minutes (or at "T minus 6 hours and 30 minutes," as NASA referred to it. In the context of a countdown, "T" refers to the launch itself, so all events occurring before launch are referred to in increments of T minus, or time left to expire before launch, while events occur-ring after launch are referred to in increments of T plus. The T plus times are also expressed as "mission elapsed time," or the time since the mission began). At that point, technicians at the launch pad made the final preparations for launching the Redstone rocket launch vehicle, a process that took 15 hours. The countdown picked up again at 11:30 P.M. on May 4. The next built-in hold took place at T minus 2 hours and 20 minutes, to allow for a final check of the spacecraft before Shepard arrived at the launch pad.

After their April 1959 selection by NASA as the first American space explorers, Shepard and his fellow Mercury Seven astronauts had trained for many months to prepare for their opportunity to fly in space for the first time. In Shepard's case, the excitement of the moment also held a larger significance, as his flight would be the first manned launch of the entire program.

Shepard's launch day began in the early morning hours of May 5. He woke up shortly after 1:00 A.M., and participated in a pre-launch ritual that would become a tradition throughout the early years of the space program when he ate a breakfast of steak and eggs. He was then outfitted with sensors that NASA medical personnel would use to monitor his condition during the various phases of the flight, and was helped into his spacesuit. At 5:15 A.M., he was delivered to the launch pad.

Looking genuinely otherworldly, wearing his silver spacesuit and space helmet and carrying a suitcase-sized box that few uninitiated observers would recognize as his portable air conditioner, Shepard was conveyed up the gantry apparatus along-side the launch vehicle by an elevator, and entered the Freedom 7 spacecraft at 5:20 A.M. He was helped into the capsule by technician Joseph Schmitt, who wished him "Happy landings" before closing the hatch.

About an hour later, the environment inside the spacecraft was replaced with pure oxygen.

The countdown then proceeded to 15 minutes before lift-off when cloud cover caused mission controllers to delay the launch again, to ensure that the start of the mission could be properly photographed.

An additional problem developed when an inverter in the launch vehicle malfunctioned and had to be replaced, which resulted in a delay of nearly an hour.

After the balky electrical component had been replaced, yet another difficulty arose: a fault in a computer at the Goddard Space Flight Center (GSFC) in Greenbelt, Maryland forced mission controllers to halt the start of the flight again.

Far from being the smooth, efficient process implied by the technical jargon and the precision of the ticking clock, the Freedom 7 countdown was a start-and-stop affair that eventually suffered an unplanned delay totaling 2 hours and 34 minutes. By the time his spacecraft finally lifted off, at 9:34 A.M. on May 5, 1961, Shepard had been strapped into his seat in the Freedom 7 capsule for 4 hours and 14 minutes.

Several of Shepard's fellow Mercury Seven astronauts played key roles throughout the launch process and during the flight. John Glenn accompanied Shepard throughout the pre-flight preparations; Gordon Cooper was with him as part of the pad support team in the blockhouse; and Donald "Deke" Slayton served as the primary spacecraft communicator ("Capcom") for the flight. Walter "Wally" Schirra was the "chase pilot" for the mission, monitoring conditions around the launch pad from the air in an F-106 aircraft and then following the launch vehicle skyward as far as possible after lift-off.

Glenn had also served as Shepard's stand-in inside the capsule during the long pre-flight tests of the spacecraft's systems. When he left Freedom 7 for the last time prior to Shepard's entering the craft, Glenn left behind a note attached to the vehicle's controls, which read "No handball playing here."

For the first 45 seconds after launch, Shepard enjoyed a relatively smooth ride. About a minute and a half into the flight, he began to feel a rattling—but expected—turbulence, which lasted a short while before the engine of the Redstone launch vehicle shut down at 2 minutes and 22 seconds. The temperature within the spacecraft reached 91 degrees Fahrenheit, but Shepard's spacesuit control system kept the temperature within his spacesuit at a maximum of 75 degrees Fahrenheit.

At the three-minute mark, the spacecraft was automatically re-oriented from a top-forward position to a bottom-first orientation, so its heat shield led the way forward for the rest of the flight.

Shepard then achieved the primary objective of the mission when he switched control of Freedom 7 from automatic flight to manual operation. The Mercury spacecraft featured separate controls for pitch, which governed the vehicle's up and down movements, yaw, which directed the craft's motion to the left and right, and roll, which controlled the capsule's ability to revolve.

Shepard was able to assume control of the spacecraft's systems without difficulty and he successfully placed Freedom 7 in the position necessary for firing its retrorockets before re-entry into the Earth's atmosphere.

Because it was the first flight in which NASA sent an astronaut into space—and only the second human spaceflight in history—the Freedom 7 mission was primarily designed to test whether or not an astronaut could function usefully in space. NASA medical personnel worried that an astronaut might become insensible or suffer unexpected physical effects brought on by the stresses of the experience.

During and after his Freedom 7 flight, Shepard decisively put such worries to rest. He communicated clearly and effectively with Slayton throughout the mission and smoothly took over control of the spacecraft as scheduled.

He also made observations of the Earth as planned, although he was hampered somewhat in that chore by a filter he had installed on the spacecraft's periscope during his long wait on the ground before the launch. While it had worked well for observations from the top of the launch stack prior to lift-off, the filter tended to cloud his view of Earth from space. Initially, Shepard tried to remove the filter, but in doing so, he bumped the control on the cockpit instrument panel that would normally be used to invoke an abort. That being a close-enough call, he decided to leave the troublesome filter in place while making the rest of his observations.

From his extraordinary perspective, Shepard was able to see the continents and recognized several landmarks, including the western coast of Florida.

Another spacecraft control exercise called for the pilot to manipulate the craft's flight by using a "fly-by-wire" technique, which involved a firing of the capsule's hydrogen peroxide jets. Shepard successfully invoked the proper mode for the maneuver at two separate points of the flight. In the first instance, just prior to the firing of the spacecraft's retrorockets before re-entry, he struggled to maintain the spacecraft at the appropriate pitch of 34 degrees. He tried again during re-entry, with better results.

Freedom 7's drogue parachute deployed at an altitude of 21,000 feet, and the main parachute popped open at 10,000 feet. Shepard splashed down in the Atlantic Ocean on May 5, 1961, after a flight of 15 minutes and 28 seconds.

He was relieved to find that his spacecraft had endured the water impact without suffering any leaks. A fluorescent dye spilled out of the craft as planned, to make it easier for recovery forces to find the capsule. As things turned out, the lead rescue helicopters had sighted Freedom 7 during its descent and were able to hook a cable onto the spacecraft just minutes after Shepard splashed down.

Shepard then climbed out of his Mercury capsule and was hoisted aboard a helicopter. He was delivered to the primary recovery ship, the aircraft carrier USS *Lake Champlain,* just 11 minutes after Freedom 7 hit the water.

MERCURY 2: LIBERTY BELL 7

(July 21, 1961)

Manned U.S. Spaceflight

NASA chose Virgil I. "Gus" Grissom to be the second American to fly in space. His mission was to confirm the positive results of the first U.S. spaceflight, in which Alan Shepard had proven that a pilot could remain alert, function coherently in space, and could actively control the flight.

In keeping with the tradition begun by Shepard, Grissom named his Mercury capsule the "Liberty Bell 7," in honor of the famous symbol of American independence and in reference to his fellow Mercury Seven astronauts.

Gus Grissom climbs into Liberty Bell 7 at the start of America's second manned space mission, July 21, 1961. [NASA/courtesy of nasa images.org]

Grissom, Shepard, and John Glenn had been informed that they were the primary candidates to pilot the first Mercury flights prior to the first mission. Glenn served as backup for both Shepard and Grissom as they prepared for the first two flights, which were sub-orbital, in that they reached an altitude beyond the Earth's atmosphere and entered space, but did not go into orbit. Glenn would make the first orbital flight, during the third Mercury mission. Shepard served as spacecraft communicator ("Capcom") at the Mercury Control Center for Grissom's Liberty Bell 7 flight.

Each of the Mercury Seven astronauts played a role in their fellow pilot's flights; in the case of Grissom's mission, Walter "Wally" Schirra joined Shepard at the Mercury Control Center; Scott Carpenter and Donald "Deke" Slayton served as part of the support team in the blockhouse at Cape Canaveral as Grissom prepared to begin the flight; and Gordon Cooper piloted the chase aircraft, observing the scene at the launch facility from the air and tracking weather conditions in the hours before lift-off.

Training together as the flight approached, Grissom and Glenn ran through the entire sequence of activities planned for the 15-minute mission approximately 100 times, in simulated space environments at Cape Canaveral in Florida and at NASA's Langley Research Center in Hampton, Virginia.

Although the basic mission profiles for the first and second Mercury flights were largely identical, several important modifications had been made to the Mercury spacecraft by the time Grissom lifted off on his flight. Engineers at the McDonnell Aircraft Corporation, which was responsible for the manufacture of the Mercury capsule, replaced the two small (10-inch), round windows that had adorned Shepard's Freedom 7 capsule with a single, large window that would make it easier for Grissom to observe Earth's terrain from space. The new triangular-shaped window was 11 inches wide at its base, 19 inches high, and 7.5 inches wide on top.

Manufactured in several layers from special materials that could survive temperatures as high as 1,800 degrees Fahrenheit, the new window was developed by the Corning Glass Works of Corning, New York. The expanded view afforded Grissom a field of vision that extended 30 degrees horizontally and 33 degrees vertically.

Another innovation, a redesigned hatch outfitted with explosive bolts to allow an astronaut to get out of the space capsule quickly, would play a major role in the frightening conclusion of the Liberty Bell 7 mission, in which the space capsule filled with water and sank as Grissom scrambled for safety.

The explosive hatch had been designed in response to the astronauts' concerns that the previous emergency exit procedure, which involved the pilot crawling out through the antenna compartment at the narrow end of the capsule, was too cumbersome and would take too much time in the event of an emergency. Ironically, Grissom's flight would be the only Mercury mission to experience an emergency that involved a quick exit from the spacecraft.

Working with their fellow engineers at the McDonnell Corporation, NASA personnel also incorporated many minor changes in the spacecraft and its systems, such as adding more padding to the pilot's head rest and changing the layout of the cockpit instrument panel, after a detailed analysis of Alan Shepard's experience during the first Mercury flight.

Originally scheduled for lift-off on Tuesday, July 18, 1961, the Liberty Bell 7 launch was delayed for one day by cloudy conditions in the area in which the capsule was scheduled to be recovered at the end of the flight.

Grissom woke in the early morning hours of July 19 to prepare for the next launch attempt, which mission managers had hoped to complete successfully by 7 A.M. He traveled to the launch pad, rode the elevator up the scaffolding apparatus aside the launch vehicle, and entered the spacecraft to wait for the countdown to run out. At T minus 10 minutes and 30 seconds, mission controllers called a hold to assess the weather, and the sequence remained stuck at that point until it was decided that the weather remained unfavorable to an unacceptable degree. The launch was postponed again, or "scrubbed," as NASA phrased it.

On July 21, the process began again with Grissom rising early, arriving at Launch Pad LC-5 after a brief van ride from his quarters, and ascending the gantry to the space capsule, where he once more slid onto the contoured seat within Liberty Bell 7 from which he would experience his first flight in space.

Again, the countdown wound toward lift-off and was once more halted—this time at the T minus 45-minute mark—while technicians checked one of the explosive bolts on the spacecraft's new hatch. Although the bolt appeared to be improperly aligned, technicians were satisfied that the impact of a single bolt (there were 70 in all) would likely be insignificant in both keeping the hatch in place and in blowing it off, so the countdown resumed after a 30-minute delay.

Two more delays—the first to turn off lights at the launch pad that interfered with the telemetry system that conveyed data to mission controllers during launch, and the second to evaluate the weather conditions again—stretched the long wait for another 50 minutes. By the time the countdown finally reached its completion, Grissom had been inside Liberty Bell 7 for 3 hours and 22 minutes.

Then, at 7:20 A.M. on July 21, 1961, Gus Grissom lifted off in Liberty Bell 7 atop a Redstone rocket from Launch Pad LC-5 at Cape Canaveral to begin the second U.S. spaceflight.

His launch experience was smooth, and although he felt vibrations similar to those that had rocked Alan Shepard in the early part of his Freedom 7 flight, Grissom did not suffer any difficulty seeing as a result. The escape tower at the top of Liberty Bell 7 was jettisoned as expected shortly after lift-off, and the main engine of the Redstone rocket cut off on schedule, at 2 minutes and 22 seconds into the flight. Ten seconds after the engine stopped, the Liberty Bell 7 capsule was separated from the Redstone launch vehicle by the firing of small rockets.

The intense preparation and training, including careful simulations of every aspect of the flight, left Grissom confident and comfortable with the sequence of events he was experiencing, as extraordinary as they were to the many observers who followed the flight from Earth. He anticipated the next milestone in his mission, when the space capsule was automatically re-oriented to fly in a bottom-first position with its heat shield leading the way for the rest of the flight.

Grissom next set to work on the primary activity of his mission profile, taking over control of the spacecraft. An exceptional pilot with long experience and a reputation for performing well under pressure, Grissom proceeded with his assigned tasks even as the remarkable sight of the Earth's horizon beckoned outside the window of his Liberty Bell 7 capsule. Maintaining his attention on the task at hand, he managed to direct the spacecraft through several planned maneuvers that included changes in the vehicle's pitch, which governed the craft's up and down movements, and its yaw, which directed the capsule's motion to the left and right.

He also successfully completed a period of observations in which he recognized a number of landmarks, including Cape Canaveral.

As the short flight neared its conclusion, Grissom positioned the craft for re-entry, and the spacecraft's retrorockets were fired. His return through the atmosphere went as planned, and as expected, he felt the force of gravity building steadily during his plunge toward the Atlantic. The spacecraft's drogue parachute deployed as expected at 21,000 feet.

Liberty Bell 7's main parachute popped open at a higher than expected 12,300 feet, and the pilot noticed with some concern that the chute was torn in several spots, but the chute worked as planned and slowed his descent to a reasonable rate. The capsule's flotation device deployed as scheduled, and the initial impact of the splashdown was relatively light. Grissom was pitched over onto his left side briefly before the capsule tipped back into proper position, and he then braced himself against the heavy rolling of the Atlantic waves as he prepared for the arrival of the recovery forces.

Helicopters closed in on the capsule while Grissom wrote down the data being displayed on the instruments in the cockpit. After the recovery forces established communication with him, Grissom armed the explosive hatch by pulling the pin on the hatch detonator, and then relaxed; he had landed close to the target splashdown point (within about three miles), and knew that the helicopters were just minutes away.

Suddenly—and for reasons that have never been satisfactorily explained—the explosive bolts that held the hatch on Grissom's Mercury capsule in place were prematurely detonated, and the hatch popped off of the spacecraft, hitting the waves about five feet away from the craft and then skipping off like a stone tossed along the surface of a lake.

Within seconds, the turbulent waves of the Atlantic were filling the tiny spacecraft with water. Thinking quickly, Grissom flipped off his helmet and climbed out of the Liberty Bell 7 capsule. He swam rapidly away from the sodden spacecraft to make sure that he was free of any part of the vehicle that might catch on his spacesuit.

The first recovery helicopter arrived at that point. Noting that Grissom appeared to be, at least at that moment, safely floating at a reasonable distance from the rapidly sinking Liberty Bell 7, the helicopter pilot focused his efforts on snaring the loop at the top of the capsule. Drawing on the experience gained in extensive simulations prior to the flight, the helicopter pilot was able to hook a line through the loop to begin the recovery process just as the spacecraft disappeared beneath the waves. His co-pilot, meanwhile, turned his attention to Grissom and prepared to throw a second line that was specially outfitted to enable the recovery of the astronaut.

A second helicopter arrived to assist with the recovery just as warning lights in the cockpit of the first helicopter began to flash an alarming warning: the added weight of the water within the capsule was stressing the engine of the aircraft to the brink of failure. As the first helicopter crew continued their struggle with the attempt to recover the capsule, they radioed the crew of the second helicopter to take over the task of rescuing Grissom.

By that point, Grissom had begun to swim back toward the floundering space capsule, with the intent of helping with its recovery. He checked to make sure that the recovery line was securely hooked through the loop at the top of Liberty Bell 7, and then moved away again to watch the salvage effort.

As he bobbed in the water beside his sinking Mercury spacecraft, Grissom gradually began to feel heavy and fatigued. His spacesuit was slowly taking on water, and he found it increasingly difficult to keep his head above the waves. Several long minutes passed as he struggled, and for an instant his survival seemed imperiled.

At just that unpleasant instant, the second helicopter crew seemed to realize that the astronaut was in trouble, and its co-pilot—a friend of Grissom's, George Cox, who had participated in recovery training and previous capsule recoveries during the testing phase of the Mercury program—threw him the "horse collar" flotation device and life line that the crew then used to pull Grissom up to and into the helicopter.

Along with his rescuers aboard the second aircraft, Grissom watched as the first helicopter crew continued to wrestle with the completely flooded Liberty Bell 7; the added weight of the water raised the total load to more than 5,000 pounds, which was well beyond the helicopter's lifting ability. The battle ended in defeat for the overloaded helicopter. With its wheels already dipping into the crest of the waves and the craft in danger of being pulled fully into the ocean swells, the crew finally cut the space capsule loose. The spacecraft sunk to the Atlantic floor, settling on the bottom of the ocean at a depth of 15,000 feet.

Grissom was safely delivered to the aircraft carrier USS *Randolph,* where he was found to be in good health despite the trauma of the difficult recovery and the loss of his spacecraft. He had experienced weightlessness for about 5 minutes of his 15 minute and 37-second sub-orbital flight, achieved an altitude of 118.3 miles, and traveled a distance of 302 miles. Despite the unsettling conclusion of the flight, Grissom successfully accomplished all major goals of his assigned mission profile.

Once Grissom's good health was positively ascertained, NASA officials turned their attention to the difficult recovery. The Space Task Group appointed a committee to investigate the incident to try to determine a cause for the premature loss of the explosive hatch. Extensive tests of the hatch detonator mechanism and all its related systems failed to duplicate the unintentional firing of the explosive bolts.

In addition to the pilot's ability to activate the hatch by pressing a plunger within the cockpit, the explosive bolts on the hatch could also be detonated from outside the spacecraft, where a rescuer could set the process in motion by pulling a lanyard. Among the modifications to the hatch system after the Liberty Bell 7 flight, the lanyard was more securely fastened in its place on the outside of the capsule, to avoid the possibility of its accidentally causing the hatch to blow.

Whatever the cause of the inadvertent loss of the hatch, Grissom was adamant in his conviction that he had not been the cause of the malfunction. The investigating committee largely agreed that it was unlikely that the astronaut had accidentally activated the hatch.

Other aspects of the post-flight investigation included study of the misaligned explosive bolt that had been removed from the hatch on the launch pad before lift-off, but the loss of the single bolt was determined to not have played a role in the malfunction. Having lost the chance to remove any memento from his spacecraft, Grissom kept the bolt as a souvenir.

Although the exact cause of the accident was not clear, mission managers sought to avoid a repeat of the event by instituting new flight rules. Under the new plan, a pilot would not arm the explosive bolts until helicopter recovery forces had already securely locked onto the capsule.

As time passed and other Mercury flights were flown with no duplication of the Liberty Bell 7 accident, the mystery of the prematurely exploding hatch was largely forgotten. A good indicator that the pilot was not at fault, however, became evident during the later flights, when each of Grissom's fellow astronauts used the plunger to activate the hatch within their capsules. At the end of each flight, every subsequent Mercury astronaut was found to have suffered a slight injury to his hand after initiating the hatch sequence; Grissom, however, had not suffered the same injury, thereby supporting his assertion that he had not inadvertently caused the hatch of his spacecraft to detach prematurely.

In a remarkable turn of events, the Liberty Bell 7 spacecraft was recovered after 38 years beneath the waves of the Atlantic Ocean, on July 20, 1999, when it was raised from its resting place 90 miles northeast of the Bahamian island of Great Abaco.

MERCURY 3: FRIENDSHIP 7

(February 20, 1962)

Manned U.S. Spaceflight; First Manned
*   U.S. Spacecraft to Orbit the Earth*

After sending two astronauts into space on sub-orbital flights, NASA prepared to place the first American in orbit with the third manned launch of the Mercury program.

Just two individuals—Soviet cosmonauts Yuri Gagarin and Gherman Titov—had flown in orbit prior to John Glenn's Friendship 7 mission.

Medical officials in both the U.S. and Soviet space programs had little data to work with when trying to determine the impact that exposure to the space environment might have on an individual's physical and mental capabilities, and simply did not know if flying around the Earth in a tiny space capsule would adversely affect a person's health in the long-term, but given the positive experiences of the first two sub-orbital Mercury flights and the lack of evidence that the first two Soviet cosmonauts in orbit had suffered any catastrophic ill effects from their missions, NASA officials felt confident that Glenn could safely fly in orbit.

The first orbital flights of the U.S. and Soviet programs demonstrated the vastly different approaches that the two nations took to the design and conduct of space missions, and the role that each program assigned to the crew members who flew aboard space vehicles. While the Soviets relied on heavily automated systems and favored a control arrangement centered on Earth-based flight control with

John Glenn inside Friendship 7 on February 20, 1962. Glenn was the first American to orbit the Earth. [NASA/ courtesy of nasaimages.org]

relatively little input from the cosmonaut-pilot, officials within NASA defined a more active role for the pilots of U.S. spacecraft.

To that end, in the years leading up to Glenn's first U.S. orbital mission, a widely representative group of NASA personnel—including the Mercury astronauts themselves—carefully studied the likely capabilities that space pilots would have during a given flight and worked to design a program of planned activities and options for pilot control of the spacecraft and its systems.

In order to prepare for both their assigned duties and potential contingencies during the first orbital flights, Glenn and Scott Carpenter, who would fly the fourth Mercury mission (and second orbital flight) engaged in an intensive period of mission-specific training that included interaction with simulated flight conditions in an altitude chamber and simulations in the air-lubricated, free-axis (ALFA) trainer at the NASA Langley Research Center (LaRC) in Hampton, Virginia. They also underwent extensive medical evaluations and made an exhaustive study of the Mercury spacecraft and systems so that they would be thoroughly familiar with the vehicle and its equipment prior to launch.

By the time of the originally scheduled launch, Glenn had participated in simulations of virtually every aspect of the mission he was about to undertake, including on-site run-throughs of the procedures he would take to arrive at and enter the spacecraft on launch day, and an elaborate simulation of the recovery process.

Ironically, it was perhaps the only eventuality for which neither the astronaut nor NASA flight planners were adequately able to prepare that initially tested the effort to put an American in orbit. After a successful orbital test flight on November 29, 1961, in which a chimpanzee, Enos, traveled around the Earth and landed safely, Glenn's Friendship 7 launch was delayed by a series of nine mostly weather-related postponements. NASA officials originally hoping for a launch in December endured a long, anxious wait as the weather pushed the orbital mission into January, and then into February.

Countdown on January 27, 1962 proceeded to 13 minutes before lift-off before weather concerns forced that launch attempt to be canceled; similar attempts—and similarly frustrating results, due to both weather concerns and a leak in the launch vehicle's fuel tank—followed on February 13, 14, 15, and 16.

Then, in the early morning hours of February 20, 1962, Glenn began his pre-flight ritual—a shower and a hearty breakfast, followed by a medical checkup—for what would turn out to be the final time before his first flight. Accompanied by physician William Douglas and spacesuit technician Joseph Schmitt, Glenn traveled to Hanger S at Cape Canaveral, where Schmitt helped him put on his spacesuit. They arrived at Launch Pad LC-14, where the Friendship 7 capsule was perched atop its Atlas launch vehicle, at 5:17 A.M.

Within 10 minutes, the possibility of yet another delay seemed ominously likely, when a fault was detected in the launch vehicle guidance system, but technicians overseeing the launch were able to correct the problem within a little more than two hours.

Glenn entered the Friendship 7 capsule a little after 6 A.M. The spacecraft's hatch was initially sealed at T-90 minutes (90 minutes before the scheduled launch), but one of the hatch's 70 bolts broke during the process, and 40 more minutes passed while it was repaired. The countdown then proceeded mostly according to plan, winding down as expected except for minor interruptions and the built-in holds, or planned delays, for tasks such as the final fueling of the launch vehicle's liquid oxygen tanks.

During the delays, Glenn chatted with his fellow Mercury 7 astronauts Alan Shepard and Scott Carpenter, who served as spacecraft communicators in the control center at Cape Canaveral. He also passed the time by studying the final details of his spacecraft systems checklist and occasionally using a bungee-like exercise device that had been installed to give the pilot an opportunity to stretch his muscles in the confined space of the small capsule.

Flight director Christopher Kraft called for one final, brief hold six and a half minutes before lift-off to make sure that NASA's Bermuda tracking station was working properly. The worldwide NASA tracking system had failed during a test in early November, but had since been brought into excellent working order.

Friendship 7 lifted off from Launch Pad LC-14 at Cape Canaveral at 9:47:39 A.M. on February 20, 1962, carrying John Glenn into space and into history as the first American to orbit the Earth.

The launch was witnessed by an estimated television audience of 100 million people in the U.S., and by a crowd of 50,000 in the area around the launch site.

In the initial moments of the flight, the launch vehicle was controlled by a guidance system that had been developed jointly by the General Electric Corporation and the Burroughs Company, and the Mercury-Atlas vehicle successfully achieved the proper orbital insertion trajectory necessary for Glenn to reach Earth orbit.

Glenn settled into an orbit of 162.2 by 100 statute miles at a velocity of 17,544 miles per hour. Upon entering weightlessness, he reported an immediate comfort with the sensation and appeared delighted to describe his view of the Earth from virtually the first moment that he could see the planet from space.

The first orbit passed comfortably and, apparently, more swiftly than the pilot would have liked, as Glenn noted aloud that he was falling behind in his scheduled tasks. He began his first major maneuver at the end of the first orbit, when he used Friendship 7's yaw attitude control to turn the capsule around, so he was facing forward as he continued the flight.

Glenn continued to describe his unique view of the Earth, noting the sunset and describing stars above and the outlines of cities below. He communicated at one point with fellow Mercury 7 astronaut Gordon Cooper, and reported that he was comfortable and enjoying the flight.

He came upon an odd sight as he approached his first sunrise in orbit: as he looked outside his space capsule, he was amazed to see thousands of tiny specks of light, which he likened to fireflies when he reported their presence to mission controllers. They seemed to flit nervously around the outside of Friendship 7, and then suddenly disappeared, much to the astronaut's surprise.

Later investigation revealed the strange phenomena to simply be drops of waste fluid expelled from the capsule, but the sudden appearance of such an odd sight was something for which Glenn could not have been prepared even in the exhaustive pre-flight simulations. To his credit, he dealt with the incident with great calm and good-natured fascination.

At the end of the first orbit, ground controllers noted a problem with the space-craft's attitude control system. One of the thrusters designed to enable further yaw motions had clogged after the first yaw maneuver.

Having already become aware of a slight drift in the capsule's flight—and the failure of the automatic control system to correct the problem—Glenn switched the spacecraft to manual control and set the vehicle back into the proper position. He switched back and forth between automatic control and the manual "fly-by-wire" mode several times to test each for the best results and least fuel consumption, and eventually decided that the manual mode, while absorbing more of his attention and disrupting some of his other assigned activities, would work best. Mission controllers agreed without hesitation, and for the rest of the flight Glenn demonstrated the value of the hands-on, astronaut-centric approach to spaceflight.

During his second orbit, Glenn was kept busy maintaining Friendship 7 in the proper attitude, taking photographs, and checking in with the various ground stations in NASA's worldwide tracking network. As he passed over Bermuda, he spoke with Virgil "Gus" Grissom, who, as the commander of the second Mercury flight the previous July, was one of just two Americans (along with Alan Shepard, the first American in space) who could truthfully say they knew how Glenn felt in the tiny capsule, experiencing the weightlessness of traveling in space.

As the flight progressed, an ominous reading flashed onto the monitors of those NASA engineers responsible for tracking the health of the Friendship 7 spacecraft. The signal appeared to indicate that the heat shield and inflatable landing apparatus on the Mercury capsule had become loose in the course of the flight and were no longer securely locked into place at the base of the vehicle. Mission managers anticipating the rapidly approaching completion of the flight anxiously considered potential responses to the problem, knowing that the integrity of the heat shield was an absolute necessity if Glenn were to return to Earth safely.

In the hope that the instrument reading might be the result of a mispositioned switch, mission controllers asked communicators at the worldwide tracking sites to query Glenn to ensure that the control for the capsule's inflatable landing bag was not engaged.

Understanding the need to formulate a quick response to the potentially tragic situation, mission managers decided the best method of keeping the heat shield attached during re-entry—if indeed it had come loose, as the instrument read-out indicated—would be to keep the spacecraft's retrorockets attached to the vehicle during re-entry. Grouped in a package, the retrorockets were strapped to the bottom of the spacecraft; they were fired to position the vehicle for re-entry, and would normally be jettisoned after they were fired. In the case of Friendship 7, they would be retained during re-entry in the hope that the strapping would keep the heat shield in place long enough to protect Glenn before it was burnt off the vehicle toward the end of the spacecraft's trip through the atmosphere.

Mission controllers conferred with Maxime Faget, who had been largely responsible for the design of the Mercury spacecraft, to make sure that the plan to maintain the spent retrorockets was the best option. Faget agreed that the solution was the best one possible, given the circumstances.

As the drama unfolded within Mission Control, Glenn was completing the second of his three orbits. Puzzled by the questions from the ground about the position of the landing bag switch, Glenn tried to remain focused on his manual control of his spacecraft's flight and to continue working on his assigned experiments. He was able to carry out a space vision test and took a series of photographs of clouds and of the Earth, but an exercise in which he was supposed to observe balloons released by a tracking ship in the Indian Ocean had to be called off because of poor weather conditions.

When Glenn entered his third orbit, mission controllers polled their engineers and scientists and received conflicting advice about whether the retrorockets package should or should not be retained. Glenn, meanwhile, had figured out the situation, and despite the danger he faced, calmly relayed information to the ground, noting that he did not hear any rattling that might have indicated that the heat shield was loose.

Ultimately, the decision about maintaining the spent Friendship 7 retrorockets was made by Mercury program director Walter Williams and flight director Christopher Kraft. They relayed their choice—to keep the package in place during re-entry—to Glenn's fellow Mercury Seven astronaut Walter "Wally" Schirra at the tracking site in California. Schirra then sent the word on to Glenn.

The retrorockets fired as planned, and re-entry proceeded according to schedule. Glenn positioned the spacecraft properly, and Friendship 7 began its return trip to Earth. A fiery torrent surrounded the craft as it passed through the atmosphere, and Glenn watched as the disintegrating retrorockets burnt up outside his window. The spacecraft shuddered violently and fuel ran out before the spacecraft's drogue parachute deployed.

The chute opened as planned, 28,000 feet from the surface of the Atlantic Ocean, followed by the main parachute, and the first American to have flown in orbit dropped into the water in his Friendship 7 spacecraft about 800 miles southeast of Bermuda, 40 miles from the targeted splashdown site. Within 21 minutes of hitting the waves, Glenn and his spacecraft were aboard the destroyer USS *Noa*. To exit from the craft, he activated the explosive hatch, cutting his hand in the process. Other than the resulting minor abrasion on his knuckles, he emerged unharmed and smiling after his superb flight and nerve-wracking re-entry. While on the Noa, he underwent his first post-flight medical examinations, which confirmed that his health was indeed as good as it appeared, and he received a congratulatory phone call from President John F. Kennedy.

Glenn's pioneering Friendship 7 flight had taken just 4 hours, 55 minutes, and 23 seconds. During his three orbits of the Earth, he traveled 76,679 statute miles, and experienced weightlessness for 4 hours, 48 minutes, and 27 seconds.

Post-flight analysis revealed that the reading indicating that the heat shield had come loose—which had caused so much consternation at the end of the flight—had been in error. Despite the difficulty it had caused, the realization that the crisis had arisen from a fault in instrumentation rather than a defect in the construction of the spacecraft was something of a relief for mission planners, as it meant that the next mission—and the Mercury program as a whole—could proceed safely.

John Glenn was celebrated in parades in Washington, D.C. and in New York City—which proclaimed March 1 "John Glenn Day" in his honor—and appeared at the United Nations headquarters in New York and in a special ceremony in his hometown of New Concord, Ohio. President Kennedy also welcomed Glenn and his family at the White House, marking the occasion of the first U.S. flight in orbit with eloquence and resolve: ". . . this is the new ocean, and I believe the United States must sail on it and be in a position second to none."

After thorough examination by NASA engineers, the Friendship 7 Mercury capsule was put on public display during a tour of 17 countries, and was viewed by millions of people. It was also exhibited at the Century 21 Expo in Seattle in August 1962, and, a year after it had carried John Glenn into orbit, it was added to the permanent collection of the Smithsonian Institution, where it was placed on display alongside the airplane that the Wright Brothers flew during their first flight and the Spirit of St. Louis airplane that Charles Lindbergh flew across the Atlantic in May 1927.

MERCURY 4: AURORA 7

(May 24, 1962)

Manned U.S. Spaceflight

Given the mission of confirming the results of the first U.S. orbital spaceflight, Malcolm Scott Carpenter trained long and hard for his turn in space. With his fellow Mercury Seven astronauts, Carpenter prepared himself physically and mentally to

Aurora 7 (the small, darker object at top of the launch vehicle) prepared for launch on May 24, 1962. [NASA/ courtesy of nasaimages.org]

fly into space during one of the first launches of the U.S. space program. As things worked out, he got his chance during the fourth Mercury flight, aboard a capsule that he named Aurora 7.

Rising in the early hours of May 24, 1962, Carpenter edged closer to his historic flight by quickly progressing through the sequence that was becoming a tradition among the first astronauts: he ate breakfast, endured a brief physical exam and the affixing of medical sensors to his skin, and was helped into his spacesuit.

Accompanied by a small horde of technicians and space agency personnel, including his predecessor in orbit, John Glenn, Carpenter arrived at Launch Pad 14 at Cape Canaveral at 3:45 A.M. EST. The group waited at the launch site for about 45 minutes before Carpenter was given the all-clear to climb aboard the elevator on the launch vehicle support structure, and then, after expressing his gratitude to his support team, he traveled the short distance up to the capsule, stopped for a final check by technicians at the top, and then entered Aurora 7 a little after 4:30 in the morning.

Several short holds delayed the flight for about 45 minutes, but the delays had no discernible effect on Carpenter's good humor; he busied himself with reviewing his flight plans and pre-flight checklist, and chatted with his wife and children while he waited calmly for lift-off.

The launch took place at 7:45 A.M. An audience estimated at 40 million watched the live broadcast of the event, as Carpenter lifted off the pad in Aurora 7 atop an Atlas-D launch vehicle. The initial phase of the flight went according to plan, and the launch vehicle performed as expected.

In contrast to the previous orbital flight, during which John Glenn's Friendship 7 capsule was oriented by the spacecraft's automatic control system at a high cost in fuel consumption, Carpenter manually oriented Aurora 7 for its insertion into orbit.

Carpenter's descriptions of his experiences during the early part of the flight were strikingly similar to Glenn's. He reported the pleasant sensation of entering weightlessness, and noted that he did not feel the intense speed at which he was traveling, even though he was intellectually aware that Aurora 7 was hurtling through space at more than 17,000 miles per hour. He also matched Glenn in the wistful, awed tone in which he described the sights and sensations of flying in orbit, particularly as he encountered the horizon for the first time.

Carpenter labored to complete his first assignment, photographing the horizon, after struggling to load film into the camera he had been given. He was able to complete the work, however, and made a series of Earth observations, calling out landmarks to mission controllers as he passed over the various sites. He communicated with fellow Mercury Seven astronaut Donald "Deke" Slayton as Aurora 7 passed over the NASA tracking site in the Perth suburb of Muchea, in Western Australia, but was unable to spot flares that were set off for him to observe near Woomera, Australia.

Medical personnel monitoring Carpenter's health during the flight became alarmed when they received readings indicating that his temperature was as high as 102 degrees Fahrenheit, but he assured them that he was quite comfortable and speculated that the readings were probably in error. Medical data and the evaluation of a pilot's ability to function properly in space were given a high priority during the Mercury program, as at that time there was little evidence about the effect that spaceflight was likely to have on human beings. A primary goal of Carpenter's flight was to confirm that an astronaut could perform useful work and communicate effectively while in orbit, as John Glenn had during his Friendship 7 flight. Carpenter's observations throughout his first orbit revealed him to be alert and in good spirits.

Carpenter continued to assert manual control over the spacecraft's flight for a number of maneuvers during the second orbit, as he took more photographs and made additional observations; in doing so, however, he made several unintentional firings of the craft's attitude control thrusters, which resulted in greater than expected expenditures of fuel.

In an innovative experiment designed to measure the amount of resistance encountered by the capsule as it sped through space, Carpenter deployed a brightly colored balloon that popped out of a compartment in the side of his spacecraft on a tether. The balloon only partially inflated, largely scuttling his attempt to secure the desired measurements, and later proved difficult to jettison, although it ultimately had no ill effect on the overall mission.

For most of his third and last orbit, Carpenter placed Aurora 7 into a long, slow drift to conserve fuel. He reported the hour-plus ride as a pleasant experience, and the exercise also provided engineers monitoring the spacecraft's flight with data that would later prove helpful in the planning of longer missions, in which an astronaut would need to set his vehicle into a similar glide while sleeping or resting.

Carpenter was not inactive during his drifting orbit, however. He made a concise description of the setting Sun, took a series of 19 photographs, and then described the airglow phenomenon—the hazy layer about 10 degrees over the horizon—in great detail. He also observed the firefly-like lights that had fascinated John Glenn during his Mercury flight; to Carpenter, the objects appeared to be frost particles peeling off the outside of the spacecraft. At the time of the second orbital flight, the strange phenomena had not yet been identified as expended drops of waste water from inside the capsule.

Also carried aboard Aurora 7 was an experiment designed to study how liquids behave in weightlessness.

As he passed over the NASA tracking station on Hawaii near the end of his third orbit, Carpenter received word that it was time to prepare for the firing of Aurora 7's retrorockets, to orient the spacecraft for re-entry into the Earth's atmosphere. The procedure called for him to switch the capsule from manual to automatic control, a process simulated often enough in Earth-based training prior to the mission to seem routine. The switch did not go smoothly, however, and the automatic system appeared to be malfunctioning. Carpenter switched back to manual mode in an attempt to diagnose the trouble, and as a result, expended additional amounts of the hydrogen peroxide fuel necessary for both systems to function properly.

The Aurora 7 pilot was able to orient the craft just as he passed out of communication with the Hawaii tracking station, and he then communicated with fellow Mercury Seven astronaut and first American in space Alan Shepard, who was monitoring his progress from NASA's tracking site in Arguello, California. Shepard reminded Carpenter to deactivate the automatic system for orienting the spacecraft, and after doing so, Carpenter prepared to manually activate the spacecraft's solid fuel retrorockets, which would maneuver him out of orbit and set him on his way back to Earth.

A lapse of about three seconds in firing the retrorockets resulted in Aurora 7 veering far off course during its return trip through the atmosphere. Given the lack of fuel remaining, Carpenter could do little to affect the vehicle's attitude during re-entry, but he was able to quell the violent shaking the craft experienced just after he passed back into the Earth's gravity field.

Serving as capsule communicator at Cape Canaveral, Virgil "Gus" Grissom—who during the Mercury Liberty Bell 7 flight had been the second American to fly in space—reminded Carpenter to engage the faceplate on his space helmet in anticipation of the jolt that would come when Aurora 7 splashed down in the Atlantic.

Carpenter endured more violent shaking of his spacecraft during the fiery trip through the atmosphere, until the capsule's drogue parachute was deployed at 25,000 feet. The craft's main parachute deployed at 9,500 feet, and Aurora 7 subsequently dropped into the waves of the Atlantic with a relatively gentle—but awkwardly slanted—impact at 12:41 P.M. on May 24, 1962 after a flight of 4 hours, 56 minutes, and five seconds, three orbits of Earth, and a total distance of 76,021 miles.

Tilted at an angle somewhere between 45 and 60 degrees, the spacecraft failed to orient itself properly as it had been designed to do, and Carpenter was faced with the dilemma of deciding whether to remain within the hot, cramped capsule for what promised to be a wait of at least an hour (the time it would take for recovery forces to arrive, as Grissom had informed him during re-entry) or to try to crawl out of the spacecraft to await rescue in a life raft in the sunlight and fresh air.

After trying and failing to make radio contact with recovery forces, Carpenter decided it was best to get out of Aurora 7. Rather than risk the possibility of allowing water to get inside the craft through the open hatch, he wriggled his way out the narrow top of the capsule, struggling mightily. Once free, he dropped into the water, inflated and climbed onto a rubber life raft, and settled down for a long wait.

Having tracked his descent by radar, recovery forces were able to predict Carpenter's splashdown site with fair accuracy, even though he had hit the water some 250 miles from the targeted site. Concern grew among those waiting for word of the recovery, however, as time passed and he remained out of radio contact. On a level commensurate with public concern over Gus Grissom's near-drowning at the end of the Liberty Bell 7 flight and the potentially faulty heat shield on John Glenn's Friendship 7 capsule, anxiety arose as millions remained close to television sets and radios in hopes of hearing the good news that Carpenter had been found alive and well.

Rescuers hurrying toward the site were in the area a little more than a half hour after Carpenter hit the water. About an hour after splashdown, two parachute rescuers—frogmen, in the vernacular—arrived at the side of the astronaut's life raft. Carpenter was initially startled by the site of his rescuers, as he had been keenly watching the aircraft circling overhead and had not seen them floating down into the water, but he was obviously in good health and still exhibiting the same good humor that he had displayed throughout his flight in space.

Grateful for the company, Carpenter cheerfully offered his rescuers some of his food rations.

Three hours passed before Carpenter was hoisted from the rubber raft by a helicopter, still clutching the camera containing the images he had captured during his flight in space. A near calamity occurred when he dipped briefly into the waves at the end of the rescue line as he was pulled onto the helicopter, but he was able to keep the camera out of the water. Finally, 4 hours and 15 minutes after splash-down, he was set down onto the deck of the aircraft carrier USS *Intrepid,* where he received a congratulatory call from a relieved President John F. Kennedy.

An equally relieved public and press celebrated the astronaut's safe return, and Carpenter was subsequently honored in his hometown of Boulder, Colorado and by a Memorial Day gathering of 300,000 in Denver, Colorado.

MERCURY 5: SIGMA 7

(October 3, 1962)
Manned U.S. Spaceflight

Walter M. "Wally" Schirra, Jr. was the fifth American astronaut to fly in space. He made his first spaceflight aboard a Mercury vehicle he named Sigma 7, in honor of the vast collective effort of the engineers whose work had made the flight possible (sigma being the engineering symbol for summation, and the flight being the summation of the massive effort of the large group of engineers).

The mission profile for Schirra's Sigma 7 flight called for doubling the duration of each of the two previous Mercury orbital missions, which had each lasted three orbits and about four hours.

With the methodical expansion of the mission profile, NASA officials established a routine of incremental improvement that would carry the agency and its astronauts all the way to the manned lunar landings of the later Apollo program. Eschewing any temptation to rush into longer flights to compete with the Soviet Union, which had sent cosmonaut Gherman Titov into space for 17 orbits in Vostok 2 in August 1961 and launched Vostok 3 and Vostok 4 in a joint flight in August 1962, NASA chose to expand its spaceflight capabilities slowly and steadily, so as to present the smallest risk possible to its astronauts while yielding the largest reasonable expectation of adding to its knowledge and capabilities.

The superb outcome of the Sigma 7 mission was probably prefigured in events leading up to the launch. Schirra exhibited a determined professionalism in his training for the flight, and even as a series of potential difficulties worried mission planners, he quickly and expertly assumed a state of readiness that increased the confidence of everyone involved with the mission.

Added to the reasonable engineering challenges such as how to reduce the high fuel expenditures of the previous, shorter orbital flights, potentially dangerous new problems included fallout—both rhetorical and literal—from an atmospheric test of a nuclear weapon that had been conducted over the Pacific Ocean in July 1962. NASA officials consulted with officials at the Atomic Energy Commission and the McDonnell Aircraft Corporation, which manufactured the Mercury spacecraft, to determine the degree of threat that might be posed to the Sigma 7 flight by radiation left over from the nuclear test, which had been known as Project Dominic. In the months following the test, several satellites had suffered battery failures, but as the months passed, radiation levels dropped to levels deemed to pose no significant risk to the fifth Mercury mission.

Even with the reassuring news that the flight would not be susceptible to any radiation threat, NASA engineers outfitted the Sigma 7 craft with a dosimeter to measure radiation levels and provided Schirra with a portable version of the instrument.

In keeping with the routine established by the first four Mercury Seven astronauts to fly in space, Schirra awoke in the early hours of his scheduled launch day, ate breakfast, endured a brief physical exam and the attaching of sensors to his skin that would supply data to medical personnel monitoring his health during the flight, and was then helped into his spacesuit by suit technician Joseph Schmitt.

A short van ride later, he arrived at Launch Pad 14 at Cape Canaveral, where he rode the elevator on the launch vehicle apparatus to arrive at his Sigma 7 spacecraft atop the Atlas launcher that would propel him into orbit. Schirra was inside Sigma 7 by 4:41 A.M. on October 3, 1962.

Except for a brief hold about 45 minutes before the scheduled launch, while technicians made repairs to a radar system at NASA's Canary Island tracking station, the countdown for Schirra's launch went smoothly. He lifted off at 7:15 A.M. and communicated his pleasure with the launch to fellow Mercury Seven astronaut Donald "Deke" Slayton, who was in the Mercury Control Center.

A minor error in the timing of the cut-off of the launch vehicle sustainer engine—it had continued to fire 10 seconds longer than planned—resulted in Sigma 7 entering a higher-than-expected orbit of 175.8 by 100 miles, which would ultimately prove to be the highest orbit of any Mercury mission.

Schirra piloted Sigma 7 through the necessary turnaround maneuver to orient the spacecraft for orbital flight with a minimal expenditure of the craft's hydrogen peroxide fuel.

About midway through his first orbit, Schirra felt unusually warm, and medical personnel monitoring his flight from the ground noticed that the instruments keeping track of his temperature were reading higher than normal. Similar readings had occurred during Scott Carpenter's flight during the previous Mercury mission, even though Carpenter reported that he had not felt any change in temperature and was in fact quite comfortable throughout his trip.

Given their previous experience, NASA's doctors decided to give Schirra more time in orbit before becoming concerned enough to give the prospect of ending his flight early any serious consideration. Flight Director Christopher Kraft was charged with making the ultimate decision about whether to allow a second orbit or not, and after listening to the medical advice, decided that the flight should continue.

For his part, Schirra reacted to the temperature dilemma with the same careful, methodical approach that he had used in solving problems during pre-flight simulations. He increased the circulation of the cooling system in his spacesuit in small increments, waiting to judge the impact of each adjustment before making another change, until he arrived at a reasonably comfortable temperature.

Because the primary goal of his mission was to evaluate the capabilities of the pilot and the spacecraft for six orbits, Schirra was assigned a relatively light program of scientific work. In one innovative, if brief, experiment championed by NASA doctor Robert B. Voas, Schirra tried to locate particular dials on his control panel with his hand while keeping his eyes closed. He was able to locate each dial in all but three tries. The majority of the flight plan, however, called for Schirra to test and monitor the Sigma 7 spacecraft's controls and systems.

Of particular importance, he was able to accurately gauge the vehicle's yaw movements—the craft's motion to the left and right—in relation to the prescribed flight plan for the spacecraft. Schirra could gauge the yaw attitude using both the craft's periscope and by viewing the Earth through the spacecraft's window. He tested the yaw controls during both day and night conditions, and achieved remarkably accurate results, given the equipment and circumstances of the flight and the fact that it was just the third orbital mission of the U.S. space program.

Schirra also displayed his characteristic good humor in the lighter moments of his trip, joking with fellow Mercury Seven astronaut Scott Carpenter as he passed over NASA's tracking station at Guaymas, Mexico, for example, that he was in "chimp configuration," meaning that the spacecraft was in the automatic flight mode used during early Mercury test flights involving chimpanzees. He also communicated with Mercury Seven spaceflight veterans Virgil "Gus" Grissom, who was monitoring the flight from the NASA tracking site on Hawaii, and John Glenn, at NASA's tracking site in Point Arguello, California.

After an orbit of drifting flight during which he shut off the spacecraft's electrical power, Schirra checked out the craft's systems and equipment when he brought the systems back online. As expected, all worked well, and he continued his flight without difficulty.

As Sigma 7 passed over the Indian Ocean, the crew of one of the ships tracking the flight reported seeing the spacecraft for a brief period.

Schirra again moved the spacecraft into drifting mode near the end of his third orbit, and made use of the time during which he was not actively controlling the flight to check the radiation dosimeter that he had been given to gauge the amount of radiation left over from the July nuclear test in the atmosphere over the Pacific. He reported to Donald "Deke" Slayton, his fellow Mercury Seven astronaut at Cape Canaveral, that the amount of radiation detected by the instrument was very small.

At that point in the flight, about halfway through his planned six orbits, the Sigma 7 pilot took care of a variety of non-engineering chores. Schirra took a series of photographs and tried but failed to observe several phenomena created for the occasion as targets to check his vision (in one case, a large balloon was launched to test his ability to locate it). He also made a live broadcast of about two minutes as he passed over the United States, chatting with John Glenn in California about the details of his flight for the benefit of the millions of people following his progress via television and radio.

Even as he passed the previous milestone of a third orbit—the longest previous flight of the U.S. program—Schirra reported that he was comfortable and that his spacecraft was performing as expected. Mission controllers monitoring the flight saw that all their available data agreed with his assessment, and the flight continued flawlessly through the fourth, fifth, and sixth orbits, accomplishing a great leap forward in experience for NASA and confirming the capabilities of the agency's spacecraft, systems, and equipment.

At the end of the flight, Schirra quickly and competently proceeded through the items on his checklist while preparing for re-entry. He received a count from Alan Shepard, his fellow Mercury Seven astronaut who was tracking the flight from a ship in the Pacific Ocean, to begin firing Sigma 7's retrorockets. Shortly thereafter, he piloted the craft into re-entry attitude and began his descent through the atmosphere.

The Sigma 7 drogue parachute and main parachute deployed as expected, and the craft splashed down fewer than five miles from its target in the Pacific Ocean. Schirra returned to Earth after six orbits and a total flight of 143,983 miles, in 9 hours, 13 minutes, and 11 seconds. As a result of the slight error that had made his orbital altitude the highest of any of the Mercury flights, his speed in orbit, 17,557 miles per hour, would ultimately be the fastest of the entire Mercury program.

Recovery of the Sigma 7 capsule, with Schirra still inside, was made within 45 minutes by the aircraft carrier USS *Kearsarge*. Activating the spacecraft's explosive hatch, Schirra emerged from the capsule on the ship's deck tired but in good health after his superb mission, which NASA would later characterize as a "textbook flight."

MERCURY 6: FAITH 7

(May 15–16, 1963)

Manned U.S. Spaceflight

Having doubled the three-orbit Mercury flights of John Glenn and Scott Carpenter with Walter "Wally" Schirra's superb six-orbit and nine-hour flight in Sigma 7 in October 1962, NASA officials decided that the final Mercury mission should try for a full day in space and at least 18 orbits.

Gordon Cooper was selected as the pilot of the final Mercury flight, and he named his spacecraft Faith 7, in recognition, he said, of "my trust in God, my country, and my teammates."

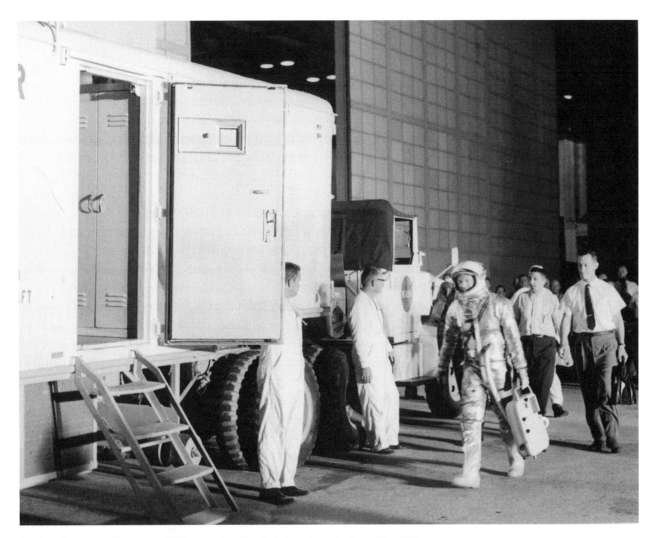

Gordon Cooper on his way to Faith 7 to start the first day-plus mission of the U.S. space program, May 15, 1963. [NASA/courtesy of nasaimages.org]

A first attempt at launch, on May 14, was postponed by a radar malfunction at NASA's backup control center in Bermuda. Cooper spent nearly six hours inside Faith 7 that morning, enduring several delays before the decision was made to put the launch off for another day, but he seemed entirely unaffected by the long, frustrating wait. He joked about not being able to get to the "real fun part" of lifting off, and spent the afternoon fishing, while waiting for another chance at launch the following day.

Cooper received good news from Walter Williams, the director of the Mercury program, in the early evening: all preparations for launch the following day were proceeding according to plan and the countdown had begun again in earnest.

Cooper lifted off in Faith 7 from Launch Pad 14 at Cape Canaveral at 8:00 A.M. on May 15, 1963, atop an Atlas launch vehicle. The initial phase of the flight went exactly according to plan, with the separation of the booster rocket and the sustainer stage of the Atlas vehicle correctly propelling the Mercury capsule into orbit at 165.9 by 100.3 statute miles. Like his predecessors, Cooper made a smooth turnaround maneuver to orient the spacecraft properly for orbital flight, and he began his mission in excellent form and high spirits.

Before he had even finished his first orbit, mission managers on the ground let him know that the superb insertion maneuver at the start of the flight and his position in orbit made it likely that he would be cleared to make at least 20 orbits. He was granted an official "go," NASA's term for approved mission goals based on the current status of the flight, for seven orbits as he passed over the NASA tracking station in Guaymas, Mexico, where his fellow Mercury Seven astronaut Virgil "Gus" Grissom was monitoring the flight.

Cooper relayed his satisfaction with both the performance of Faith 7 and with the pleasant nature of the flight to Grissom and, a short while later, to fellow Mercury veterans Wally Schirra and Alan Shepard at Cape Canaveral. They in turn assured him that all data being collected on the ground also indicated that both he and the spacecraft were in excellent condition and were performing well.

The first two orbits passed peacefully enough for Cooper to take a brief nap when not engaged in Earth observations. During the third orbit, he began to work on the program of experiments that he had been assigned. Because his flight was the longest of the Mercury program, he was given scientific tests greater in number and complexity than those carried out on earlier flights. The 11 experiments he would conduct during his day in space placed a high priority on medical research, including the collection of samples and the monitoring of his body temperature and blood pressure.

A particularly innovative test involved the release of a small beacon equipped with xenon strobe lights, which Cooper was to try to observe in orbit after it was released. Cooper was initially disappointed in his efforts to track the device after he deployed it during his third trip around the Earth, but was subsequently elated when he recognized the flashing light during his fourth, fifth, and sixth orbits. In keeping with his good humor and amply demonstrating his joy at the experiment's success, he joked with his fellow Mercury astronaut Scott Carpenter, who was monitoring the flight from NASA's tracking station on Hawaii: "I was with the little rascal all night."

Cooper was also the first Mercury astronaut to see Earth-based phenomena that had been deployed on Earth for his benefit, when he recognized a bright light set up in a small town in South Africa.

An experiment designed to measure the amount of resistance encountered by the spacecraft as it sped through space, and which required Cooper to deploy a brightly colored balloon, failed when the balloon would not eject from its compartment in the side of the capsule. The same experiment had been flown aboard Carpenter's Aurora 7 mission in May 1962, with moderately better results.

Cooper fared much better with his photography and Earth observation assignments, taking a remarkable series of photographs and describing in great detail the clarity with which he could see even relatively obscure landmarks such as rivers and roads and even houses in small villages on the high ground of the Tibetan plateau.

From his tracking position aboard a ship off the coast of Japan, fellow Mercury astronaut John Glenn reminded Cooper that he was scheduled to sleep during the seven or so hours that would elapse between his 9th and 13th orbits. After a dinner of powdered roast beef, Cooper dozed off during his 10th pass over the Earth. He slept off and on for about six hours in the drifting Faith 7, with the spacecraft's electricity powered down.

During his 14th orbit, after a careful check of his spacecraft and its systems and gratefully aware of his own continued high spirits and good health, Cooper activated the craft's voice recorder and spoke a heartfelt prayer of gratitude and hope. Recorded onto tape for posterity, Cooper's simple faith distilled all of the enormous effort, worry, and true belief of the early years of the U.S. space program into a concise posture of gratitude, when he said in part: "Thank You for the privilege of being able to be in this position, to be up in this wondrous place, seeing all these many startling, wondrous things that You've created."

Cooper continued onward in his orbital journey, turning his attention again to engineering tasks and checking the spacecraft's systems and equipment. During the 16th orbit, he took a series of photographs of the zodiacal light and the night airglow layer over the Earth's horizon, and in later orbits, he took photos that would subsequently be used in the design of navigation systems for the Apollo lunar landing program and a series of infrared photographs of weather phenomena.

In lighthearted moments during his 18th and 19th passes over the Earth, Cooper sang and joyously described the remarkable experience of seeing large portions of the United States all at once.

His mood—and that of all those tracking his flight on Earth—took a serious turn near the end of the 19th orbit, however, when a warning light indicated that Faith 7 was slowing down. Although Cooper initially reckoned the warning to be the result of faulty instrumentation, it proved to be a genuine problem during the next orbit, and was immediately compounded by a second, even more serious difficulty when a short circuit caused the spacecraft's automatic control system to fail.

Even as he prepared to take over manual control of the spacecraft as it neared the end of its final orbit, Cooper was faced with yet another cause for concern when the level of carbon dioxide in his spacesuit and space capsule began to rise. Maintaining an admirable calm amid the suddenly tumultuous circumstances, he noted the multiple problems in characteristically concise terms: "Things are beginning to stack up a little."

Despite the multiple failures in his vehicle and its systems, Cooper took over the spacecraft's controls and gamely oriented the craft for re-entry. He fired the retrorockets on John Glenn's count as precisely as he had in pre-flight Earth-bound simulations, and then piloted Faith 7 to a remarkably precise splashdown in the Pacific Ocean just four miles from the primary recovery ship, the aircraft carrier USS *Kearsarge*.

Cooper hit the water on May 16, 1963, after 22 and a half orbits and a flight of 1 day, 10 hours, 19 minutes, and 49 seconds. His journey had taken him a total distance of 546,167 miles.

Observing protocol to a degree that would likely be appreciated by even the most fastidious military officer, Cooper—an officer in the U.S. Air Force—requested permission to board the Kearsarge, a U.S. Navy ship, before being recovered.

Aboard the ship 40 minutes after splashdown, Cooper blew the explosive hatch on his spacecraft and was helped to exit Faith 7 after a brief initial medical checkup. Unsteady, dehydrated, and tired, he was nonetheless in good health overall and showed no emotional distress from the harrowing end to his excellent flight, during which he had overcome the failings of his spacecraft's systems to successfully pilot his day-long mission to a successful conclusion.

Cooper was subsequently honored in a series of celebrations, including a parade of historic proportions in New York City and a heartfelt ceremony in his hometown of Shawnee, Oklahoma. He also appeared before a joint session of the U.S. Congress.

His remarkable flight confirmed the validity of the astronaut-centric approach of the U.S. space program, which placed a high value on the abilities and observations of the individual within the spacecraft as well as on the vehicle and its systems. In the case of Cooper's Faith 7 odyssey, his capabilities as a pilot and his ability to remain calm and focused in the face of multiple systems failures were key to the success of the flight, and to his safe return to Earth.

MESSENGER

(2004–)

Mercury Exploration Mission

The Mercury Surface, Space Environment, Geochemistry, and Ranging (MESSENGER) mission began on August 3, 2004.

MESSENGER is equipped with five instruments, including the Mercury Dual Imaging System, a gamma-ray and neutron spectrometer, an X-ray spectrometer, the Mercury Laser Altimeter, and an atmospheric and surface composition spectrometer.

MESSENGER made its first flyby of Mercury on January 14, 2008. In October 2008, the spacecraft transmitted the highest resolution color image of Mercury's surface yet captured, and a series of images that together display a close-up mosaic of the planet's surface.

MEXICO: FIRST CITIZEN IN SPACE.

See Neri Vela, Rodolfo

MICHEL, F. CURTIS

(1934–)

U.S. Astronaut

Curtis Michel was born in La Crosse, Wisconsin on June 5, 1934. He graduated from C. K. McClatchey High School in Sacramento, California and then attended the California Institute of Technology, where he received a Bachelor of Science degree, with honors, in physics in 1955.

As an undergraduate, Michel participated in the Air Force ROTC program. He briefly worked as an engineer with the Guided Missile Division of the Firestone Tire and Rubber Company, and then joined the U.S. Air Force. He received his initial training as a pilot at Marana Air Force Base in Arizona and in Texas, at Laredo Air Force Base and Perrin Air Force Base.

Michel served three years of active duty in the Air Force, flying F-86D Interceptor aircraft. During his military service, he accumulated 1,000 hours of flying time, including 900 hours in jet aircraft.

After completing his Air Force service, he returned to the California Institute of Technology, where served as a research fellow and earned a Doctorate degree in physics in 1962. In 1963, he became a member of the faculty of Rice University in Houston, teaching space sciences.

NASA selected him for training as an astronaut in June 1965, as one of its Group Four selection of scientist astronauts. After several years with the space agency, he left NASA in September 1969 to return to his academic career.

Michel subsequently served as chairman of the Space Physics and Astronomy department at Rice University, and in 1974 he became the university's Andrew Hays Buchanan Professor of Astrophysics. In 1979, he was awarded a Guggenheim Fellowship and spent a year at the University of Paris; in 1982, he received the Humboldt Senior U.S. Scientist Award to study in Heidelberg, Germany.

MIR SPACE STATION

(February 20, 1986–March 23, 2001)
Soviet/Russian Space Station

Successfully operated in Earth orbit for more than 15 years and ultimately serving as a destination for 28 long-duration crews, the Mir space station outlived the totalitarian regime responsible for its development and launch to become a symbol of international cooperation and a gateway to the modern age of cooperative space exploration.

After years of preparation and careful engineering that expanded the capabilities of the Salyut series of space stations, the Soviet Union launched its next-generation space station, Mir, on February, 20, 1986, just 23 days after the United States suffered the loss of the space shuttle Challenger and its seven-member crew.

Befitting both the significance of the event and the emerging climate of glasnost ("openness," in English) introduced by Soviet leader Mikhail Gorbachev, the unmanned Mir lifted off from the Baikonur Cosmodrome in Kazakhstan with television cameras broadcasting its progress live. Media coverage also provided details of the station's structure, systems, and mission—all of which would have been, in previous times, unthinkable, given the tight government control of information related to the nation's space program.

The Mir space station (top) docked to the space shuttle Atlantis, July 4, 1995. Mir orbited the Earth for 15 years and housed 28 long-duration crews. [NASA/courtesy of nasaimages.org]

Mir represented a major advance over the first Soviet space stations, the Salyuts, and the military Almaz series. Intended as a long-term orbiting structure that would play continuous host to crew members over the span of many years, Mir was designed with six docking ports to accommodate crews arriving in Soyuz spacecraft, unmanned Progress resupply vehicles, and several large expansion modules that provided specialized work space for scientific investigations. The ability to gradually increase the station's size and functionality over the span of many years represented the first modular approach to space station construction, a method that was later adopted by the planners of the International Space Station (ISS).

Mir Expansion Modules

In its first decade in space, Mir was expanded by the addition of six modules, growing from the initial weight of its core, which launched at 20.4 tons in 1986, to more than 100 tons when fully expanded with all of its modules attached. In its completed configuration and with a Soyuz and Progress spacecraft attached, Mir measured 107 feet long and 90 feet wide.

The Mir core module provided the station's primary living area. It was first augmented by the arrival of the Kvant 1 module, an astronomy laboratory that was launched from the Baikonur Cosmodrome atop a Proton rocket on March 31, 1987. Equipped with a battery of X-ray telescopes supplied by a variety of international interests, including the European Space Agency (ESA), Great Britain, the Netherlands, and West Germany, Kvant 1 represented the first major test of the Soviet's modular approach to building Mir in orbit.

A first attempt to automatically dock Kvant 1 at Mir, on April 5, 1987, failed to achieve a "hard dock" (the module would not securely link with the station). Another attempt four days later also failed; finally, mission managers called upon Mir 2 crew members Yuri Romanenko and Aleksandr Laveykin to make a spacewalk to determine the source of the docking trouble. The cosmonauts performed their EVA task on April 11 and were surprised to find plastic debris wedged in Mir's rear docking port—apparently left over from a Progress resupply spacecraft that had visited the station previously. With the unexpected barrier removed from the docking port, Kvant 1 docked successfully on April 12.

The efficiency of the modular approach to space station construction particularly proved its worth in the case of the Kvant 2 module, which delivered upgraded life support systems to the station in 1989. A series of frustrating delays preceded the November 26 launch of Kvant 2, and the module suffered several malfunctions en route to the station, including a problem with its solar arrays that mission controllers were able to overcome fairly quickly. A computer problem required a more thorough analysis by engineers on the ground, in consultation with Mir 5 crew members Aleksandr Viktorenko and Aleksandr Serebrov, but the difficulty was overcome and the module was successfully docked at the space station on December 6 1989.

On May 31, 1990, the Mir Kristall module lifted off unmanned from Baikonur to travel to Mir, where it would dock with the station on June 10 to provide more room for scientific experiments. The Kristall module was equipped with a docking port that was intended to accommodate the Soviet space shuttle Buran, but in the aftermath of the dissolution of the Soviet Union, the Soviet shuttle program was canceled after a single, unmanned test flight. As had been the case with the two previous Mir expansion modules, Kristall also required several docking attempts, with mission controllers eventually relying on the module's backup propulsion system to enable it to dock with the station.

In anticipation of the arrival of the Spektr module, which was launched on May 20, 1995 to accommodate a program of scientific work to be carried out by U.S. researchers on the Russian space station, Mir 18 Commander Vladimir Dezhurov and Flight Engineer Gennady Strekalov made a series of five remarkable spacewalks to reconfigure the station for the arrival of Spektr and for the station's first docking with a U.S. space shuttle.

The Spektr module arrived at Mir on June 1, 1995 and was at first automatically docked at the station's front port, while Dezhurov and Strekalov continued their EVA work to prepare the module for its final docking position. Spektr was successfully moved to its appropriate port, but a test of the module's solar arrays four days later proved frustrating, as one array became stuck in a half-open position, resulting in a loss of about 20 percent of the electrical power mission planners had expected to have available for the scientific work the module was intended to support. The problem turned out to be less debilitating than originally thought, however, and a careful study of the station's overall power budget made it clear that the scheduled first space shuttle docking could proceed as planned despite the solar array malfunction.

With the arrival of the U.S. space shuttle Atlantis during STS-74 in November 1995, Mir experienced its fifth expansion, as the STS-74 crew delivered a Docking Module designed to make it easier for shuttles to dock at the station in the future. The Docking Module was attached to Mir on November 15, 1995.

Designed to provide capabilities for the remote sensing of Earth resources, the final Mir expansion module, Priroda, was launched from the Baikonur Cosmodrome on April 23, 1996. Priroda was automatically docked at Mir three days later, on April 26.

Russian—U.S. Cooperation Aboard Mir

In 1994, Mir served as the centerpiece of Russia's first post-Soviet program of cooperative spaceflight with the United States. Anticipating their future joint participation in the development of the International Space Station (ISS), the two former superpower rivals embarked on the Cooperation in Human Space Flight Program, which included dockings between the U.S. space shuttle and Mir and a series of long-duration visits to the space station by U.S. astronauts that resulted in a string of 907 days of continuous U.S. presence aboard Mir.

The cooperative program began with the first flight of a Russian cosmonaut aboard a U.S. space shuttle, which was achieved by Sergei Krikalyov during the STS-60 flight of the shuttle Discovery in February 1994.

In February 1995, the STS-63 Discovery crew achieved the first shuttle-Mir rendezvous, and a month later, on March 14, NASA astronaut Norman Thagard became the first U.S. astronaut to visit Mir when he launched in Soyuz TM-21 as part of the Mir 18 crew, with Russian cosmonauts Vladimir Dezhurov and Gennady Strekalov. The three crew mates lived and worked aboard Mir, primarily conducting a program of medical research, for nearly four months. They returned to Earth with the STS-71 crew of the shuttle Atlantis on July 7, 1995.

A total of nine shuttle dockings were achieved between the STS-71 Atlantis first docking on June 29, 1995 and the final docking in June 1998 by the STS-91 Discovery crew. The shuttle/Mir dockings were remarkable for their complete lack of significant problems. Although the station itself experienced a number of difficulties during the period in which U.S. astronauts were in residence, the docking procedures and the transfer of crew members, equipment, and supplies created no difficulties throughout the cooperative program.

The period of continuous U.S. presence aboard the station began with the arrival of NASA astronaut Shannon Lucid during the March 1996 STS-76 flight of Atlantis and concluded with the departure of Andrew Thomas during STS-91.

A series of accidents and equipment failures occurred during the stays of several of the U.S. crew members, including a fire on board the then 11-year-old station in February 1997 and a collision with an unmanned Progress resupply spacecraft four months later that nearly cost the crew members their lives and ultimately rendered the Spektr module useless.

In the first incident, which occurred on February 24, 1997, a fire broke out in the Kvant 1 module as a result of an accident involving an oxygen-generating device, putting the six crew members aboard—Mir 22 crew members Valery Korzun, Alexander Kaleri and NASA astronaut Jerry Linenger, Mir 23 cosmonauts Vasili Tsibliyev and Aleksandr Lazutkin, and visiting cosmonaut Reinhold Ewald of Germany—in immediate danger. The crew reacted quickly and was able to extinguish the fire, but Kvant 1 sustained damage and heavy smoke filled the station for several minutes. The crew members were fortunate to escape injury.

Following the fire, Mir was plagued by a series of other safety hazards, including a leak in the station's coolant system, an electrical outage, and a failure of the station's attitude control system.

A positive interlude offered a nice change of pace when Tsibliyev and Linenger made a landmark spacewalk—the first to involve a Russian cosmonaut and an American astronaut, during a 4 hour and 57-minute EVA on April 29, 1997.

Just two months later, however, on June 25, the unmanned Progress M-34 collided with Mir's Spektr module, causing the atmosphere within Spektr to escape and sending the crew—which at that time included Mir 23 cosmonauts Tsibliyev and Lazutkin and NASA astronaut Michael Foale—scrambling to close off the hatch to the damaged module to avoid the possibility that the life-sustaining atmosphere of the entire station would be vented out into space. The crew was fortunate to be able to seal off Spektr so quickly, but the incident signaled the end of the useful lifespan of the module, which had served as crew quarters and laboratory space for visiting U.S. astronauts during the period of continued U.S. presence aboard the station.

Mir Crew Members

During its 15 years in orbit, Mir supported 28 long-duration crews. Mir crew members conducted extensive scientific and medical research and greatly expanded long-duration spaceflight experience and the understanding of the physiological impacts of living and working in orbit. The station also initially furthered the political goals of the Soviet Union by providing spaceflight opportunities for visitors from other countries, and later served as a priceless asset for the emerging Russian Republic as the nation made its first steps toward cooperating with the United States on a program of shared spaceflight and exploration.

One of the most amazing achievements of the long Mir odyssey was the station's near-continuous habitation. Beginning with Soviet cosmonauts, including international visitors, and then featuring Russian crew members from the transformed Russian nation and finally also counting among their number long-term visitors from the United States, cosmonauts were aboard Mir for a total of 4,594 of the total 5,511 days the space station was in orbit.

By the time it was brought back to Earth in 2001, Mir had completed a remarkable 89,067 orbits, and hosted 40 Soviet and Russian cosmonauts (15 of which visited more than once), 18 international visitors, including 5 French and 4 German cosmonauts, and 7 U.S. astronauts who lived and worked aboard the station for extended periods—in addition to the crew members of nine space shuttles who visited briefly during shuttle/Mir docking operations.

Among the many exceptional achievements of Mir crew members, several stand out as particularly remarkable. Anatoly Solovyov made the most visits to the legendary station, with five trips, first visiting aboard Soyuz TM-5 in 1988, and then living and working aboard the station as a member of the Mir 6, Mir 12, Mir 19, and Mir 24 long-duration crews.

Many cosmonauts served significantly long missions aboard Mir; as part of a long-term program of medical research, Valeri Polyakov spent two exceptionally long spans aboard Mir, the first lasting more than 240 days, from August 1988 to April 1989, and the second a remarkable 437 days, from January 1994 to March 1995. During these two missions, Polyakov accumulated 678 days, 16 hours, and 33 minutes in space.

Other notable long-duration stays at Mir include Sergei Avdeyev's Mir 26 and Mir 27 missions, during which he remained aboard the station for a second tour of duty after concluding his first assignment and as a result accumulated a total of 379 days, 14 hours, and 52 minutes in space; Sergei Krikalyov's 311 days, 20 hours, and 1 minute from launch to landing during the Mir 9 and Mir 10 missions, which he also served consecutively; the Mir 3 mission, during which Vladimir Titov and Musa Manarov became the first individuals to spend an entire year in space, logging 365 days, 22 hours, and 39 minutes from launch in December 1987 to their return to Earth in December 1988; the 207 days, 12 hours, and 50 minutes of Mir 25, which featured Talgat Musabayev of Kazakhstan as commander, with Flight Engineer Nikolai Budarin; and the Mir 20 long-duration flight of Thomas Reiter of Germany, during which he represented the European Space Agency (ESA) for the EuroMir '95 science mission while serving 179 days, 1 hour, and 42 minutes in space with crew mates Sergei Avdeyev and Yuri Gidzenko.

A total of 78 Extravehicular Activities (EVAs)—each involving two crew members—were conducted from Mir. Because EVA work requires specialized training and the Russian approach to working in EVA relies heavily on teamwork, many of the 35 individuals who participated in Mir-based spacewalks did so more than once, and often with the same partner. Yuri Romanenko and Aleksandr Laveykin carried out the first Mir EVA on April 11, 1987; the final spacewalk from Mir was conducted by Sergei Zalyotin and Alexander Kaleri, on May 12, 2000.

Mir 1 Commander Leonid Kizim and Flight Engineer Vladimir Solovyov had the honor of being the first crew to occupy Mir after its launch. They traveled to the station on March 13, 1986, aboard Soyuz T-15, and during an odyssey in which they became the first crew ever to visit two space stations during a single mission, they also became the final crew to occupy the Salyut 7 space station, where they spent 51 of their total 125 days in space during the Mir 1 long-duration mission.

When Kizim and Solovyov left Mir at the end of their long flight, the station was left unoccupied for one of just three brief periods in its existence. A main goal for the station was that it was to be manned continuously; thus, only circumstances of major technical or political upheaval interrupted the station's otherwise constant habitation. The initial period of vacancy occurred while the first expansion module, Kvant 1, was being readied for launch; it lasted from the departure of the Mir 1 crew, on July 16, 1986, to the arrival of the second long-duration crew, in February 1987.

The station was also unoccupied from April to September 1989—the interval between the departure of the Mir 4 crew and the arrival of the fifth resident crew—during a period of political upheaval while Soviet officials pondered the future course of the nation's space program. One final, unintended period in which Mir hosted no crew members was the span between the final official crew, Mir 27, and the actual last visitors to the station, cosmonauts Sergei Zalyotin and Alexander Kaleri, who traveled to Mir in 2000 to help evaluate its potential for development as a tourist destination operated by commercial space developer MirCorp. The station was kept in orbit without a crew from August 1999 to April 2000 while the private development plans were evaluated; plans to privatize the station ultimately fell through, however, and Zalyotin and Kaleri were the final inhabitants of Mir. They returned to Earth on June 16, 2000, and the legendary space station was de-orbited on March, 23, 2001.

MISHIN, VASILY.

See Russian Federal Space Agency

MITCHELL, EDGAR D.

(1930–)

U.S. Astronaut and Sixth Human Being
to Walk on the Moon

Edgar Mitchell, the sixth person to walk on the Moon, was born in Hereford, Texas on September 17, 1930. He received his initial education in Roswell, New Mexico and graduated from Artesia High School. He then attended the Carnegie Institute of Technology, where he earned a Bachelor of Science degree in industrial management in 1952.

He joined the U.S. Navy that same year, and in 1953 he graduated from the Navy Officer's Candidate School in Newport, Rhode Island and was commissioned as an Ensign. After receiving flight training in Hutchinson, Kansas, he was deployed on Okinawa while attached to Patrol Squadron 29.

Edgar Mitchell on the surface of the Moon during Apollo 14, February 6, 1971. [NASA/ courtesy of nasaimages.org]

Edgar Mitchell on the surface of the Moon during Apollo 14, February 6, 1971. [NASA/ courtesy of nasaimages.org]

In 1957, he was assigned to Heavy Attack Squadron Two, where he flew A3 aircraft while serving on the USS *Bonhomme Richard* and USS *Ticonderoga*. He then served with Air Development Squadron Five as a research project pilot.

Mitchell graduated first in his class from the U.S. Air Force Aerospace Research Pilot School at Edwards Air Force Base in California and subsequently served as an instructor at the school.

In 1964, he became chief of the project management division of the Navy's field office for the Air Force Manned Orbiting Laboratory (MOL) program. The MOL project was designed to place members of the U.S. military in orbit for manned surveillance missions. The program resulted in a single, unmanned test flight before it was canceled in 1969.

Continuing his education concurrently with his military career, Mitchell attended the U.S. Naval Postgraduate School, where he earned a Bachelor of Science degree in aeronautical engineering in 1961, and the Massachusetts Institute of Technology, where he earned a Doctorate degree in aeronautics and astronautics in 1964.

During his outstanding military career, Mitchell logged 4,000 hours of flying time, including 1,900 hours in jet aircraft, and had risen to the rank of captain in the U.S. Navy by the time of his retirement from the service.

NASA chose him for training as an astronaut as a member of its Group Five selection of pilot astronauts in April 1966. His initial assignments at the space agency included service as a member of the astronaut support crew for the Apollo 9 mission, and he also served as the backup lunar module pilot for the Apollo 10 lunar orbit flight.

On January 31, 1971, Mitchell lifted off as lunar module pilot during Apollo 14, along with commander Alan Shepard and command module pilot Stuart Roosa. Once they arrived at the Moon, Mitchell and Shepard entered the Antares lunar module and began their descent to the lunar surface. Then, just as they were about to make their final preparations for landing on the Moon, a malfunction within Antares nearly caused the craft's abort program to engage.

As Mitchell and Shepard circled the Moon, awaiting word from Mission Control about how best to deal with the malfunctioning abort switch, NASA engineers devised a solution: the software logged into Antares's on-board computer would need to be reprogrammed. To put the fix into operation, Mitchell carefully entered 60 new lines of instructions into the lunar module's software program without making a single error—despite the long distance over which the new code had to travel and the tense circumstances in which the success of his work would determine whether or not he and Shepard could continue with their flight and land on the surface.

After Mitchell successfully reprogrammed Antares to work around the abort switch problem, Shepard brought the lunar module to a soft—if uneven—landing in the Fra Mauro Highlands of the Moon, where the spacecraft settled onto the surface on an angle.

More drama ensued when difficulties arose with the spacecraft's communications system just before the scheduled start of the first EVA; the Moonwalkers waited anxiously within the lunar module for 49 minutes before they were given the go-ahead to begin their first EVA.

Mitchell and Shepard took their first steps onto the lunar surface on February 5, 1971. Their first excursion outside of Antares was just the fourth Moonwalk in history.

Their initial duties included the gathering of a 42.9-pound sample of surface materials, and they also set up a television camera and an S-band antenna, deployed experiments, and planted the American flag on the surface. They spent a total of 4 hours and 49 minutes in EVA during their first trip outside the lunar module.

On February 6, Mitchell and Shepard made their second Moonwalk when they explored the rim of the 1,000-foot wide Cone Crater. To make the journey from Antares to Cone easier, they utilized the Modularized Equipment Transporter (MET), a lightweight cart with wheels that carried their EVA tools and containers for collecting samples.

The astronauts reached the base of the crater about an hour and a half into their EVA. The 2,800-foot climb upward to the rim proved an arduous task, and both men struggled to keep their footing in the loose, dusty soil. They were eventually forced to carry the MET after its wheels got bogged down in the boulders strewn around the crater.

Mission controllers worried about the astronauts' well being as the difficult climb continued, but granted a half hour extension to the EVA in the hope that Mitchell and Shepard might be able to reach the crater rim without rushing. The goal of reaching the rim ultimately had to be abandoned to ensure their safety, but Mitchell and Shepard were able to secure samples from the lower elevations of the crater before they returned to the lunar module, exhausted, after 4 hours and 46 minutes on the lunar surface.

After their strenuous second Moonwalk, Mitchell and Shepard completed their stay on the surface and then rejoined Roosa in the Apollo 14 command module Kitty Hawk. The three crew mates returned to Earth on February 9, 1971.

During their time on the lunar surface, Mitchell and Shepard gathered nearly 100 pounds of rocks and soil samples for return to Earth.

Edgar Mitchell spent 9 days and 48 minutes in space during his remarkable Apollo 14 Moon landing mission and walked on the surface of the Moon for a total of 9 hours and 35 minutes.

Mitchell subsequently served as the backup lunar module pilot for the Apollo 16 mission.

Among the many honors he received for his achievements as an astronaut, Mitchell was awarded the Presidential Medal of Freedom, the John F. Kennedy award of the Arnold Air Society, and the City of New York Gold Medal.

MOHRI, MAMORU

(1948–)

Japanese Astronaut

A veteran of two spaceflights, Mamoru Mohri was born in Yoichi, Hokkaido, Japan on January 29, 1948. In 1966, he graduated from Hokkaido Yoichi High School and then attended Hokkaido University, where he earned a Bachelor of Science degree in chemistry in 1970 and a Master of Science degree in chemistry in 1972.

Mohri then attended Flinders University of South Australia, where he earned a Doctorate degree in chemistry in 1976.

He began his career as a member of the nuclear engineering faculty at Hokkaido University, where he conducted research in a wide variety of fields and subjects, including chemistry, ceramic and semiconductor thin films, environmental pollution, and physics.

As a member of a research team working on a Japanese study of nuclear fusion, Mohri was involved in the development and use of systems designed to confine plasma. His work led to his being selected in 1980 to serve as an exchange scientist during the first nuclear fusion collaboration program of Japan and the United States.

During his outstanding academic career as a scientist, researcher, and instructor, Mohri has published more than 100 technical papers based on his research in materials science and vacuum science.

In 1980, he was honored with the prestigious Fifth Kumagai Memorial Award for the best science paper in the field of vacuum science, and he has also received awards from the Prime Minister of Japan, Japan's Minister of Science and Technology, the government of Hokkaido, the Japan Society of Microgravity Application, the Japanese Society for Space Biology, and the Japan Society of Mechanical Engineering.

Mohri was chosen by the National Space Development Agency of Japan (NASDA) in 1985 for training as a payload specialist astronaut to be assigned to the Spacelab-J program of scientific research aboard the U.S. space shuttle.

Mohri joined the physics faculty of the University of Alabama in Huntsville, Alabama in 1987 as an adjunct professor, and also worked at the university's Center for Microgravity and Materials Research. While conducting research related to the Spacelab-J experiments at the center, he also carried out microgravity experiments at NASA's Marshall Space Flight Center (MSFC) in Huntsville, and in the agency's KC-135 aircraft.

Mohri began his first spaceflight on September 12, 1992 when he lifted off from the Kennedy Space Center (KSC) in Florida as primary payload specialist for the Spacelab-J scientific program aboard the space shuttle Endeavour during STS-47. STS-47, the 50th mission of the space shuttle program, was commanded by Robert Gibson, and the crew also included pilot Curtis Brown, payload commander Mark Lee, and mission specialists Jan Davis, Jay Apt, and Mae Jemison.

Mohri was the first Japanese citizen to fly on an American space shuttle; the journalist Toyohiro Akiyama had been the first Japanese citizen to fly in space, during a visit to the Russian space station Mir in 1990.

During STS-47, the crew conducted 43 experiments in life sciences and materials science as part of the Spacelab-J program.

The life sciences work included experiments in animal and human physiology and behavior, biological rhythms, cell and developmental biology, human health, and space radiation, and was carried out with test subjects that included crew members, animal and plant cells, chicken embryos, frogs and their eggs, fruit flies, Japanese koi, and seeds.

Among the 24 materials science investigations included in Spacelab-J, the crew conducted experiments in biotechnology, electronics, fluid dynamics and transport phenomena, glass and ceramics, and metals and alloys. The scientists also made acceleration measurements.

As part of his work as the primary payload specialist for Spacelab-J, Mohri conducted live educational broadcasts during the mission to transmit the details and progress of the scientific work to the Japanese public, which followed his exploits with enormous enthusiasm.

The large amount of scientific work being accomplished onboard led NASA mission controllers to extend the flight by one day to allow the crew more time to finish their investigations. Endeavour returned to Earth at KSC on September 20, 1992 after a flight of 7 days, 22 hours, 30 minutes, and 23 seconds.

After his first flight in space, Mohri became the first general manager of NASDA's Astronaut Office, in October 1992.

He returned to the United States in August 1996 to begin training at NASA's Johnson Space Center (JSC) in Houston, Texas for assignment to future space shuttle missions as a mission specialist. After completing his training, he served in the payloads and habitability branch of NASA's Astronaut Office, where he helped to develop plans for the integration of the Kibo Japanese Experiment Module for the International Space Station (ISS).

On February 11, 2000, Mohri launched on his second flight in space, this time as a mission specialist aboard the space shuttle Endeavour during STS-99, the Shuttle Radar Topography Mission.

Mohri's crew mates for his second space mission included STS-99 commander Kevin Kregel, pilot Dominic Gorie, and mission specialists Janet Kavandi and Janice Voss of NASA and European Space Agency (ESA) astronaut Gerhardt Thiele of Germany.

Utilizing a 13-ton radar mapping instrument that extended 200 feet from the shuttle, the STS-99 crew was able to facilitate the capture of high quality, three dimensional images of the surface area of Earth that is home to nearly 95 percent of the planet's inhabitants. Crew members worked around the clock in alternating shifts to map the Earth between 60 degrees north latitude and 56 degrees south latitude.

Most of the areas mapped during the flight were covered twice, and when the radar mast was reeled into the shuttle on February 21, the crew had captured high quality images of all but 80,000 square miles of the Earth's surface. Endeavour returned to Earth at KSC on February 22, 2000, after traveling a total of 4.1 million miles in 11 days, 5 hours, 39 minutes, and 41 seconds.

Mamoru Mohri spent more than 19 days in space during two space shuttle missions.

MONGOLIA: FIRST CITIZEN IN SPACE.

See Intercosmos Program

MOON: MANNED AND UNMANNED EXPLORATION.

See Apollo Program; Luna Program; Lunar Orbiter Program; Surveyor Lunar Exploration Program; Zond Program

MUKAI, CHIAKI

(1952–)

Japanese Astronaut; First Japanese Woman to Fly in Space

The first Japanese woman to fly in space, Chiaki Mukai was born in Tategbayashi, in Gunma Prefecture, Japan on May 6, 1952. In 1971, she graduated from Keio

Chiaki Mukai, the first Japanese woman to fly in space, aboard the shuttle Columbia during STS-65, July 8, 1994. [NASA/courtesy of nasa images.org]

Girls High School in Tokyo and then attended the Keio University School of Medicine, where she earned a Doctorate degree in medicine in 1977.

She completed a general surgery residency at Keio University Hospital in Tokyo in 1978, and subsequently worked as a surgeon at Shimizu General Hospital in Shizuoka Prefecture and at Saiseikai Kanagawa Hospital in Kanawaga Prefecture. In 1980, she began a residency in cardiovascular surgery at Keio University Hospital and also served at Saiseikai Utsunomiya Hospital in Tochigi Prefecture as a cardiovascular surgeon before returning to Keio University Hospital in 1983 as chief resident in cardiovascular surgery.

In 1988, Mukai earned a Ph.D. in physiology from the Keio University School of Medicine. She was board certified by the Japan Surgical Society as a cardiovascular surgeon in 1989.

During her outstanding medical and academic career, she has published over 60 papers and articles based on her research.

Mukai began her association with NASA in 1987, when she worked at the Space Biomedical Research Institute at the Johnson Space Center (JSC) in Houston for a year as a visiting scientist representing the National Space Development Agency of Japan (NASDA).

In 1992, she became a research instructor in the department of surgery at Baylor College of Medicine in Houston, and also began teaching in the department of surgery at Keio University School of Medicine in Tokyo that same year as a visiting associate professor. She was promoted to visiting professor at Keio in 1999.

NASA chose Mukai for training as a payload specialist astronaut in 1985, as one of three Japanese candidates for the STS-47 Spacelab-J mission. She served as a backup payload specialist for STS-47.

On July 8, 1994, Mukai became the first Japanese woman to fly in space when she lifted off from the Kennedy Space Center (KSC) in Florida during the STS-65 flight of the space shuttle Columbia.

Robert Cabana served as commander of STS-65, and the crew also included pilot James Halsell and mission specialists Leroy Chiao, Carl Walz, and Donald Thomas.

The second flight of the International Microgravity Laboratory (IML-2), STS-65 featured 82 experiments in life science and microgravity science. The crew worked in a pressurized Spacelab module in two alternating teams to conduct scientific research around the clock. The STS-65 scientific program was designed by scientists from six space agencies, and included studies in biology, bioprocessing, human physiology, protein crystal growth, and radiation biology.

Using the Biorack facility developed by the European Space Agency (ESA), the crew made observations of biological and chemical samples that included bacteria, eggs, fruit flies, human and mammalian cells, sea urchin larvae, and seedlings. Another IML-2 component, the Nizemi centrifuge, allowed the astronaut-researchers to study the impact of different gravity levels on jellyfish and plants. The Nizemi was designed by the German space agency Deutsches Zentrum für Luft- und Raumfahrt e.V. (DARA), which also provided the Tempus, an instrument for studying the process of how materials transform from a liquid into a solid state.

During the flight, researchers at various locations on Earth monitored the experiments in real time, and mission controllers at the Spacelab Mission Operations Control in Huntsville, Alabama sent over 25,000 commands to the Spacelab instruments—more than during any previous flight.

At the conclusion of the mission, poor weather forecasted for the scheduled landing site at KSC in Florida caused the mission to be extended by one day. As a result, Mukai, who was already making a milestone flight as the first Japanese woman to fly in space, also set a record for the longest shuttle flight by a female astronaut. Columbia landed at KSC on July 23, 1994, after making 236 orbits in 14 days, 17 hours, and 55 minutes, which was also a new record for the longest shuttle mission up to that time.

After her landmark first spaceflight, Mukai served as a backup payload specialist for the STS-90 Neurolab mission—the final flight of the Spacelab module—in 1998.

Mukai lifted off on her second space mission on October 29, 1998 aboard the space shuttle Discovery, during STS-95, John Glenn's return to space.

The STS-95 mission was commanded by Curtis Brown, and in addition to payload specialists Mukai and Glenn, the crew also included pilot Steven Lindsey and mission specialists Scott Parazynski and Stephen Robinson of NASA, and European Space Agency (ESA) astronaut Pedro Duque of Spain.

Returning to spaceflight 36 years after his pioneering 1962 Mercury mission, during which he had become the first U.S. astronaut to fly in orbit, John Glenn, at the age of 77, set a new record during STS-95 as the oldest individual to be launched into space. He was the subject of a series of experiments designed to evaluate various aspects of the aging process.

The STS-95 crew also released and retrieved the SPARTAN 201 free-flying satellite to gather information about the solar wind and deployed the Petite Amateur Naval Satellite (PANSAT), which was designed to capture and transmit very faint or noisy radio signals that are not detectable by standard satellites.

Using a pressurized SPACEHAB module, crew members also conducted more than 80 experiments in areas including astronomy, human physiology, materials science, and physical sciences.

Discovery returned to Earth at KSC on November 7, 1998 after 134 orbits and a flight of 9 days, 19 hours, 54 minutes, and 2 seconds.

Chiaki Mukai spent more than 24 days in space during her two space shuttle missions.

Since her second spaceflight, she has continued her astronaut career as a payload specialist for the Japan Aerospace Exploration Agency (JAXA).

MULLANE, RICHARD M.

(1945–)
U.S. Astronaut

Richard Mullane was born in Wichita Falls, Texas on September 10, 1945. In 1963, he graduated from St. Pius X Catholic High School in Albuquerque, New Mexico and then attended the United States Military Academy at West Point, where he received a Bachelor of Science degree in military engineering in 1967.

Mullane was stationed at Tan Son Nhut Air Base in Vietnam from January to November 1969, and flew 150 combat missions during the Vietnam War as an RF-4C weapon system operator.

He then served in England for four years before attending the U.S. Air Force Institute of Technology, where he was recognized as a distinguished graduate and earned a Master of Science degree in aeronautical engineering in 1975. He also attended the U.S. Air Force Navigator Training School, where he received the Commander's Trophy as the outstanding graduate of his class, and the U.S. Air Force Test Pilot School, where he was also recognized as a distinguished graduate when he graduated in 1976. He subsequently served with the 3,246th Test Wing at Eglin Air Force Base in Florida as a flight test weapon system operator.

After 23 years of service during his outstanding military career, Mullane retired from the U.S. Air Force in 1990, with the rank of colonel.

Mullane became an astronaut in 1979 and first flew in space as a mission specialist during STS-41D, the first flight of the space shuttle Discovery. The eventful mission began with the first pad abort of the space shuttle program when a fault in one of Discovery's Space Shuttle Main Engines (SSMEs) caused a launch attempt on June 26, 1984 to be stopped just nine seconds before the scheduled lift-off.

Eventually launched on August 30, 1984, STS-41D was a resounding success. Judith Resnik became the second American woman to fly in space, and the crew launched three commercial satellites: Satellite Business System SBS-D, SYNCOM IV-2 (also known as LEASAT 2) and TELSTAR 3-C.

The STS-41D crew also tested a solar wing designed by the Office of Application and Space Technology (OAST-1). Equipped with a variety of solar arrays, the large (102 feet long by 13 feet wide) solar wing was designed to prove that large solar arrays could be used to provide power for future space facilities, including the International Space Station (ISS).

Other payloads included the Continuous Flow Electrophoresis System (CFES) III, which was operated in space by Charles Walker, an employee of its manufacturer, McDonnell Douglas, for the first time, an IMAX camera, and a crystal growth experiment that had been designed as part of the Shuttle Student Involvement Program (SSIP).

In a unique exercise, the crew used the Remote Manipulator System (RMS), the shuttle's robotic arm, to remove ice particles from Discovery's skin, leading mission controllers to dub the astronauts "icebusters." The shuttle landed on September 5, 1984, at Edwards Air Force Base in California.

Mullane made his second flight in space as a mission specialist during STS-27, which launched aboard the space shuttle Atlantis on December 2, 1988. STS-27 was the second shuttle flight following the Challenger accident in January 1986 and the third shuttle mission devoted to the classified activities of the U.S. Department of Defense. During the launch, insulating material from one of the shuttle's SRBs came loose and hit Atlantis about a minute and a half after the shuttle left the launch pad; similar damage years later, in 2003, led to the loss of the STS-107 crew and the shuttle Columbia. Despite the mishap at launch, Atlantis and her STS-27 crew returned safely after a little more than four days in space, landing at Edwards Air Force Base on December 6, 1988.

On February 28, 1990, Mullane lifted off on a remarkable third spaceflight, again serving as a mission specialist aboard Atlantis during a classified mission. The sixth shuttle flight dedicated to the classified activities of the U.S. Department of Defense, STS-36 launched after a week's delay due to the illness of the mission's commander, John Creighton. The mission was flown at an inclination of 62 degrees—the highest inclination flown during a manned U.S. space mission up to that time. Atlantis made 72 orbits of the Earth before landing on March 4, 1990.

Richard Mullane spent more than 14 days in space during his career as an astronaut.

He retired from NASA in 1990 and has since pursued a career as a professional speaker and writer. His first novel, *Red Sky: A Novel of Love, Space, & War,* was published in June 1993.

MUSABAYEV, TALGAT A.

(1951–)

Kazakh Cosmonaut

A veteran of three spaceflights, including two long-duration missions and eight spacewalks, Talgat Musabayev was born in Karaganda, Kazakhstan on January 7,

1951. In 1974, he graduated from Riga Civil Aviation Engineers Institute with a specialty in the operation of aviation radio equipment.

He began his career as an engineer with the Burundaisk Combined Aviation Detachment of the Air Service of Civil Aviation in 1974. In 1975, he became secretary of the committee of the Komsomol of the detachment in Alma-Ata, and the following year he was assigned to the Kazakh Administration of Civil Aviation's political education department as an instructor. He was subsequently elevated to the position of senior instructor, and in 1979 he became deputy commander of the 240th flying detachment.

In 1983 and 1984, Musabayev was a member of the championship aerosports team in the Soviet Union and was recognized as a master of sport in aerobatics and competitive aerial gymnastics.

He received training at the Alma-Ata aeroclub, as a participant in the Soviet Voluntary Society for Cooperation with the Army, Air Force, and Navy, and was also trained in the 30th flying detachment of civil aviation. In 1986, he received his civil aviation pilot's certificate.

In 1989, he completed training at the Ulyanovsk Flight, Dispatcher, and Engineering Personnel Training Center of Civil Aviation, and that same year became a commander in the Burundaisk combined aviation detachment. He also flew the Tu-134 aircraft, as second pilot, in the First Flying Detachment of the Alma-Ata Combined Aviation Detachment.

Musabayev was chosen by Soviet space officials for training as a cosmonaut as the sole member of the Kazakhstan Group One selection in May 1990. He underwent a period of intensive training at the Gagarin Cosmonaut Training Center (GCTC) in Star City, Russia and was qualified for assignment to future spaceflights as a cosmonaut researcher.

He was originally scheduled to fly in space as part of an all-Kazakh crew that was to visit the Mir space station, but that mission was later canceled amidst financial difficulties in Russia as the Soviet Union dissolved.

In addition to his cosmonaut training, Musabayev also attended Aktubinsk Civil Aviation High Flying School, where he earned a pilot-engineer diploma in February 1993.

Musabayev first flew in space as a flight engineer during the Soyuz TM-19/Mir 16 long-duration mission. Launched from the Baikonur Cosmodrome in Kazakhstan on July 1, 1994 with commander Yuri Malenchenko, Musabayev spent more than four months aboard the space station.

The two cosmonauts joined Valeri Polyakov, who was already aboard Mir when they arrived, for their long odyssey in space. Polyakov was in the midst of a record-setting long-duration mission of more than 437 days.

During their Mir 16 mission, Musabayev and Malenchenko made two spacewalks. On September 9, 1994, they spent more than five hours in EVA while inspecting and repairing damage that had been done to the station during two previous docking maneuvers. They inspected the station's Kristall module and were pleased to find less damage than they expected from the impact that Soyuz TM-17 had made when it docked in January 1993. Musabayev and Malenchenko fixed what damage they found and then moved on to the spot where the unmanned Progress-M 24 supply spacecraft had hit Mir while docking in August 1994. Another pleasant surprise awaited them at the second site: they found that the station had not suffered any damage from the second impact.

During their second EVA, on September 14, they inspected Mir's solar arrays in preparation for the later repositioning of the arrays before the first docking of the U.S. space shuttle Atlantis at Mir, which was scheduled for June 1995. They next retrieved experiments from the outside of the station and installed an amateur radio antenna, which Polyakov tested from inside of Mir. During their second trip outside the station, the two spacewalking cosmonauts added just over six hours to their EVA total for the flight.

Musabayev and Malenchenko completed their long stay at Mir on November 4, 1994, when they returned to Earth in Soyuz TM-19 with European Space Agency (ESA) astronaut Ulf Merbold of Germany, who had arrived on October 3 with the Mir 17 crew.

On January 29, 1998, Musabayev began his second space mission, this time as the commander of Soyuz TM-27/Mir 25. He traveled to Mir with Russian flight engineer Nikolai Budarin and guest cosmonaut Léopold Eyharts of France, who returned to Earth with the Mir 24 crew on February 18. Musabayev and Budarin remained at Mir to begin their long stay with NASA astronaut Andrew Thomas, who was already in residence at the station as the seventh and last U.S. astronaut to serve on Mir as part of the joint Russian-American first phase of preparation for the International Space Station (ISS) program.

Musabayev, Budarin, and Thomas conducted an intensive program of scientific research and performed the engineering and space station maintenance tasks necessary for keeping the station in good repair during their stay. Musabayev and Budarin also made a series of long spacewalks to make repairs to Mir.

They began their work outside the station on April 1, when they spent 6 hours and 40 minutes fixing Mir's solar array panels, work they continued on April 6 during a second EVA, which lasted 4 hours and 15 minutes.

On April 11, they turned their attention to Mir's main engine, during a spacewalk that lasted more than 6 hours. They continued their work on the station's propulsion system on April 17, adding 6 hours and 33 minutes to their EVA total, and again on April 22, when they finished the work in a final EVA, which lasted about 6.5 hours.

The U.S. space shuttle Discovery arrived at Mir on June 4, 1998, during STS-91, the ninth and final docking of an American shuttle and the Russian space station. The STS-91 crew collected Andrew Thomas for his return to Earth, and, as the final flight in the cooperative program of human spaceflight between the United States and Russia—and the start of the two nations' joint development of the ISS—the Discovery STS-91 flight also celebrated the good will of the successful step forward into the modern era of international cooperation in space.

Musabayev's role as commander of Mir during the successful conclusion of the Russian-American program was imbued with a particular poignancy, as his own career as a cosmonaut of Kazakh descent with dual Russian citizenship epitomized the promise of the new era of post-Soviet, post-Cold War opportunities in space for gifted pilots from a wide variety of nations.

After the space shuttle undocked and returned to Earth, Musabayev and Budarin remained aboard Mir for several more months, carrying out their assigned scientific research and performing the necessary maintenance and engineering tasks to keep the station operating safely. They returned to Earth in Soyuz TM-27 with Yuri Baturin, who arrived at the station a short while earlier with the Mir 26 crew.

In 2000, Musabayeve received the degree of candidate of technical sciences.

He next flew in space with Yuri Baturin when the two cosmonauts lifted off aboard Soyuz TM-32 on April 28, 2001 to travel to the ISS with U.S. businessman Dennis Tito, who had arranged for his flight by paying the Russian space agency $20 million, to become the first space tourist. As commander during his third spaceflight, Musabayev delivered both Tito and the Soyuz TM-32 spacecraft to the ISS, where the spacecraft was left to replace the Soyuz that had been docked at the station for several months and was nearing the end of its safe-operations limit.

At the ISS, Musabayev, Baturin, and Tito visited with ISS Expedition 2 crew members Yury Usachev of Rosaviakosmos, and flight engineers James Voss and Susan Helms of NASA. The two crews visited for about a week, and Musabayev, Baturin, and Tito then returned to Earth aboard Soyuz TM-31 on May 6, 2001, after a total flight of just under eight days.

Talgat Musabayev spent more than 341 days in space, including more than 37 hours in EVA, during his three space missions.

Among the many awards he has received for his remarkable achievements as a cosmonaut, Musabayev was honored as a Hero of the Russian Federation in 1994, a National Hero of Kazakhstan in 1995, and NASA's Space Flight Medal in 1998.

MUSGRAVE, STORY

(1935–)

U.S. Astronaut

Story Musgrave was born in Stockbridge, Massachusetts on August 19, 1935. He attended St. Mark's School in Southborough, Massachusetts but left school before receiving his high school diploma, and in 1953 he joined the U.S. Marine Corps, in which he served as an aviation electrician, instrument technician, and aircraft crew chief. During his time in the Marine Corps, he served in Korea, Japan, and Hawaii, and aboard the aircraft carrier USS *Wasp*. For his service, he was awarded the National Defense Service Medal and shared in an Outstanding Unit Citation as a member of Marine Corps Squadron VMA-212 in 1954.

He next attended Syracuse University, where he received a Bachelor of Science degree in mathematics and statistics in 1958. He briefly worked for the Eastman Kodak Company in Rochester, New York as a mathematician and operations analyst, and then moved to Los Angeles to pursue his graduate education. In 1959, he received a Master of Business Administration degree in operations analysis and computer programming from the University of California at Los Angeles, and he then attended Marietta College in Marietta, Ohio, where, in 1960, he received a Bachelor of Arts degree in chemistry.

Continuing his remarkable academic career at Columbia University in New York, Musgrave earned a Doctorate degree in medicine in 1964 and then completed a one-year surgical internship at the University of Kentucky Medical Center in Lexington, Kentucky. In 1965, he was granted a post-doctoral fellowship by the U.S. Air Force to continue his work at the center, where he focused on aerospace medicine and physiology. In 1966, he received a Master of Science degree in physiology and biophysics from the University of Kentucky and was the recipient of a post-doctoral fellowship from the National Heart Institute in support of his teaching and research in cardiovascular and exercise physiology.

He continued to work at the University of Kentucky Medical Center as a part-time professor of physiology and biophysics from 1967 to 1989, while simultaneously working part-time as a surgeon at the Denver General Hospital.

An accomplished pilot, Musgrave holds FAA ratings as an instructor, instrument instructor, glider instructor, and airline transport pilot, and has earned U.S. Air Force Wings. He has accumulated more than 18,000 hours of flying time in 160 different types of civilian and military aircraft, including 7,500 hours in jet aircraft, and has also made more than 800 free fall parachute jumps, including 100 experimental descents as part of his research into human aerodynamics. At the time of his retirement from NASA, he had accumulated more time flying the Northrop Grumman T-38 jet aircraft than any other pilot.

Musgrave has also authored 25 scientific papers in a variety of areas, including aerospace medicine and physiology, temperature regulation, exercise physiology, and clinical surgery. He is a member of the Civil Aviation Medical Association, and was honored with the Outstanding Airman of the Year Award by the Flying Physicians Association in 1974 and 1983.

NASA selected him for astronaut training in 1967. He was initially assigned to work on the Skylab program, using his skills as a scientist-astronaut to help develop the systems and procedures that would successfully launch and maintain America's first space station. He served as a member of the backup crew for the first Skylab mission, and worked as a spacecraft communicator (Capcom) during the second and third Skylab flights.

Musgrave also played a key role in the development of all the EVA equipment that would soon become ubiquitous on space shuttle spacewalks, including spacesuits, life support systems, airlocks, and manned maneuvering units.

He first flew in space as a mission specialist during the first flight of the space shuttle Challenger, during STS-6 in April 1983. The crew launched the first Tracking and Data Relay Satellite (TDRS-1), and Musgrave and Donald Peterson made the first spacewalk of the shuttle program, testing new spacesuits and EVA equipment and procedures for 4 hours and 10 minutes in the shuttle's cargo bay. The five-day mission ended with Challenger's safe landing at Edwards Air Force Base in California on April 9, 1983.

Musgrave's second space mission, the STS-51F Spacelab-2 flight, began with a harrowing start, as the crew experienced the first abort-to-orbit of the shuttle program. The mission actually suffered two separate abort procedures, each as a result of the shutdown of one or more Space Shuttle Main Engines (SSMEs). The originally scheduled launch on July 12, 1985 ended in a pad abort when a faulty coolant valve caused the spacecraft's three SSMEs to shut down just three seconds before ignition of the shuttle's solid rocket boosters (SRBs). On launch day, July 29, 1985, the number one SSME shut down prematurely, which meant that the shuttle would be unable to reach its planned orbit. NASA mission controllers called for an Abort-to-Orbit (ATO), which meant that the shuttle would enter a temporary orbit lower than the one called for in the original mission profile by using the thrust from its two properly functioning SSMEs.

Once in the orbit dictated by the ATO trajectory, the crew used the shuttle's Orbital Maneuvering System (OMS) thrusters to move Challenger into its originally intended orbit. The combination of the ATO procedure and the on-orbit maneuvering saved the mission, but resulted in a need for extensive alteration of the original flight plan in order to accomplish the mission's objectives. The flight was extended by one day to accommodate the changes.

Musgrave served as systems engineer for the launch and re-entry portions of the STS-51F flight and took a turn as pilot while the shuttle was in orbit.

With the mission safely underway, the crew deployed the Spacelab-2 experiments in the shuttle's cargo bay. Spacelab-2 consisted of experiments in astronomy, astrophysics, and life sciences, mounted on three pallets that were exposed to space, and utilizing a pressurized support module known as the "igloo." Despite the mission's scary start, the STS-51F crew was able to successfully achieve most of their objectives during 127 orbits, in a flight of 7 days, 22 hours, 45 minutes, and 26 seconds. They landed on August 6, 1985.

Continuing his education concurrently with his astronaut career, Musgrave attended the University of Houston, where he received a Master of Arts degree in literature in 1987.

In November 1989, he completed a third space mission, STS-33. Lifting off in the dark on November 22, STS-33 was the first shuttle mission in the period after the Challenger accident to begin with a nighttime launch, and the fifth flight of the space shuttle program to be devoted to the classified activities of the U.S. Department of Defense. Discovery landed at Edwards Air Force Base in California on November 27, 1989, after a flight of 5 days, 6 minutes, and 48 seconds and 79 orbits.

Musgrave's fourth flight featured classified payloads for the U.S. Department of Defense and also included unclassified payloads. The 10th flight of the space shuttle Atlantis, STS-44 launched after dark on November 24, 1991. The crew deployed a satellite and conducted other tests, including the Military Man in Space experiment. They had completed much of the planned mission profile before the scheduled 10-day flight had to be cut short when one of the shuttle's three orbiter inertial measurement units failed, endangering the shuttle's ability to navigate.

Plans for landing at the Kennedy Space Center (KSC) in Florida also had to be changed because of the malfunction, and Atlantis landed—safely—at Edwards Air Force Base in California on December 1, 1991.

In a spectacular nighttime launch on December 2, 1993, Musgrave launched aboard the space shuttle Endeavour for his fifth trip into space, during one of the most complex and ambitious missions of the entire shuttle program, the STS-61 servicing of the Hubble Space Telescope (HST).

The first servicing of the HST required a record five EVAs in which crew members spent more than 35 hours working in space. Musgrave was payload commander and lead spacewalker for STS-61; Richard Covey served as commander of the flight, and the crew included pilot Kenneth Bowersox and mission specialists Kathryn Thornton, Jeffrey Hoffman, Thomas Akers, and European Space Agency (ESA) astronaut Claude Nicollier of Switzerland.

Endeavour caught up with the HST on the third day of the mission, and Nicollier used the shuttle's Remote Manipulator System (RMS) robotic arm to grapple the instrument into the shuttle's cargo bay.

Then, working in teams during an amazing series of spacewalks over the course of five days, Musgrave and Hoffman alternated with Akers and Thornton to service and repair the telescope.

Musgrave and Hoffman made the first EVA, working on the HST for 7 hours and 54 minutes while they replaced the instrument's Rate Sensing Units (RSUs)—the devices containing the HST's gyroscopes—and their underlying electronics and electrical hardware.

Akers and Thornton made their first trip outside the shuttle the following day, replacing the HST's solar array panels during an EVA of 6 hours and 35 minutes.

Musgrave and Hoffman replaced the telescope's Wide Field/Planetary Camera and installed two new magnetometers on the telescope during their next EVA, which took 6 hours and 48 minutes.

Akers and Thornton made their second spacewalk of the flight—the fourth overall—on the seventh day of the mission. In an EVA of 6 hours and 50 minutes, they removed another of the HST's instruments, the High-Speed Photometer, and replaced it with a corrective device designed to help overcome the flaws in the telescope's primary mirror. They also installed a co-processor in the computer that controls the HST.

On flight day eight, Musgrave and Hoffman made the fifth spacewalk of the mission, installing the electronics for the new solar arrays and working on the HST for a total 7 hours and 21 minutes, raising the total spacewalking time for the flight to a remarkable 35 hours and 28 minutes. The HST was redeployed the following day, and Endeavour returned to Earth at the Kennedy Space Center in Florida in the early hours of December 13, 1993.

On November 19, 1996, Musgrave lifted off on a remarkable sixth space mission, this time aboard the shuttle Columbia during STS-80. With his launch as a member of the STS-80 crew, he became the first person to fly aboard the space shuttle six times, and at 61, he was the oldest person to fly in space up to that time (a mark later eclipsed by John Glenn, who was 77 at the time of his STS-95 flight in 1998). STS-80 also set a new record for the longest shuttle flight, at 17 days, 15 hours, 53 minutes, and 18 seconds.

Commanded by Kenneth Cockrell, the STS-80 crew also included Kent Rominger, who served as pilot of the shuttle Columbia, and Musgrave's fellow mission specialists Thomas Jones and Tamara Jernigan.

During the flight, the crew launched and retrieved two free-flying spacecraft: the Wake Shield Facility-3 (WSF-3) and the Orbiting and Retrievable Far and Extreme Ultraviolet Spectrometer-Shuttle Pallet Satellite II (ORFEUS-SPAS II). Musgrave was responsible for the operation of the WSF-3, which successfully cultivated seven thin film semiconductor materials for use in advanced electronics, and the ORFEUS-SPAS spacecraft made 422 studies of about 150 astronomical bodies during its period of free flight.

The record-setting STS-80 flight ended with Columbia's landing on December 7, 1996 at KSC.

Story Musgrave spent more than 1,281 hours in space, including 26 hours and 13 minutes in EVA, during his career as an astronaut. He left NASA in 1997 to pursue private interests.

N

NAGEL, STEVEN R.

(1946–)

U.S. Astronaut

Steven Nagel was born in Canton, Illinois on October 27, 1946. In 1964, he graduated from Canton Senior High School and then attended the University of Illinois, where he received a Bachelor of Science degree in aeronautical and astronautical engineering in 1969 with high honors.

During his undergraduate years at the University of Illinois, Nagel participated in the university's U.S. Air Force Reserve Officer Training Corps (AFROTC) program, and he was commissioned in the Air Force upon graduation. He received undergraduate pilot training at Laredo Air Force Base in Texas and was honored for his exceptional performance with the Commander's Trophy, the Flying Trophy, the Academic Trophy, and the Orville Wright Achievement Award. After completing his undergraduate pilot training, he received additional training in F-100 aircraft at Luke Air Force Base in Arizona.

He was subsequently assigned to the 68th Tactical Fighter Squadron at England Air Force Base in Louisiana, where he served until July 1971, when he began a one-year tour of duty in Udorn, Thailand as an instructor tasked with teaching operation of T-28 aircraft to Laotian Air Force pilots.

Nagel returned to England Air Force Base in Louisiana in October 1972 to serve as an instructor pilot and flight examiner. In 1975, he was selected to attend the U.S. Air Force Test Pilot School at Edwards Air Force Base in California. Upon graduation, he was assigned to the 6512th Test Squadron, where he flew F-4 and A-7D aircraft as a test pilot.

Continuing his education concurrently with his military career, Nagel attended California State University at Fresno, where he earned a Master of Science degree in mechanical engineering in 1978.

During his outstanding military career, Nagel accumulated 9,400 hours of flying time, including 6,650 hours in jet aircraft. He had risen to the rank of colonel in the U.S. Air Force by the time of his retirement from the service in 1995.

Nagel became an astronaut in 1979. His initial technical assignments at NASA included serving as a backup chase pilot for the first space shuttle mission, STS-1, and as a member of the support crew for several shuttle flights. He has also worked in the Shuttle Avionics Integration Laboratory (SAIL) and the Flight Simulation Laboratory (FSL), and served as acting chief of the Astronaut Office.

Nagel first flew in space as a mission specialist during STS-51G, which launched aboard the shuttle Discovery on June 17, 1985. The STS-51G crew deployed three communications satellites during the busy flight, including the Mexican MORELOS-A satellite and the ARABSAT-A spacecraft of the Arab Satellite Communications Organization. The international crew included payload specialists Patrick Baudry of France and Sultan Salman Al-Saud, the first citizen of Saudi Arabia to fly in space. The mission also featured the High Precision Tracking Experiment (HPTE), a laser tracking test related to the proposed Strategic Defense Initiative (SDI). The seven-day flight ended with Discovery's landing at Edwards Air Force Base in California on June 24, 1985.

Nagel's second space mission was STS-61A, aboard the space shuttle Challenger, in 1985. He served as pilot for the flight, which launched on October 30. Devoted to the German scientific module Spacelab D-1, STS-61A also marked the first time that eight astronauts were launched into space aboard a single spacecraft.

The international cooperation at the heart of the flight was an early sign of the coming era of routine multinational space exploration; Nagel's crew mates included payload specialists Reinhard Furrer and Ernst Messerschmid of West Germany (Germany was at the time still divided into East and West, prior to the country's reunification after the fall of the Soviet Union) and Wubbo Ockels, the first citizen of The Netherlands to fly in space. Although the shuttle's flight operations were overseen in the usual manner by mission controllers at the Johnson Space Center, German space officials at the German Space Operations Center at Oberpfaffenhofen, West Germany oversaw the Spacelab portion of the flight. Challenger landed at Edwards Air Force Base in California on November 6, 1985, following a flight of just over seven days.

Nagel served as commander of his third space mission, STS-37, which launched aboard the shuttle Atlantis on April 5, 1991. The STS-37 crew successfully deployed the Compton Gamma Ray Observatory (GRO) during the flight, but the observatory's antenna became stuck and would not extend, putting the GRO's entire mission at risk. To correct the problem, mission specialists Jay Apt and Jerry Ross made an unscheduled spacewalk of 5 hours and 32 minutes on April 7 to force the antenna loose, and they then deployed it manually. The effort was a success; the observatory returned scientific data for almost nine years, through June 2000. Atlantis returned to Earth on April 11, 1991.

On April 26, 1993, Nagel lifted off on a remarkable fourth spaceflight as commander of the STS-55 Spacelab mission, the second Spacelab flight dedicated to German scientific research, aboard the space shuttle Columbia. The crew endured a pad abort on March 22, when, just three seconds before the shuttle was scheduled to lift off, Columbia's computers detected an incomplete ignition in the number three Space Shuttle Main Engine (SSME). The fault was later traced to a liquid oxygen leak; at the time, however, the T-3 abort was a nerve-wracking ordeal for crew and mission controllers alike.

Fortunately, the launch on April 26 proceeded without incident. Once in orbit, the crew worked around the clock in alternating shifts to complete 88 experiments, and the flight was extended by one day to give the astronauts more time to finish their work. The experiment program included investigations in a wide range of disciplines, including astronomy, the Earth's atmosphere, life sciences, materials science, physics, and robotics. Columbia landed on May 6, 1993.

Steven Nagel spent more than 30 days in space during his career as an astronaut.

He retired from the Astronaut Office in 1995 to become deputy director for Operations Development in the Johnson Space Center (JSC) Safety, Reliability, and Quality Assurance Office, and subsequently transferred to NASA's Aircraft Operations Division to work as a Research Pilot.

NATIONAL AERONAUTICS AND SPACE ADMINISTRATION (NASA)

(October 1, 1958–)

U.S. Space Agency

From its first documented use, when the Chinese used a black powder version to fend off attacking Mongols, to the World War II V-2 research of Nazi Germany, rocket technology had been associated with warfare for centuries. Thus it was

NASA's first two groups of astronauts: the Mercury Seven (front, left to right): Gordon Cooper, Virgil "Gus" Grissom, Scott Carpenter, Walter Schirra, John Glenn, Alan Shepard, Donald Slayton; and Group Two astronauts (back, left to right): Edward White, James McDivitt, John Young, Elliot See, Charles "Pete" Conrad, Frank Borman, Neil Armstrong, Thomas Stafford, and James Lovell. [NASA/courtesy of nasaimages.org]

A panoramic view of Mission Control during the flight of Gemini 5, August 21, 1965. [NASA/courtesy of nasaimages.org]

with a certain irony that the United States and Soviet Union embarked in the late 1950s on the most intense development of rocket technology in history in order to achieve the peaceful goal of sending a human being into space and, eventually, to the surface of the Moon.

The origins of the "space race" between the two superpowers can be traced to the end of their partnership as World War II allies. Although they had worked together to defeat Germany and the Axis Powers, the two nations' vastly different political, social, military, and philosophical approaches, to each other, the rest of the world, and toward the future, soon made it clear that the United States and the Soviet Union would compete for the hearts and minds of the rest of the world's peoples.

By the end of the war, the initial stakes of space research were clear to both countries. German V-2 rockets had terrorized the citizens of London for nearly a year by the time the German rocket center of Peenemunde was captured by rapidly advancing Russian forces in May 1945. Correctly gauging the importance of securing the personnel, data, and hardware of the V-2 program, the Soviet leadership marked Peenemunde for priority capture. What neither the Soviets nor the Americans knew, however, was that the V-2 program and its scientists had relocated their operations to a complex of mines at Nordhausen, leaving behind little more than a gutted wreck at Peenemunde. The U.S. Army took possession of tons of documents and hundreds of V-1 and V-2 rockets, and captured Wernher von Braun and more than 100 of his engineers at Nordhausen while the Soviets scrambled to secure whatever they could find of the remnants of the program at Peenemunde and elsewhere.

The American edge in space-related research at the end of World War II was not without precedent; in 1926, American scientist Robert H. Goddard had launched the first liquid-fueled rocket in history. He continued to launch rudimentary rockets in the following decades, but received relatively little attention or acclaim for his experiments until years after his death in 1944.

A far more concentrated effort was made to capitalize on the V-2 cache. From March 1946 to September 1952, the United States launched 63 of the captured V-2 rockets at the first launch facility of the modern era, in White Sands, New Mexico. Among the major milestones of those early days were the launch of the entirely American-made WAC Corporal rocket and the launch of a combined WAC Corporal and a V-2, an early precursor of the multi-stage rockets that would later propel manned missions into space.

Even as their potential use expanded, from weapons development to scientific experimentation, the context for America's space rockets in the first decade after the war was primarily a military one. The U.S. Naval Research Laboratory was developing the Viking rocket as a potential launch vehicle for a small, unmanned satellite the nation hoped to send into space during the 1957–58 International Geophysical Year. The satellite idea had been the most intoxicating challenge put forth by the scientists of 40 countries when they had first proposed the year-long effort to focus on space during a milestone conference in Rome in 1954.

Competing with the Navy's Viking in the American race to launch a satellite were the U.S. Army's Redstone rocket and the Air Force-designed Atlas rocket. Given the early identification of the Army with the captured V-2 and the German scientists, led by von Braun and Hermann Oberth, the Redstone seemed to have an obvious edge, but it was the Navy's Viking launcher, soon re-named Vanguard, that was given the first opportunity to put an American satellite into space.

Before the United States could even make its first attempt, however, the Soviet Union shocked the world with the successful launch of Sputnik 1 on October 4, 1957. Emitting a series of "beeps" that could be heard by ham radio operators on Earth as it orbited above, the 185-pound Sputnik was greeted with great alarm in the United States. To many Americans, the Soviets' technological success with Sputnik 1 implied the frightening possibility that the Russians were ahead of the United States in everything from weapons development to scientific achievement. Although U.S. President Dwight Eisenhower tried to calm the fears of the American people, the press, and the scientific community by citing the progress of the ongoing U.S. program, it soon became apparent that nothing less than a successful satellite launch would serve as an appropriate answer to the Soviets' Sputnik.

As if the launch of Sputnik 1 were not enough to alarm the average U.S. citizen, on November 3, 1957 the Soviets launched the larger, more complex Sputnik 2, with a female dog, a husky named Laika, on board.

In the face of the two successful Sputniks, U.S. efforts to put a satellite in orbit intensified—initially with disappointing results. The attempted launch of Vanguard Test Vehicle-3 (TV-3) on December 6, 1957 ended in flames when the rocket lifted only a short distance off its launch pad and then toppled over.

Perhaps without the challenge of the Sputniks, the Viking (Vanguard) rocket might still have launched the first U.S. satellite, but the rush for America to get into space quickly was won by von Braun and the Army space team, albeit not with a Redstone rocket, but with a launch vehicle known as the Jupiter, which was in essence the next generation of the V-2 the Germans had devised during the war. The Jupiter successfully propelled the Explorer 1 satellite into orbit on February 1, 1958, officially launching the United States into the space age and its long competition with the Soviet Union for ever-greater "firsts" in space exploration.

Explorer 1 was outfitted with a Geiger counter at the insistence of Dr. James Van Allen, for whom the Earth's Van Allen radiation belts were named when the satellite became the first to confirm their existence, thanks to the onboard radiation-monitoring instrument.

What is sometimes lost in accounts of the history of the period is that most of the "firsts" of the early space programs of the two superpowers were in most cases even more important as political events than as technological achievements. As forward-looking and fascinating as the first satellites were, it was their symbolic value as products of American democracy or Soviet communism that meant the most to the political leaders of both nations.

The importance attached to being the first to achieve a particular space milestone led to distorted perspectives on both sides of the space race. On one hand, the amazing string of early successes on the Soviet side obscured the fact that the American and Russian programs were, in the larger context of developing entirely new technology for entirely new applications, very closely matched in management skills, engineering abilities, and both specific and general aims and objectives. At the same time, the intense emphasis on the race to be first to land a man on the Moon has, over the years, tended to lessen the impact of the Russian space achievements both before and after the American Apollo lunar landings while also blurring the truly remarkable nature of the Soviet feats of the late 1950s and early 1960s and the very real fear and tension they elicited in the United States.

A string of American setbacks, in retrospect, a not unlikely series of events, given how new the technology and applications were at the time, gave a general impression of futility in the U.S. space program. And yet, even as five consecutive Pioneer probes aimed at but failed to achieve lunar orbit, Vanguard 1 joined Explorer 1 in Earth orbit. Launched March 17, 1958 from Cape Canaveral in Florida, Vanguard 1 remains in orbit 51 years later; but at the time of its launch, the Soviets had already placed Sputnik 1 and Sputnik 2 in orbit, and would soon follow (on May 15, 1958) with Sputnik 3. On January 2, 1959 the Soviets upped the ante even further with the launch of Luna 1, which became the first spacecraft to escape the gravitational pull of the Earth.

In contrast to the intensely focused approach of the totalitarian Soviet regime, which amassed vast resources around the SS-6 Intercontinental Ballistic Missile (ICBM), a dizzying variety of ideas for spaceflight emerged in the United States based on an equally bewildering array of rockets supplied and championed by different branches of the U.S. military. The multifaceted approach was a logical outgrowth of the innovation and technical expertise of military planners who had developed the weapons potential of rocket technology out of necessity and who suddenly found themselves faced with the challenge of adapting its use to spaceflight in the post-World War II world.

Bolstered by its employment of Wernher von Braun and his V-2 scientists, the U.S. Army called for manned space missions based on its Jupiter Intermediate Range Ballistic Missile (IRBM), the same launch vehicle that had launched America's first successful satellite into orbit.

The U.S. Air Force went much further in its long-range plans for putting Americans into space, into orbit, onto manned space stations, and onto the surface of the Moon. For their time—both the Army and the Air Force plans were presented in 1958—these bold visions were remarkable in their optimism and earnestness and reflected the considerable research and expertise in space-related technology that had already been accumulated within the U.S. military by the mid-1950s.

Several aspects of the Air Force proposal proved particularly forward-looking as America's space program developed by the decade's end. The early call for a manned lunar landing was famously echoed in President John F. Kennedy's special message on urgent national needs three years later, and the space station idea later took shape first as the Air Force's own Manned Orbiting Laboratory (MOL) project and eventually found its ultimate expression in the Skylab program. Perhaps most importantly, however, was the Air Force plan to partner with the National Advisory Committee for Aeronautics (NACA).

Since its inception in 1915, NACA had been the U.S. government's primary nonmilitary resource for aircraft research and experimentation. By the 1950s, the organization had amassed a large number of the country's brightest thinkers in the areas of aircraft design and flight; the sudden national passion for space exploration had long since been a natural interest for NACA's more than 8,000 scientists and engineers.

Given his long, distinguished military career, culminating in the supreme command of all allied forces in Europe during World War II, the course that President Dwight D. Eisenhower chose for the future of America's space program might seem ironic, but at about the same time that the various military paths to space were beginning to take shape, Eisenhower, in April 1958, expressed his strong preference that the U.S. space program should be directed by civilian authorities. Brilliantly surmising that a new civilian agency would draw upon and benefit from the considerable space expertise of the nation's military while at the same time advancing the space program as a nonmilitary venture aimed at the higher goal of expanding humanity's knowledge and understanding, Eisenhower asked the U.S. Congress to create the National Aeronautics and Space Administration (NASA), which it did on October 1, 1958. At its inception, NASA was essentially NACA re-tasked for the space age; the older organization was completely subsumed into the new, and its former director, Hugh Dryden, became deputy administrator of NASA, reporting to the new agency's first administrator, T. Keith Glennan.

The new organization quickly brought order to the American space effort. The director of NACA's Langley Laboratory in Virginia, Robert Gilruth, was given the task of leading the Space Task Group, a project whose goal was to put an American in space. Beginning as such initiatives often do, with meetings and plans, presentations, approximations, and estimates, it was a project that would soon emerge from the complex thicket of scientific and technical challenges to capture the fascinated awe of the nation as Project Mercury.

As those responsible for the first of America's programs to send human begins into space, the Mercury administrators, astronauts, engineers, and scientists were faced with the difficult job of creating from scratch all of the systems, equipment, and procedures they would need to achieve the country's first spaceflights. The unprecedented nature of their work was made to seem even stranger by the intense time pressures imposed by the race to keep up with their counterparts in the Soviet program, who were clearly out-performing the United States in space. With the advent of the Mercury program, NASA had to quickly—and yet carefully—learn how to design and build (or more accurately, have built by its long list of contractors and subcontractors) spacecraft, spacesuits, rockets, communications systems, procedures, tools, and equipment for every phase of a typical space mission, a computer network for use both on Earth and in space, a media relations operation, and a means of recruiting, evaluating, and training the people who would fly in space. In addition to these concerns, the agency also needed to formulate detailed contingency plans for alternate courses of action in the event that the original plans encountered unforeseen difficulties. What was more, from the beginning, all these elements had to be developed concurrently by a host of teams working at the same time so the pieces would coalesce at the precise moment that mission planners set for each milestone flight.

Of course, military planners had already begun some of the work. Development of the rocket technology necessary to send a human being into space was already well underway in the United States by the late 1950s. The U.S. Army's Redstone rocket was chosen for the first planned Mercury flights, which would send an astronaut just beyond Earth's atmosphere to an altitude of 100 miles or more, but not into orbit. The Redstone's thrust was rated at 75,000 pounds, enough for the first sub-orbital Mercury flights. A more powerful rocket would be needed for the Mercury spacecraft to develop enough speed to travel into orbit. For that task, NASA turned to the Atlas Intercontinental Ballistic Missile (ICBM) developed by the U.S. Air Force. The thrust of the Atlas was 360,000 pounds. Anticipating six manned flights and a number of test launches, NASA ordered eight Redstone rockets, which cost $1 million each, and nine Atlas rockets, at a cost of $2.5 million apiece, at the outset of the Mercury program.

The Mercury spacecraft presented an entirely different set of challenges. In the course of its design and development, it gradually took on the shape of a cone, with the round cylinder at its top housing a set of parachutes designed to deploy and slow the craft's descent at the end of its flight. In the bulging middle of the spacecraft, soon more accurately dubbed a capsule, as it was very much an abbreviated environment for each of its successive inhabitants, there was the astronaut's tiny chamber, packed with miles of electric works and a sort of cosmic dashboard with switches, gauges, and lights that represented, in the late 1950s, the highest expression of advanced computing and electronic technology. At the bottom of the Mercury capsule was the smooth, flat, delicately curved ablative heat shield, a truly brilliant bit of innovation that would blunt the impact of the intense 3,000 degree heat the spacecraft would endure during its re-entry from space into the Earth's atmosphere. NASA chose the McDonnell Aircraft Corporation to manufacture the Mercury capsule.

Although the computers used throughout the first era of space exploration (from the late 1950s to the 1970s) were primitive by later standards, the overall communications infrastructure built to track the spacecraft and to communicate with astronauts in space was far from rudimentary. The original $40 million network that was built to support Project Mercury involved construction at 18 sites around the globe, and its operation required carefully coordinated activities on the part of a far-flung team of well-trained professionals, all devoted to the single goal of ensuring a safe mission for the astronaut in the spacecraft overhead.

Even with all the carefully developed technology and science, it was the astronauts who were at the heart of America's fascination with space exploration from the very beginning. NASA chose its first group of astronauts, the Mercury Seven, on April 9, 1959. The procedure began a long-held NASA tradition of selecting astronaut trainees in groups and identifying individuals according to the group to which they belonged (i.e., as a "Group One" or "Group Four" astronaut, for example).

President Eisenhower, who sought to avoid the mad rush of astronaut candidates that would likely have arisen from an open call, had largely proscribed the original selection. He asked NASA to limit its initial pool of space flyers to military test pilots with at least 1,500 hours of flying time. The original astronauts also had to be under the age of 40, hold a college degree, be certified to fly in jet aircraft, be in excellent health, and be less than five feet eleven inches tall, this last requirement made necessary by the cramped interior of the Mercury capsule, which simply would not accommodate a taller astronaut.

Given President Eisenhower's key contributions to the early years of the America's space program, it is perhaps not surprising that his successor, John F. Kennedy, also recognized and supported the importance of the space program and of NASA. However, President Kennedy's interest in America's fledgling space agency and its ambitious plans went far beyond that which could reasonably have been foreseen in the heated campaign of 1960, even as he and his Republican opponent, Vice President Richard M. Nixon, debated the "missile gap" between the U.S. and Soviet Union.

With a unique sense of the historical moment and a great deal of political courage, on May 25, 1961 President Kennedy set the goal for the United States in space as no less than "landing a man on the Moon and returning him safely to Earth" before the end of the 1960s. He also noted that such a feat would be more difficult and more expensive than any other space project of its time, and he said all this at a moment when the United States had only flown in space once, for just 15 minutes and 28 seconds (thanks to Alan Shepard, during his Freedom 7 Mercury flight on May 5, 1961).

Asking Congress to support his lunar landing vision with $148 million in immediate funds and $531 million for 1962, Kennedy also knew well that the space age he was helping to create by suggesting the Moon as a target might well take the entire decade to develop. As such, he knew he was committing the nation to a goal that might not be reached while he was still in office, as the most he could hope to serve as president was mandated by law to be a maximum of eight years.

In addition to extending beyond the term limits of his own presidency, Kennedy's vision of space exploration also broadened beyond the circumstances of the U.S.-Soviet Cold War. He publicly suggested the possibility that the superpower rivals might work together on a joint project to reach the Moon, and even broached the subject directly with Soviet leader Nikita Khrushchev when they met in Vienna in June 1961. Given the Soviet Union's advantage in the space race at the time, as well as the overall intransigence of the totalitarian Soviet regime, the suggestion was not seriously considered.

The impetus of presidential support, the larger goal of the decade-long reach for the Moon, the intense public fascination with the first astronauts, and the frightening implications of the Soviet Union's early successes all served to spur the rapid development of America's expertise during Project Mercury.

The Mercury program featured six manned launches, including the first flight of an American in space and the first American in orbit. NASA used 12 primary contractors to develop the Mercury hardware, rocketry, systems, and equipment, as well as 75 major and 7,200 lesser subcontractors, and in one way or another utilized the talents of about two million people, including 1,300 space agency employees and 18,000 military personnel for each mission during the nation's first great push into space.

As it developed its capability for human spaceflight, NASA also continued to pursue unmanned exploration of other planets. Begun in July 1962, the Mariner program of unmanned Venus and Mars probes grew in sophistication and success as the agency's knowledge and expertise grew over the years. The second vehicle in the series, Mariner 2, achieved NASA's first flyby of another planet when it passed by Venus, and Mariner 4 succeeded in flying by Mars and returning the first photos of the Martian landscape in July 1965. The Mariner program also gave NASA the honor of being the first space agency to place a spacecraft into orbit around another planet, when Mariner 9 began circling Mars in 1971. Perhaps most impressive of all, the program would eventually result in the remarkable flight of Mariner 10, which flew past both Venus and Mercury in 1974.

The unmanned successes of the 1970s were still far in the future as NASA embarked on Project Gemini, however. Much had changed on Earth between the last Mercury flight, in May 1963, and the first manned launch of the Gemini program, in March 1965. Most significantly, both of the space superpowers had new leaders: in the United States, Vice President Lyndon B. Johnson had succeeded President Kennedy when Kennedy was assassinated on November 22, 1963, and in the Soviet Union, less than a year later, Nikita Khrushchev lost an internal power struggle and was replaced by Leonid Brezhnev.

While the impact of President Kennedy's assassination was deeply felt in virtually every aspect of American society and gave rise to a profound sense of sadness and anger in the nation's public discourse, the impact of the changes in leadership of the two space rivals was even more acutely felt in the space program of the Soviet Union than in the U.S. program. Kennedy's vision and challenge to reach the Moon by the end of the decade had already become the central motivating force for NASA and the U.S. program by the time of his death, and remained a solemn commitment at the agency and in the space policy of his successor, Lyndon Johnson, who had himself been a long-time supporter of the space program. In the years following Khrushchev's removal, however, the Soviet program lost its drive and direction, and, in January 1966, it also lost its primary architect, Sergei Korolev. The Soviets re-focused their efforts to reach the Moon after Korolev's untimely death, but ultimately fell short of their attempt to beat the United States to a manned lunar landing and subsequently gave up the attempt entirely.

In the United States, NASA's long-range planners carefully constructed a series of incrementally more difficult space launches as they built their systems and expertise toward the eventual Moon landing. Confident of the continued support of the nation's political leadership and the American public, NASA officials designed the Gemini program as the next logical step in the expanding U.S. program. With the enormous advances of Project Mercury already in place, the Gemini teams of administrators, astronauts, engineers, and scientists were able to concentrate on methodically expanding the agency's knowledge and experience.

The main goals of the Gemini program were to launch spaceflights with crews of two, to extend the length of time that astronauts spent in space to as long as two weeks, to gather data about the effect that longer-duration stays in space might have on the astronauts, the spacecraft, and its systems, and to acquire expertise in rendezvous, docking, and extravehicular activity—all of which would become particularly important during the later Apollo lunar landing program.

Gemini flights would be controlled from the new state-of-the-art Manned Spacecraft Center (MSC) in Houston, which would soon be almost universally referred to by the name it assumed while flights were in progress: Mission Control. The facility would later be renamed the Johnson Space Center in honor of President Johnson, in recognition of his support for the space program.

With two crew members and equipped with a host of new electronics and communications equipment, the Gemini capsule required a new, more powerful rocket in place of the Redstone and Atlas launch vehicles that had powered the Mercury launches. The Gemini launch vehicle was the multi-stage Titan II, an ICBM that had been developed by the Martin Company for the United States Air Force. The Titan II featured a thrust of 430,000 pounds, compared to the 360,000-pound thrust of the Atlas and the 75,000-pound thrust of the Redstone, which had been used for the Mercury sub-orbital flights. Unlike the earlier rockets, the Titan II used hypergolic propellants, which were considered more reliable and safer than those previously used.

To carry out the rendezvous and docking tests that were at the heart of the Gemini program, NASA officials also had to develop an unmanned target spacecraft that the manned Gemini vehicles could use during rendezvous and docking exercises. For the target craft, the space agency turned to the Lockheed Company, which had developed the Agena B rocket for the U.S. Air Force. After modifications necessary to customize the basic Agena for its intended use during Gemini flights, the rocket would be launched as the upper stage of a combined Atlas-Agena rocket stack and then separated from the Agena in orbit.

Although the Gemini spacecraft itself was technically an expanded version of the Mercury capsule, it was outfitted with many improvements over the course of its development and could rightly be said to have evolved into a completely new vehicle by the time it was first used on a manned flight, during Gemini 3 in March 1965. The McDonnell Aircraft Corporation manufactured both the Mercury and Gemini capsules.

A key proponent of many of the changes incorporated in the Gemini craft was James Chamberlin, chief of NASA's Space Task Group engineering division. Chamberlin worked closely with the McDonnell staff to expand the capabilities of the Gemini spacecraft in a manner that kept pace with the broad vision of NASA mission planners, who would assign increasingly complex tasks to the vehicle as the Gemini program took shape in the early- and mid-1960s.

The astronauts who flew the 10 manned Gemini flights greatly expanded the knowledge and capabilities of the U.S. space program, and carefully, methodically moved the nation closer to the ultimate goal of landing an American on the surface of the Moon by the end of the decade. Frequently faced with difficult circumstances in space, the fuel cell failure of Gemini 5, for example, or the harrowing docking mishap of Gemini 8, the Gemini astronauts and their support teams on Earth performed exceptionally well and proved that even dire challenges could be successfully met with precision, concentration, and courage.

Virtually all the skills that would prove vital to the Apollo lunar landings, including the crucial rendezvous, docking, EVA, and extended duration flight capabilities, were successfully tested and made commonplace during the Gemini program, and many of the individuals who would make the Apollo flights a reality from their positions on Earth gathered their skills for launching, tracking, and safely returning the astronauts from the Gemini program.

With the successful Mercury and Gemini flights behind them, NASA officials entered the Apollo era optimistic about the United States' chances of meeting the late President Kennedy's goal of landing a man on the Moon before the end of the decade.

The culmination of the decade-long space race between the United States and the Soviet Union and the fulfillment of President Kennedy's lunar landing challenge, the Apollo program successfully landed 12 astronauts on the Moon, and returned them—and 838 pounds of carefully documented samples of lunar materials—to Earth.

In anticipation of the need for a large, multi-stage rocket for use during the Apollo Moon flights, NASA began development of the Saturn V launcher in January 1962. At the time, the agency had achieved only the first two manned U.S. spaceflights, the suborbital Freedom 7 and Liberty Bell 7 missions, and yet, with a forward-looking vision that would soon become typical of the organization during the 1960s, NASA's large complement of engineers and scientists began their work on the launch vehicle necessary for the Moon missions even with the knowledge that the flights might never occur if any of the missions in between ended in serious failure.

Led by Wernher von Braun, the engineers at NASA's Marshall Space Flight Center (MSFC) in Huntsville, Alabama spearheaded the development of the Saturn V in cooperation with three leading aerospace companies.

The Boeing Corporation was given the job of creating the first stage of the large launch vehicle, the S-IC. Capable of a thrust of 7.5 million pounds, the S-IC was fueled with a mix of liquid oxygen and kerosene.

For the second stage of the Saturn V, the S-II, the Rocketdyne division of North American Aviation developed an innovative liquid hydrogen engine known as the J-2; five J-2 engines were included in the S-II.

At the top of the stack of vehicles that made up the Saturn V, the S-IVB stage, which was built by the Douglas Aircraft Company, was fueled by some 60,000 gallons of liquid hydrogen.

Recognizing a need for more data about the Moon's orbital and surface environment and specific information about the suitability of potential landing sites, NASA's forward-looking mission planners also designed a series of unmanned lunar probes, which were sent to the Moon in the mid-1960s as part of two separate programs: the Lunar Orbiter Program (August 1966 to January 1968) and the Surveyor Lunar Landing Program (May 1966 to January 1968).

The various stages of the Saturn V also underwent a series of test launches while the first Apollo crew prepared for the first manned flight. Commanded by Mercury veteran Virgil "Gus" Grissom, whose crew included Edward White, who had been the first U.S. astronaut to walk in space, and rookie astronaut Roger Chaffee, the Apollo 1 mission progressed through a period of intensive training while the Apollo spacecraft and its equipment and systems were still being developed.

In addition to the multi-stage Saturn V launcher and its smaller counterparts, the complex system for sending astronauts to the Moon also included the Command Module (CM), Service Module (SM), and the Lunar Module (LM).

The CM was the crew cabin, which would carry crews of three to and from lunar orbit; the SM contained the equipment necessary to maneuver the three docked vehicles during their trip to and from the Moon and in lunar orbit; the

LM was the lunar landing vehicle that two of the three space travelers would use to descend to the surface of the Moon, while the CM pilot continued to circle in lunar orbit until the lunar explorers returned. North American Aviation built both the CM and SM; the Grumman Aircraft Engineering Company built the LM.

The mission profile for the first planned Apollo flight was understandably limited primarily to a test of an early version of the CM and the SM. Sadly, a tragic fire occurred during a pre-launch test of the Apollo 1 spacecraft on January 27, 1967, killing the three crew members and sending NASA into a period of deep mourning and scathing self-analysis. The agency's leadership appointed a commission to study the circumstances in which the test was conducted and to attempt to find the cause of the fire; ultimately, however, an exact cause was not established.

The results of the fire, however, were clear and far-reaching. Some 1,300 changes were made in the Apollo spacecraft, its equipment, and systems, and perhaps most importantly, a renewed and deepened sense of commitment to the safety of future crews took hold throughout NASA, contributing directly to the success of subsequent missions.

Mercury veteran Walter "Wally" Schirra led the United States back into space after the Apollo 1 accident when he commanded Apollo 7 in October 1968. The remarkable Apollo 8 mission, which resulted in crew mates Frank Borman, James Lovell, and William Anders becoming the first human beings ever to orbit another celestial body when they entered orbit around the Moon, followed in December 1968.

Two more manned test flights—the Apollo 9 Earth orbit test of the LM, and the lunar landing dress rehearsal Apollo 10—took place during the first half of 1969 in anticipation of the historic first lunar landing flight, Apollo 11.

On July 20, 1969, Apollo 11 astronauts Neil Armstrong and Edwin "Buzz" Aldrin became the first human beings to walk on the Moon. They would be followed by Charles "Pete" Conrad and Alan Bean during Apollo 12 in November 1969, Alan Shepard and Edgar Mitchell (Apollo 14) and David Scott and James Irwin (Apollo 15) in 1971, and John Young and Charles Duke (Apollo 16) and Eugene Cernan and Harrison Schmitt (Apollo 17) in 1972.

Following the success of the first two lunar landing flights, the Apollo Moon program suffered a major setback with Apollo 13. A faulty oxygen tank exploded during the Moon-bound phase of the mission, imperiling the crew and necessitating a remarkable rescue effort that successfully engineered a safe return for crew members James Lovell, Fred Haise, and John Swigert.

At a total cost of more than $20 billion, the Apollo program represented a major national investment for the United States, and its six successful lunar landing flights (and the remarkable Apollo 13 rescue) collectively represented a major achievement of engineering, scientific ability, technology, and national will.

With the conclusion of the final Apollo lunar flight, Apollo 17, in December 1972, NASA turned to the immediate development of Skylab, the first U.S. space station, and to the initial phases of two longer-term projects: the Apollo-Soyuz Test Project (ASTP) and the "space transportation system," which would in the ensuing decade become popularly known as the space shuttle program.

The space agency also continued its program of unmanned exploration of the cosmos with the launch of two of its most successful planetary programs, the Viking and Voyager probes.

The two identical Viking spacecraft, Viking 1 and Viking 2, were launched in August and September 1975 and began their exploration of the planet Mars in the summer of 1976 at the height of the United States' Bicentennial celebration. Each spacecraft featured an orbiter and a lander, and the two landing craft made the first two entirely successful landings on Mars. Returning amazing photographs of the landscape of the "Red Planet," the Viking landers also conducted tests of small samples of the Martian soil in the hope of finding some evidence of life and returned confusing results that initially gave rise to conflicting interpretations within the scientific community about the possibility that the planet might have once supported some rudimentary life form. Subsequent analysis of the experiments led NASA scientists to conclude that the initial results had most likely been incorrect.

Launched in August and September 1977, Voyager 1 and Voyager 2 each made remarkable explorations of Jupiter, Saturn, Uranus, and Neptune, and then continued on trajectories that were designed to send them entirely out of the solar system, with the hope that they would eventually become the first human-made objects to enter interstellar space. At a cost of about $700 million, the two Voyager spacecraft added a wealth of priceless data to human understanding of Earth's planetary neighbors, returning amazing photographs and discovering a total of 22 new moons around the planets they visited.

The agency's post-Apollo manned spaceflight program began with the Skylab space station program. Skylab resulted from a series of visionary plans that had circulated within NASA and its large community of supporters in academia and the aerospace industry for many years. Wernher von Braun had adopted the development of a U.S. space station as one of his primary interests when he became director of the MSFC in 1964; in the ensuing years, the program hinged to an ever-greater degree on the course of the Apollo program. In its final incarnation, Skylab would as much as possible make use of the vehicles and equipment that had been developed for Apollo.

The Skylab station was, in fact, adapted from the third stage of a Saturn V rocket; its crews traveled to the orbital outpost in Apollo command modules, and Apollo 12 veterans Pete Conrad and Alan Bean commanded the first two (of three) Skylab crews.

Skylab was sent into orbit unmanned on May 14, 1973. The launch was far shakier than planned, and as a result excessive oscillation of the Saturn V launcher caused serious damage to the station As a result, the first mission, which had originally been designed primarily as an exercise in medical and scientific study, was transformed into an effort to salvage the damaged station. NASA concentrated its engineering resources on the station's repair, and the first crew successfully returned Skylab to an operational condition.

Despite its difficult start, Skylab proved a remarkable success, if perhaps a less well-publicized one than the Apollo lunar explorations. The three crews who lived and worked aboard the space station accumulated more than 171 days in space and completed a combined 2,203 orbits of the Earth. They made extensive studies of the Sun and of Earth resources and conducted extensive medical research. They also made 10 spacewalks that amounted to more than 41 hours in EVA.

As a sort of embarrassing bookend to the problems the agency faced during the launch of the station, Skylab's demise in 1979 created enormous difficulties for NASA. Concerns about the possibility that the station might survive re-entry and cause damage if it were to strike a populated area had surfaced in the media periodically throughout 1978 as NASA's engineers worked to ensure a safe re-entry for whatever portions of the station might survive.

After months of criticism, the agency was able to effect a partially controlled descent for Skylab on July 12, 1979; for the most part, the surviving pieces of the station fell into the Indian Ocean, although some landed in remote parts of Western Australia, where they caused no reported damage.

Developed concurrently with the Skylab program, the unique Apollo-Soyuz Test Project (ASTP) mission achieved a feat whose political and social ramifications nearly rivaled the historic and scientific achievements of Project Apollo. After years of intense competition that pushed U.S. astronauts and Soviet cosmonauts to the cutting edge of what each nation's technology would allow in the effort of each country's space program to better the achievements of its rival, the ASTP flight brought the American and Russian space programs together to plan and carry out their first joint space mission.

A year of unprecedented cooperation preceded the flight, which took place in July 1975. Plans for ASTP were set in motion by a May 1972 agreement signed by U.S. President Richard Nixon and Soviet leader Leonid Brezhnev, and the administrators and engineers of the two nation's space agencies worked together in the ensuing three years to construct a flight plan and build a docking unit that would allow the U.S. Apollo and Soviet Soyuz spacecraft to dock in Earth orbit.

NASA turned to Gemini and Apollo veteran Thomas Stafford to command the final flight of an Apollo vehicle in space, and assigned Mercury veteran Donald "Deke" Slayton and Vance Brand as his crew. The Soviets tapped Alexei Leonov, the first person in history to conduct a spacewalk, and Valeri Kubasov to command Soyuz 19 during ASTP.

The two crews launched separately on July 15, 1975, and on July 17, they met in orbit and docked. The docking module worked exceptionally well, proving the value of the Earth-based cooperation that had preceded the flight as well as the skills and good will of the astronauts and cosmonauts who were carrying out the mission in Earth orbit. Over the course of 47 hours of docked operations, the five crew members shared several meals together, exchanged symbolic gifts, and participated in a news conference that was aired live in both countries.

Leonov and Kubasov returned to Earth on July 21, and Stafford, Slayton, and Brand splashed down on July 24, 1975, slightly the worse for wear after a malfunction during re-entry caused them to be exposed to toxic rocket fuel fumes. Although they required five days of care in a hospital as a result of the incident, they emerged without lasting damage to their health and subsequently embarked with their Soviet counterparts on extensive tours of the United States and the Soviet Union.

As the United States entered the latter part of the 1970s, NASA was busy preparing for the first flight of the U.S. space shuttle program, which would require the agency to concentrate all of its years of spaceflight experience and collective expertise to launch, orbit, and return the first reusable spacecraft, a vehicle that its creators would proudly identify as "the most complex machine ever built."

Featuring an airplane-like spacecraft that is sent into orbit with the assistance of liquid oxygen and liquid hydrogen supplied by a huge external fuel tank and the solid propellants of two Solid Rocket Boosters (SRBs), the space shuttle launches like a rocket, maneuvers in Earth orbit as a spacecraft, and lands like an aircraft on a runway. Only the External Tank is discarded after each launch; the SRBs are jettisoned shortly after launch and then recovered and refurbished for use on subsequent missions, and the shuttle is designed to return intact, to be reflown again and again.

The development of the powerful Space Shuttle Main Engines (SSMEs) was a key component in the success of the shuttle program. Three SSMEs are installed on each shuttle for each mission, and, in concert with the SRBs, they help the shuttle to reach a speed of 17,000 miles per hour (27,538 kilometers per hour) as the spacecraft makes its way toward orbit.

In contrast to NASA's original space exploits, which were designed to build the agency's capabilities and plans toward the eventual exploration of the Moon, the shuttle program was designed to provide frequent, cost-effective access to Earth orbit. As a result, shuttle flights were designed to be conducted in Earth orbits ranging from 115 miles (185 kilometers) to 400 miles (643 kilometers) high.

A prototype shuttle, the Enterprise (named for the spacecraft in the popular 1960s television show "Star Trek"), was used in a series of test flights known collectively as the Approach and Landing Test program prior to the start of the first shuttle flight.

Columbia was the first fully operational spacecraft in NASA's shuttle fleet; prime shuttle contractor Rockwell International delivered it to the space agency in March 1979. It was followed by Challenger (July 1982), Discovery (November 1983), and Atlantis (April 1985); the shuttle Endeavour was built as a replacement for Challenger, and was completed in May 1991.

NASA inaugurated the space shuttle age with Columbia's first launch, on April 12, 1981. Gemini and Apollo veteran John Young, who had walked on the Moon during the Apollo 16 mission in 1972, served as commander of STS-1 (the designation STS, shorthand for "space transportation system," was assigned as a prefix to every space shuttle mission). He shared the flight with Robert Crippen, who had served in the U.S. Air Force Manned Orbiting Laboratory program in the 1960s before transferring to NASA in 1969. They successfully completed the first orbital test flight of the shuttle program in a little more than two days; the flight marked the first instance in which NASA sent an entirely new vehicle into space with a crew on its maiden flight.

After several more test flights, the space shuttle program began to fulfill its original promise of frequent access to Earth orbit—and frequent deployments of satellites and experiments for a variety of commercial and national clients—in earnest. After the first launch, 24 successful flights were conducted during the program's first half decade, and NASA's ability to launch and safely return the reusable shuttles appeared poised to make spaceflight a routine, albeit expensive, exercise.

A remarkable string of historic moments served as highlights of the initial shuttle flights, including the first spacewalk of the shuttle era, which was carried out by Donald Peterson and Story Musgrave during STS-6 in April 1983, the first spaceflight of an American woman, which was achieved by Sally Ride during STS-7 in June 1983, and the first spaceflight by an African American astronaut, Guion Bluford, who first flew in space during STS-8 in 1983.

Marking the culmination of a major program of cooperation between NASA and the European Space Agency (ESA), the ESA-developed Spacelab orbital laboratory first flew in space aboard the shuttle Columbia during STS-9 in 1983. During that same flight, Ulf Merbold of Germany became the first ESA astronaut to fly in space.

Spacelab would enjoy a long and storied career aboard NASA shuttles; the modular laboratory would eventually fly aboard 16 shuttle missions, making its final flight in 1998 aboard the space shuttle Columbia during STS-90.

During STS-41B in February 1984, Bruce McCandless and Robert Stewart spent more than four hours dashing around outside the shuttle Challenger while demonstrating the futuristic Manned Maneuvering Unit (MMU) EVA backpacks; for their efforts, a fascinated media likened them to science fiction space heroes Flash Gordon and Buck Rogers.

George Nelson and James van Hoften would use the devices again during the next shuttle mission, STS-41C, to make an equally well-covered and dramatic repair of the Solar Maximum satellite in orbit. The remarkable repair was the first of its kind.

The 15th flight of the shuttle program, STS-51C in January 1985, was the first shuttle mission devoted entirely to the classified activities of the U.S. Department of Defense; the next flight, STS-51D in April 1985, included Utah Senator Jake Garn among its crew. Garn was the first sitting public official to fly in space.

Florida congressman Bill Nelson became the second elected official to travel into orbit when he lifted off with the STS-61C crew of the shuttle Columbia in January 1986. His flight capped the remarkable string of highlights that had made the shuttle program seem deceptively routine to many members of the public.

The complacency of the shuttle program, and the nation as a whole, would suffer a terrible shock with the launch of the 25th shuttle flight, STS-51L, on January 28, 1986. As a result of what was subsequently found to be the weather-induced failure of an "O-ring"—a gasket used to seal the joint between sections of an SRB—the space shuttle Challenger was lost in an explosion 73 seconds after launch, and her seven crew members perished. Stunned television viewers witnessed the scene live and in frequently repeated airings; the mission had attracted wider than usual attention for the inclusion of teacher Christa McAuliffe among its crew members.

She had been selected from a wide pool of applicants to become the first teacher to conduct lessons from space.

President Ronald Reagan memorialized the Challenger crew in a memorable address that pledged the nation's eventual return to space, and the U.S. Congress approved funds for the construction of a new shuttle, which led to the construction of Endeavour. A presidential panel chaired by former U.S. Secretary of State William Rogers presented a report to President Reagan in June 1986 that was intensely critical of NASA and of the decision-making process that had allowed the Challenger launch to proceed despite concerns about the impact of the cold weather.

Many changes followed within NASA and in the agency's relationships with its primary contractors. After a long, painful period of recovery, NASA returned to spaceflight on September 29, 1988 with the STS-26 launch of the shuttle Discovery.

A series of remarkable flights followed in the post-Challenger era. Shuttle crews were responsible for the deployment of planetary probes such as the Magellan Venus radar mapper (STS-30, May 1989), the Galileo Jupiter probe (STS-34, October 1989), and the ESA Ulysses spacecraft (deployed by the STS-41 crew in October 1990), which was designed to study the polar areas of the Sun.

Several shuttle flights were designed to demonstrate the importance of space science to everyday life on Earth. The STS-40 flight of the shuttle Columbia in June 1991, for example, featured important studies in aerospace medicine and biomedical research as part of the first Spacelab flight devoted to life sciences. And NASA's "Mission to Planet Earth," which was designed to focus the attention and resources of the shuttle on the Earth's environment and resources, began with STS-48 in September 1991.

NASA also relied on the shuttle for the deployment of three of its four "Great Observatories," including the Hubble Space Telescope, which was deployed during STS-31 in April 1990 and was subsequently serviced by the crews of STS-61 (December 1993), STS-82 (February 1997), STS-103 (December 1999), and STS-109 (March, 2002); the Compton Gamma Ray Observatory, deployed by the crew of STS-37 in April 1991; and the Chandra X-ray Observatory, which was deployed during STS-93 in July 1999—in a flight that was also remarkable for its commander, Eileen Collins, who, with STS-93, became the first woman to command a space mission.

Collins had previously been the first female astronaut to serve as pilot of a space shuttle when she served aboard the shuttle Discovery during STS-63 in February 1995. That flight had been the first in which a shuttle made a rendezvous with the Russian space station Mir.

Among many highlights, the first flight of the new shuttle Endeavour provided several important milestones, including new records for the longest spacewalk and the longest spacewalk by a female astronaut, STS-49 mission specialist Kathryn Thornton, and the first three-person spacewalk, which took place during the on-orbit retrieval and repair of the Intelsat VI satellite.

During the shuttle's 50th flight, STS-47 in September 1992, Mae Jemison became the first African American woman to fly in space. A medical doctor, she served as the science mission specialist for the flight.

NASA also continued to make the shuttle available to the U.S. Department of Defense for a number of missions that featured classified payloads; the last flight scheduled to include such activities was STS-53, in December 1992.

The STS-80 crew established a new record for the longest shuttle flight when they spent more than 17 days in orbit in 1996.

On October 29, 1998, the shuttle program celebrated the nostalgic return to space of one America's original astronaut heroes when John Glenn launched as a mission specialist aboard the shuttle Discovery during STS-95. At 77, Glenn set a new record as the oldest individual to be launched into space. He served as a test subject for a series of experiments aimed at gauging various aspects of the aging process.

Throughout the mid-1990s, the shuttle served an important role in the development of a cooperative program of human spaceflight between the United States and Russia after the dissolution of the Soviet Union.

The shuttle-Mir program featured nine flights in which shuttles were docked with the legendary Russian space station to deliver equipment and supplies and to transfer crew members. The program resulted in the first flights of Russian cosmonauts aboard the U.S. shuttle (STS-60 in February 1994 and STS-63 in February 1995), and a remarkable period in which U.S. astronauts maintained a continuous presence aboard Mir for 907 consecutive days.

With the emergence of routine international cooperation in space, the shuttle program also played an important role in the construction of the International Space Station (ISS), beginning with the first ISS assembly flight, STS-88, in December 1998. The STS-88 crew delivered and installed the Unity Node to the nascent ISS, providing the station's initial electrical power and communications and propulsion systems.

During the 1990s, NASA also entered a new phase of unmanned exploration of Mars, with often spectacular results. Beginning with the frustrating failure of the Mars Observer, which launched on September 25, 1992 and was lost on August 21, 1993, just days before it was to enter orbit around the planet, NASA achieved a string of important successes with unmanned Mars probes. Designed to collect high resolution images of the surface and to gather data about the Martian atmosphere, the Mars Global Surveyor entered orbit around Mars in September 1997.

Launched in December 1996, the Mars Pathfinder spacecraft landed in the Ares Vallis region of Mars on July 4, 1997, and two days later released the Sojourner Rover, a small remote-controlled vehicle that was operated with a 10-minute delay by mission controllers on Earth. Sojourner rolled around the Martian surface for several months, returning stunning images of the surface and conducting analyses of rock and soil samples until September 27, 1997, when contact with the craft was lost.

Following the loss of the Mars Surveyor '98 Polar Lander and the Mars Climate Orbiter, the Mars Odyssey Orbiter arrived at Mars in October 2001 and successfully entered orbit around the planet, where it would serve as a communications relay for the Mars Exploration Rovers Spirit and Opportunity, which landed on the surface in January 2004.

Appearing to the uninitiated like two tiny cosmic dune buggies, the Spirit and Opportunity rovers proved the highlight of NASA's modern Mars explorations to date. Given the task of exploring the Martian surface to determine if the planet could likely ever have supported life of any sort, the rovers were launched in June and July 2003. In landings cushioned by an innovative system of protective airbags, Spirit landed first, in Gusev Crater, on January 4, 2004, followed by Opportunity, which touched down at Terra Meridiani on January 25, 2004.

The two spacecraft returned remarkable high resolution images of the surface, stoking intense public interest in their mission and resulting in NASA's decision to extend their activities into 2005 and 2006, as the rovers continued to return valuable data.

As part of the agency's ongoing effort to gather data about Mars, the Mars Reconnaissance Orbiter (MRO) was launched in August 2005 with the intent of searching for evidence of water in the planet's atmosphere and beneath its surface. The MRO spacecraft successfully entered orbit around Mars on March 10, 2006.

Shuttle crews also continued to carry out important scientific missions, such as the STS-99 Shuttle Radar Topography Mission, which gathered high quality images of an area of the Earth's surface that is home to nearly 95 percent of the planet's inhabitants.

Even as the ongoing success of the space shuttle program and the emergence of the ISS again promised to make spaceflight routine, another tragic accident cast a shadow over NASA's plans for future missions. An accident at launch in which insulation shed from the shuttle's external tank struck the vehicle's wing caused the space shuttle Columbia to break apart over the southern United States just minutes before the STS-107 crew was to have returned to Earth at the end of their mission on February 1, 2003. The seven STS-107 crew members were killed in the accident.

NASA engineers devised an elaborate monitoring system to track subsequent shuttle launches and worked to find ways to minimize the potential for errant insulation causing another tragedy. They also experimented with procedures and materials that could potentially be used to locate and repair damage to a shuttle while the vehicle was in orbit to maximize a crew's chance of returning safely even if their spacecraft had been damaged at launch.

The first shuttle flight after the Columbia accident, STS-114, launched aboard Discovery on July 26, 2005. Careful examination of data gathered during the shuttle's lift-off led NASA officials to conclude that the insulation problem still remained a serious hazard, and as a result they grounded all planned future shuttle flights even while the STS-114 mission was still in progress. The move led to heightened concern for the safety of the crew, but Discovery and her crew landed safely on August 9, 2005.

After a long delay and further attempts to ensure the safety of future missions, the STS-121 crew successfully visited the ISS in July 2006, followed by STS-115 in September 2006, giving rise to fond hopes that the shuttle would be able to complete its role in the construction of the ISS before the shuttle fleet is retired.

NASA's participation in the ISS is projected to cost the United States some $50 billion during the station's projected life span of 30 years—excluding the cost of ISS-related space shuttle flights. While the expenditure of such a large sum gives rise to the same sort of criticism of the agency, and space exploration in general, that was common even in the glory days of the 1960s, the money expended on peaceful adventures in space is, in the modern era as it was then, a defensible expense when compared to the massive war-related expenditures that were characteristic of American involvement in Vietnam in the 1960s and in Iraq in the early years of the 21st century.

In a milestone of international cooperation, NASA's William Shepherd served as commander of the first ISS crew, which also included Russian cosmonauts Yuri Gidzenko and Sergei Krikalyov. The United States and Russia provided all of the resident crew members for the station's first crews, each of which lived and worked aboard the ISS for about six months, until the arrival of ESA astronaut Thomas Reiter of Germany, who began his stay as the first European to serve as a resident ISS crew member in July 2006. By that time, more than 120 astronauts from 12 countries and three space tourists had visited the station.

Bolstered by the support of President George W. Bush, who championed a return to crewed lunar exploration and envisioned future missions which would send astronauts to the surface of Mars, in 2006 NASA announced its plans for the Orion spacecraft and Ares launch vehicles. Crucial parts of the agency's Constellation Program of future space exploration missions, the vehicles were subjected to their first review in November 2006, the first system requirements review that NASA has undertaken for a human spacecraft program since an August 1973 review of the space shuttle program.

NATIONAL SPACE AGENCY OF UKRAINE (NSAU)

(1991)

Ukrainian Space Agency

The Ukrainian people have long played an important role in the exploration of space, initially as part of the space program of the Soviet Union and, since the nation's establishment as an independent republic, through the auspices of the National Space Agency of Ukraine (NSAU).

At the dawn of the space age, when the Soviet Union chose 20 candidates for training as members of its first group of cosmonauts, three of the pilots chosen for the new program—Valentin Bondarenko, Pavel Popovich, and Georgi Shonin—were natives of the Ukraine. Two went on to fly in space, with Popovich flying during the pioneering Vostok program, aboard Vostok 4, and again in Soyuz 14 in July 1974. Shonin flew aboard Soyuz 6 during the historic group flight with Soyuz 7 and Soyuz 8 in October 1969.

Bondarenko was preparing for his first flight when he was killed in an accident during a simulation on March 23, 1961, less than three weeks before Yuri Gagarin lifted off in Vostok 1 to become the first person in history to fly in space.

Various reports in the years since have indicated that Gagarin accompanied Bondarenko to the hospital following the accident; in any case, the entire incident, including the young cosmonaut's death and subsequent burial, was suppressed by Soviet officials and not revealed until a quarter of a century later, when the Soviet government embarked on its policy of glasnost (openness, in English) in the mid-1980s.

In addition to finally allowing proper acknowledgment of his service and sacrifice in the larger cause of space exploration, the revelation of Bondarenko's involvement in the early Soviet space program also served as tacit acknowledgment of the contributions of the Ukrainian people to the success of the Soviet program.

As the space age entered its fifth decade, a total of 18 Ukrainian-born cosmonauts had flown in space. In addition to Popovich and Shonin, 15 flew as representatives of the Soviet or Russian space programs: Anatoli Artsebarsky, Georgi Beregovoi, Georgy Dobrovolsky, Yuri Gidzenko, Leonid Kizim, Anatoli Levchenko, Vladimir Lyakhov, Yuri Malenchenko, Yuri Onufrienko, Leonid Popov, Vasili Tsibliyev, Vladimir Vasyutin, Igor Volk, Alexander Volkov, and Vitali Zholobov.

Zholobov is generally recognized as the first Ukrainian cosmonaut to fly in space, as he possessed dual Ukrainian and Soviet citizenship. All of the other Ukrainian-born cosmonauts were citizens of the Soviet Union.

Leonid Kadenyuk, who began his career as a member of the Soviet Air Force in the mid-1970s and was selected and trained as a member of the Soviet cosmonaut corps, flew in space for the first time in 1997, during the STS-87 flight of the U.S. space shuttle Columbia. He was the first astronaut to fly in space as a representative of the NSAU.

Since the dissolution of the Soviet Union and the emergence of an independent Ukraine, the NSAU has carefully developed the nation's civilian space initiatives over the course of three distinct periods of development, beginning with the mid-1990s "First Program," which integrated space development efforts in the national economy and defined the role of the Ukrainian industrial sector in the design and production of space program applications and components.

A second initiative followed in the late 1990s with the goals of opening the country's access to the global market for space-related products and services, and in 2002 the Verkhovna Rada, the national parliamentary body, approved a national space program that promises to expand the nation's role in space science and in the development of communications and remote sensing satellites.

In its pursuit of satellite design and development, the NSAU worked closely with the Russian Space Agency (RSA) to create and launch spacecraft that provide scientific data of value to both nations.

The two agencies cooperated on the development of the CORONAS-I solar observatory, which was launched on March 2, 1994 from the Plesetsk Cosmodrome in Russia atop a Tskyklon 3 launch vehicle. Equipped with plasma monitors and ultraviolet and X-ray instruments, CORONAS-I failed after several months in orbit.

The Okean O1 satellite, which was also developed jointly by the Russian and Ukrainian agencies, lifted off from Plesetsk on November 11, 1994, to conduct meteorological and oceanic studies from orbit.

The following year, the NSAU launched its first "home grown" satellite, Sich 1 (the Russian and Ukrainian word *sich* translates as owl in English). A remote sensing satellite designed for oceanic studies, Sich 1 was launched from Plesetsk on August 31, 1995. In addition to its own scientific instruments, it also carried a small experimental satellite for Chile that was known as FASAT-Alpha. The Chilean spacecraft was designed to be deployed separately once Sich 1 reached orbit, but a malfunction prevented the planned separation even though the Ukrainian satellite successfully achieved orbit and returned data as intended.

On July 17, 1999, the NSAU and RSA launched Okean O, a large remote sensing satellite designed to conduct studies of ocean waves and ice and measure the salinity of ocean waters. It was launched from the Baikonur Cosmodrome atop a Zenit 2 launch vehicle on July 17, 1999 and achieved an initial orbit of 662 kilometers perigee by 664 kilometers apogee.

In 2004, the two agencies collaborated on the launch of Sich 1M. Launched from Plesetsk on December 24, 2004, Sich 1M was designed to gather images of the Earth's surface to aid in the study of natural disasters.

In addition to its close association with the Russian space program, the Ukraine also actively pursues cooperative space development projects with other space agencies, including NASA and the European Space Agency (ESA).

The NSAU operates its own mission management and tracking center in Eupatoria, in the Crimea.

The Ukraine also participates in the Sea Launch program with partners from the United States, Russia, and Norway. The Sea Launch service provides an equatorial launch platform in the Pacific Ocean for the launch of commercial payloads that are propelled into orbit atop Zenit 3SL launch vehicles. In its first decade of operation (Sea Launch was organized in 1995 and conducted its first launch four years later), the program successfully launched 19 communications satellites for a variety of commercial clients.

NEAR SHOEMAKER

(February 17, 1996–February 28, 2001)
U.S. Asteroid Exploration Spacecraft

The first mission of NASA's cost-effective Discovery initiative, the Near Earth Asteroid Rendezvous (NEAR) Shoemaker spacecraft, was launched on February 17, 1996. During its successful mission, NEAR Shoemaker became the first spacecraft to orbit and land on an asteroid.

It made careful studies of two asteroids, utilizing onboard instruments that included an X-ray/gamma ray spectrometer, a near-infrared imaging spectrograph, a multispectral camera fitted with a CCD imaging detector, a laser altimeter, and a magnetometer.

NEAR Shoemaker encountered the C-class asteroid 253 Mathilde on June 27, 1997 and captured high resolution, color images of the asteroid while also measuring its size, volume, mass, density, and craters. It also searched the asteroid for satellites, but found none.

Following its encounter with Mathilde, the craft continued on to its primary task of becoming the first spacecraft to orbit an asteroid. Originally scheduled for January 1999, NEAR Shoemaker's rendezvous with the S-class asteroid 433 Eros was delayed until February 14, 2000, when it successfully achieved orbit around the asteroid and began to collect data about its composition, size, shape, mass, magnetic field. and surface and internal structure.

Although the spacecraft was not designed for landing, NEAR Shoemaker successfully touched down on the surface of Eros on February 12, 2001. During its descent, it captured 69 high resolution images, and its gamma-ray spectrometer gathered information on the asteroid's surface until the spacecraft's final transmission, on February 28, 2001.

NEAR Shoemaker was named in honor of Gene Shoemaker (1928–1997), the American geologist who founded the astrogeology branch of the U.S. Geological Survey in 1961. Widely recognized as the originator of the field of planetary science, he was the first to apply geologic principles to the mapping of the planets.

NELSON, CLARENCE W.

(1942–)
U.S. Congressman (D-FL) Who Flew in Space (STS-61C, January 1986)

As a member of the U.S. Congress, Clarence William "Bill" Nelson flew in space during the January 1986 STS-61C flight of the space shuttle Columbia, which ended with a safe landing just 10 days before the explosion of the space shuttle Challenger.

Nelson was born in Miami, Florida on September 29, 1942 and graduated from the public school system in Melbourne. He received his initial undergraduate education at the University of Florida, and then attended Yale University, where he received a Bachelor of Arts degree in 1965. Upon graduation he served in the U.S. Army Reserve while concurrently continuing his education at the University of Virginia, from which he received a law degree in 1968.

He practiced law before entering political life with a successful run as a Democratic candidate for the Florida House of Representatives in 1972. He served in that capacity until 1978, when he was elected for the first time to the U.S. House of Representatives.

With his STS-61C flight he became America's second elected official to fly in space; Utah Senator Jake Garn had been the first, during the April 1985 STS-51D flight of the shuttle Discovery.

Originally scheduled for launch on December 18, 1985, the STS-61C flight experienced a series of frustrating delays (including poor weather, faulty instruments, and equipment difficulties), and finally lifted off on January 12, 1986, its seventh launch attempt.

Nelson flew as a payload specialist during the STS-61C mission, with commander Robert Gibson and pilot Charles Bolden, mission specialists Franklin Chang-Diaz, Steven Hawley, and George Nelson, and fellow payload specialist Robert Cenker.

Once in orbit, the crew successfully deployed the RCA Americom satellite SATCOM KU-1, but faulty batteries scuttled a 35 mm camera that had been designed to photograph the comet Halley, in an experiment known as the Comet Halley Active Monitoring Program (CHAMP).

At its end, the mission ran into delays similar to the ones that had troubled the flight at its start. Landing was originally scheduled for the Kennedy Space Center (KSC) in Florida on January 17, but it was re-scheduled for January 16 to save time in preparing the shuttle for its next flight. That landing attempt was then postponed by bad weather, as was another attempt on the originally-scheduled target of January 17. The continuing foul weather in Florida eventually led mission controllers to divert the landing to Edwards Air Force Base in California, where the shuttle finally returned to Earth during a night landing on January 18, 1986, after 98 total orbits.

As someone experienced in actual spaceflight and as a respected Democratic congressman from the state of Florida, Nelson was an important supporter of NASA during the agency's difficult recovery from the January 1986 Challenger accident.

Nelson served six terms in the U.S. House of Representatives, until 1990, when he made an unsuccessful attempt to secure the Democratic nomination for the Florida gubernatorial race. He subsequently continued his public service career with his 1994 election to the Florida Cabinet as state Treasurer, Insurance Commissioner, and Fire Marshal, a post he held until 2000.

In 2000, he was elected to the United States Senate, for a six-year term, defeating a Republican opponent in a race to replace a retiring Republican incumbent. Since his election, he has served on the Senate Commerce, Armed Service, Budget, Foreign Relations, and Aging committees, and is recognized as the Senate's leading expert on NASA.

Bill Nelson spent 6 days, 2 hours, 3 minutes, and 51 seconds in space during his STS-61C flight.

NELSON, GEORGE D.

(1950–)

U.S. Astronaut

George "Pinky" Nelson was born in Charles City, Iowa on July 13, 1950. In 1968, he graduated from Willmar Senior High School in Willmar, Minnesota and then attended Harvey Mudd College, where he received a Bachelor of Science degree in physics in 1972.

Nelson went on to attend the University of Washington, where he earned a Master of Science degree in astronomy in 1974 and a Doctorate degree in astronomy in 1978. He has performed astronomical research at the Sacramento Peak Solar Observatory in New Mexico, the Astronomical Institute at Utrecht in the Netherlands, the University of Gottingen Observatory in Germany, and the Joint Institute for Laboratory Astrophysics in Colorado.

NASA selected him for training as an astronaut in 1978. In his varied technical assignments for the agency, he has conducted scientific experiments aboard the NASA WB 57-F Earth resources aircraft, helped to develop spacesuits, and served as a photographer in the main chase plane during the first space shuttle flight, STS-1, in 1981. He has also headed the Astronaut Office Mission Development Group.

Nelson first flew in space aboard the shuttle Challenger during STS-41C, which launched on April 6, 1984. Commanded by Robert Crippen, the STS-41C crew also included pilot Francis Scobee and Nelson's fellow mission specialists James van Hoften and Terry Hart. The crew achieved a landmark on-orbit salvage mission when they retrieved and repaired the Solar Maximum satellite (which was whimsically nicknamed "Solar Max").

The STS-41C launch featured the first direct ascent trajectory of the shuttle program, meaning that the shuttle Challenger achieved orbit by a longer-than-usual firing of its Space Shuttle Main Engines (SSMEs), and just one (rather than the usual two) firings of its Orbital Maneuvering System (OMS) engines. In the typical sequence, one OMS burn maneuvers the shuttle into an elliptical orbit and a second OMS firing places the vehicle in an almost-circular orbit of the Earth. In the case of STS-41C, the direct ascent trajectory was utilized to propel Challenger directly into a higher-than-usual orbit (nearly 300 miles from Earth) so the shuttle could rendezvous with the Solar Max satellite.

After the Challenger crew located and achieved rendezvous with the satellite, Nelson and van Hoften retrieved the Solar Max craft on April 8, using the nitrogen-propelled Manned Maneuvering Units (MMUs) to sidle up to the spacecraft and wrestle it into the shuttle's cargo bay. The task proved more difficult than anticipated, and mission controllers extended the flight by one day to make sure that the astronauts would have enough time to effect the planned repairs. Nelson and van Hoften worked for 2 hours and 38 minutes in EVA while retrieving the satellite.

Once they had managed to firmly fasten Solar Max to Challenger's cargo bay, Nelson and van Hoften made a second spacewalk on April 11, accumulating another 6 hours and 44 minutes in EVA to replace the satellite's malfunctioning altitude control system and its coronagraph/polarimeter, and then returned the craft to separate flight. A remarkable achievement, the repair mission was the first of its kind, the first-ever planned fix to be made on a spacecraft while it was in orbit.

The crew also successfully achieved its other primary objective for STS-41C with the deployment of the Long Duration Exposure Facility (LDEF), an orbiting cylinder containing 57 experiments designed to test the impact of long stays in space on a variety of items, including various materials and seeds.

Challenger returned to Earth on April 13, 1984, after a flight of 6 days, 23 hours, 40 minutes, and 7 seconds and 108 orbits.

Nelson's second spaceflight was STS-61C, which lifted off on January 12, 1986 aboard the shuttle Columbia after six attempted launches and a long series of frustrating delays and postponements due to poor weather and technical problems.

The STS-61C crew, which included Congressman Bill Nelson of Florida, the second public official to fly in space (Utah Senator Jake Garn had been first, in 1985), deployed the RCA Americom satellite SATCOM KU-1 during the flight and performed a number of scientific experiments.

More poor weather in Florida forced mission controllers to divert Columbia's return to Earth to Edwards Air Force Base in California, where the shuttle made a night landing on January 18, 1986. As fate would have it, STS-61C was the last successful shuttle flight for nearly three years; 10 days after Columbia landed in California, the space shuttle Challenger was lost during launch at the Kennedy Space Center in Florida.

On September 29, 1988, Nelson lifted off on a remarkable third spaceflight, aboard the shuttle Discovery during STS-26, the first shuttle mission to be flown after the Challenger accident. Following the long, wrenching recovery after the accident, STS-26 brought America back into space after a hiatus of two years and eight months. The crew deployed the Tracking and Data Relay Satellite-3 (TDRS-3) and conducted 11 mid-deck scientific experiments during the flight. Discovery returned to Earth at Edwards Air Force Base in California on October 3, 1988.

George Nelson spent more than 17 days in space, including more than 9 hours in EVA, during his career as an astronaut.

NEPTUNE: MAJOR FLIGHTS.

See Voyager 1 and Voyager 2

NERI VELA, RODOLFO

(1952–)

First Citizen of Mexico to Fly in Space

Born on February 19, 1952 in Chilpaningo, Guerrero, Mexico, Rudolfo Neri Vela was the first citizen of Mexico to fly in space.

He attended the University of Mexico, where he received a Bachelors Degree in mechanical and electronic engineering in 1975. He began his graduate studies at the University of Essex in England, and received a Ph.D. in electromagnetic radiation from the University of Birmingham in 1979. He also completed a year of postdoctoral research in waveguides at the University of Birmingham.

As a member of the Radiocommunications Group of the Institute of Electrical Research in Mexico, Neri Vela has done research and system planning involving antennas and satellite communications systems.

He has also led the Department of Planning and Engineering of the Morelos Satellite program at the Mexican Ministry of Communications and Transportation.

The first citizen of Mexico to fly in space, Dr. Rodolfo Neri Vela (standing, on right) is pictured here with STS-61 B crew mates (front, left to right): Bryan O'Connor, Brewster Shaw; (back, left to right): Charles Walker, Jerry Ross, Mary Cleave, and Sherwood Spring. [NASA/ courtesy of nasaimages.org]

During his flight aboard the shuttle Atlantis during STS-61B (November 26 to December 3, 1985), Neri Vela was payload specialist for the deployment of the Mexican MORELOS-B communications satellite.

Rudolfo Neri Vela spent more than 165 hours in space during the STS-61B mission.

Following his flight in space, Neri Vela devoted himself to full-time research and lecturing in the post-graduate program of the National University of Mexico, specializing in antenna theory and design, satellite communications systems, and Earth station technology.

NEW HORIZONS

(January 19, 1996–)
U.S. Pluto Exploration Mission

Designed to explore Pluto and its moon Charon and then enter the Kuiper Belt, the New Horizons spacecraft was launched by NASA on January 19, 2006 atop a multi-stage Atlas launch vehicle.

Carrying seven instruments to explore Pluto's atmosphere, energetic particle environment, geology, ionosphere, surface composition, and interaction with the solar wind, New Horizons is also equipped to search for an atmosphere around Charon and to locate and study at least one Kuiper Belt Object.

In addition to the Long Range Reconnaissance Imager, which is intended to capture high resolution images of Pluto and Charon, the spacecraft is also equipped with two instruments, a CCD imager/near-infrared imaging spectrometer and an ultraviolet imaging spectrometer, that were whimsically named "Ralph" and "Alice," in honor of the main characters of the 1950s television program "The Honeymooners," who were portrayed by Jackie Gleason and Audrey Meadows.

Other instruments include the Radio Science Experiment and a student-designed dust counter intended to measure dust in the outer solar system.

The New Horizons spacecraft made its closest approach to Jupiter in late February 2007, returning stunning color images of the planet's moons Io and Europa on its way to Pluto, where it is scheduled to arrive in July 2015. The mission profile calls for the spacecraft to conduct long range imaging of Pluto and Charon during the vehicle's closest approach to Pluto and its moon.

After its planned encounter with Pluto and Charon, New Horizons is expected to continue its mission for another 5 to 10 years, while it encounters at least one Kuiper Belt Object measuring at least 35 kilometers and returns data similar to that it is expected to collect during the first phase of the mission.

The total cost of the New Horizons mission is expected to be less than $550 million.

NEWMAN, JAMES H.

(1956–)

U.S. Astronaut

James Newman was born in the Trust Territory of the Pacific Islands, which has since become the Federated States of Micronesia, on October 16, 1956. In 1974, he graduated from La Jolla High School in San Diego, California and then attended Dartmouth University, where he graduated cum laude in 1978, earning a Bachelor of Arts degree in physics. He was awarded a citation for his Senior Thesis Research by Dartmouth upon graduation.

He pursued his graduate education at Rice University, where he received a Master of Arts in physics in 1982 and a Ph.D. in physics in 1984. In recognition of his superior academic performance, he was awarded the 1982–83 Texaco Fellowship, and he received the Sigma Xi Graduate Merit award in 1985.

Newman continued his research in atomic and molecular physics during a year of post-doctoral work at Rice, during which he designed, built, and tested a new position-sensitive detection system for measuring differential cross sections of collisions of atoms and molecules. He became an adjunct professor in the university's department of physics and astronomy in 1985.

He also began his association with NASA in 1985, as an instructor responsible for conducting flight crew and flight control team training at the Johnson Space Center (JSC) in Houston. In 1988, the agency awarded him its Superior Achievement Award, and in 1989 he was chosen to attend the summer session of the International Space University in Strasbourg, France. NASA selected him for training as an astronaut in 1990.

Newman's technical assignments at NASA have included service in the Astronaut Office Mission Support Branch and Mission Development Branch, and he has also been chief of the Computer Support Branch, where he was responsible for the integration of laptop computers in space shuttle and space station missions. He also served in the Space Shuttle Program Office, where he helped to develop the Space Vision System, an effort for which he shared in the 2001 Team Award of the Rotary National Award for Space Achievement Foundation and the 2002 NASA Group Achievement Award.

He first flew in space as a mission specialist aboard the space shuttle Discovery during STS-51 in September 1993. Launched on September 12, the STS-51 crew deployed the Advanced Communications Technology Satellite (ACTS) on the first day of the flight and then deployed the Orbiting Retrievable Far and Extreme Ultraviolet Spectrograph-Shuttle Pallet Satellite (ORFEUS-SPAS) the following day. The ACTS spacecraft was propelled into geosynchronous orbit by the Transfer Orbit Stage (TOS) booster rocket, which was used for the first time on STS-51.

Newman was responsible for overseeing the operation of the ORFEUS-SPAS, which was a joint project of the United States and German space programs. The first of a series of ASTRO-SPAS missions designed to make astronomical observations from space, the ORFEUS-SPAS instrument was controlled by managers at the SPAS Payload Operations Control Center at KSC, the first time that a shuttle payload was managed from the Center. The ORFEUS-SPAS craft was retrieved after six days of free flight and returned to Earth aboard the shuttle at the end of the flight.

On September 16, 1993, Newman and fellow mission specialist Carl Walz made a spacewalk of 7 hours, 5 minutes, and 28 seconds while testing tools and techniques that would later be used in the servicing of the Hubble Space Telescope (HST).

During the busy, productive STS-51 flight, the crew also tested a global positioning system (GPS) receiver and completed an exercise involving the routing of flight data to on-board laptop computers. The mission was extended one day because of weather concerns in Florida, and then, on September 22, 1993, the flight came to a close when Discovery touched down on Runway 15 at the KSC in Florida at 3:56 A.M., in the first nighttime landing of a shuttle at KSC.

On his second space mission, Newman served as a mission specialist aboard the shuttle Endeavour during STS-69. Launched on September 7, 1995 the STS-69 crew released and retrieved two free-flying satellites during 11 days in space.

The crew released the SPARTAN (Shuttle Pointed Autonomous Research Tool for Astronomy) 201–03 satellite on the second day of the flight, and retrieved it on flight day four. The SPARTAN 201–03 was the third in a series of four satellites designed to study the Sun's outer atmosphere and the solar wind.

The Wake Shield Facility-2 (WSF-2) made its second space shuttle flight during STS-69. Deployed by the crew on the fifth day of the mission, the WSF was designed to grow thin film semiconductor materials for use in advanced electronics.

During STS-69, the WSF was able to generate four thin films (of a planned seven) in three days of free flight before being retrieved on the eighth day of the mission.

As the crew member responsible for the spacecraft's scientific operations, Newman played a key role in the WSF deployment, and he also operated Endeavour's Remote Manipulator System (RMS) robotic arm to release and retrieve the WSF at the beginning and end of its flight. His other activities during the flight included tests of the Ku-band Communications Adaptor, the RMS Manipulator Positioning Display, and he also conducted the Relative GPS experiment, an accomplishment for which he was awarded the 1995 Superior Achievement Award of the Institute of Navigation, for "outstanding accomplishments as a Practical Navigator." The STS-69 flight came to a close with Endeavour's landing on September 18, 1995, after a flight of 10 days, 20 hours, and 28 minutes in space.

On his third spaceflight, aboard the shuttle Endeavour during STS-88 in December 1998, Newman and fellow mission specialist Jerry Ross made three remarkable spacewalks while they connected the first expansion module to the International Space Station (ISS).

The main goal for the first ISS assembly flight was to deliver the ISS Unity Node and connect it to the previously launched Zarya Control Module.

Built in Russia with U.S. financial backing, the Zarya module had been launched earlier from the Baikonur Cosmodrome in Kazakhstan, and it provided the station's electrical power, communications, and propulsion systems during the initial stage of ISS operations. The Unity module provided the necessary connection points for the later attachment of additional ISS modules and components.

Newman and Ross made the first of their three STS-88 spacewalks on December 7 when they attached the new module's cables to the existing Zarya Control Module during an EVA of 7 hours and 21 minutes. Two days later, on December 9, they spent another 7 hours and 2 minutes working in space while they installed the new module's antennas. Then, on December 12, they finished their amazing orbital installation work by placing a sunshade over an external computer, installing translation aids, and attaching tools and hardware to the Unity module, adding an additional 6 hours and 59 minutes to their total EVA time for the flight. In their three sessions working outside the ISS, Newman and Ross accumulated a remarkable 21 hours and 22 minutes in EVA.

The STS-88 crew also deployed two satellites, Mighty Sat 1 and SAC-A, during their busy, productive flight. Then, with their space construction chores successfully completed, they returned to Earth on December 15, 1998, after a flight of just under 12 days.

On March 1, 2002, Newman launched on a remarkable fourth space mission, aboard the shuttle Columbia during STS-109, the fourth servicing of the HST.

Among the most complex and ambitious spaceflights of the entire shuttle program, the HST servicing missions stretched the expertise and abilities of the astronauts and mission controllers to the utmost. Deployed during STS-31 in April 1990, the HST had previously been serviced during STS-61 in December 1993, STS-82 in February 1997, and STS-103, in December 1999.

Commanded by Scott Altman, the STS-109 crew included pilot Duane Carey and mission specialists Newman, Nancy Currie, John Grunsfeld, Rick Linnehan, and Mike Massimino.

In five remarkable spacewalks over the course of five consecutive days, Newman, Massimino, Grunsfeld, and Linnehan accumulated a total of 35 hours and 55 minutes in EVA while outfitting the HST with new solar arrays, a new camera, a new power control unit, and an experimental cooling system.

Grunsfeld and Linnehan made the flight's first EVA on March 4, replacing the telescope's starboard solar array in 7 hours and 1 minute.

On March 5, Newman and Massimino made their first spacewalk, spending 7 hours and 16 minutes in EVA while replacing the HST's port solar array and outfitting the instrument with a new Reaction Wheel Assembly.

The following day, March 6, Grunsfeld and Linnehan added another 6 hours and 48 minutes to the EVA total while replacing the Hubble's Power Control Unit.

Newman and Massimino then replaced the HST Faint Object Camera with the new Advanced Camera For Surveys (ACS) during a 7 hour and 30-minute EVA on March 7.

Then, in the remarkable flight's final spacewalk, Grunsfeld and Linnehan installed a new computer chip in the telescope's experimental cooling system, during an EVA of 7 hours and 20 minutes.

In their two spacewalks during the mission, Newman and Massimino logged a total of 14 hours and 46 minutes in EVA.

With the completion of the STS-109 fourth servicing of the Hubble Space Telescope, the total number of spacewalks devoted to servicing the HST had been increased to 18, representing a total EVA time of 5 days, 9 hours, and 10 minutes of spacewalking by 14 different astronauts. Columbia returned to Earth on March 12, 2002, after 165 orbits and a flight of 10 days, 22 hours, and 11 minutes in space.

James Newman has spent more than 43 days in space, including 43 hours and 13 minutes in EVA, during four space shuttle flights.

In December 2002 he was assigned to the International Space Station Program Office, to oversee the NASA's Russian human spaceflight operations as the agency's lead representative to the Russian Federal Space Agency.

NICOLLIER, CLAUDE

(1944–)

ESA Astronaut; First Citizen of Switzerland to Fly in Space

The first citizen of Switzerland to fly in space, Claude Nicollier, was born in Vevey, Switzerland on September 2, 1944. In 1962, he graduated from Gymnase de Lausanne in Lausanne, Switzerland, and then joined the Swiss Air Force, where he completed his training as a pilot in 1966.

He then attended the University of Lausanne, where he received a Bachelor of Science degree in physics in 1970. He worked at the university's Institute of

Astronomy and at the Geneva Observatory as a graduate scientist from 1970 to 1973 and attended the Swiss Air Transport School in Zurich, where he completed training as a commercial airline pilot in 1974. Continuing his education concurrently with his research and his work

as a DC-9 pilot for Swissair, he attended the University of Geneva, where he earned a Master of Science degree in astrophysics in 1975.

During his outstanding career as a military pilot, Nicollier has accumulated 5,600 hours of flying time, including 4,000 hours in jet aircraft. He is a captain in the Swiss Air Force.

Nicollier began his association with the European Space Agency (ESA) in 1976 when he received a fellowship to work on airborne infrared astronomy projects as a research scientist at the agency's Space Science department at Noordwijk, in the Netherlands.

In 1978, the ESA selected Nicollier as a member of its first group of astronauts. With the start of the joint ESA-NASA astronaut development program in 1980, he was assigned to the Johnson Space Center (JSC) in Houston, Texas, where he received training to fly aboard the American space shuttle as a mission specialist.

His initial technical assignments in the Astronaut Office at JSC included service in the Shuttle Avionics Integration Laboratory (SAIL), and he also helped to develop the space shuttle Remote Manipulator System (RMS) and the Tethered Satellite System (TSS), which flew on STS-46 in 1992 and STS-75 in 1996.

In 1988, Nicollier attended the Empire Test Pilot School at Boscombe Down in the United Kingdom, where he was trained as a test pilot.

On July 31, 1992, he became the first citizen of Switzerland to fly in space when he launched aboard the space shuttle Atlantis during STS-46. The STS-46 crew also included Franco Malerba, who was the first citizen of Italy to fly in space.

Nicollier deployed one of the flight's two primary payloads, the ESA's European Retrievable Carrier (EURECA), which would remain in orbit until its retrieval during STS-57 in 1993.

The flight also featured the first major test of tethered spaceflight, the TSS, which had been developed over the course of a decade by teams of engineers from NASA and the Italian space agency Agenzia Spaziale Italiana (ASI) in a program in which Nicollier had participated as a European astronaut assigned to the American space agency.

The concept of joining two spacecraft with a tether for linked flight had been tested in space as early as the Gemini program in 1966. The TSS exercise on STS-46 called for linked flight between the shuttle and a satellite that would be deployed at a different orbital altitude from Atlantis, with the tethered flight being conducted at a distance of as much as 12.5 miles.

Despite all the careful preparations, a mechanical hitch foiled the test, when the tether jammed after it had been unwound to a distance of just 840 feet. The crew attempted for several days to release the stuck line, but the experiment finally had to be written off as a failure. The TSS satellite was retrieved and returned to Earth. Atlantis landed at KSC on August 8, 1992.

In a spectacular nighttime launch on December 2, 1993, Nicollier launched aboard the space shuttle Endeavour for his second trip into space, during one of the most complex and ambitious missions of the entire shuttle program, the STS-61 servicing of the Hubble Space Telescope (HST).

Nicollier served as a mission specialist responsible for operating the shuttle's Remote Manipulator System (RMS) robotic arm during the landmark flight, which required a record five EVAs during which crew members spent more than 35 hours working in space. Richard Covey was commander of STS-61, and the crew also included pilot Kenneth Bowersox, payload commander Story Musgrave, and Nicollier's fellow mission specialists Kathryn Thornton, Jeffrey Hoffman, and Thomas Akers.

Endeavour caught up with the HST on the third day of the mission, and Nicollier used Endeavour's RMS robotic arm to grapple the instrument into the shuttle's cargo bay.

Then, working in teams during an amazing series of spacewalks over the course of five days, Musgrave and Hoffman alternated with Akers and Thornton to service and repair the telescope.

Musgrave and Hoffman made the first EVA, working on the HST for 7 hours and 54 minutes while they replaced the instrument's Rate Sensing Units (RSUs)—the devices containing the HST's gyroscopes—and their underlying electronics and electrical hardware.

Akers and Thornton made their first trip outside the shuttle the following day, replacing the HST's solar array panels during an EVA of 6 hours and 35 minutes.

Musgrave and Hoffman replaced the telescope's Wide Field/Planetary Camera and installed two new magnetometers on the telescope in their next EVA, which took 6 hours and 48 minutes.

Akers and Thornton then made their second spacewalk of the flight—the fourth overall—on the seventh day of the mission. In an EVA of 6 hours and 50 minutes, they removed another of the HST's instruments, the high-speed photometer, and replaced it with a corrective device designed to help overcome the flaws in the telescope's primary mirror. They also installed a co-processor in the computer that controls the HST.

On flight day eight, Musgrave and Hoffman made the fifth spacewalk of the mission, installing the electronics for the new solar arrays and working on the HST for a total 7 hours and 21 minutes, raising the total spacewalking time for the flight to a remarkable 35 hours and 28 minutes.

Nicollier again used the RMS to re-deploy the fully restored HST the following day. Endeavour landed at the Kennedy Space Center in Florida in the early hours of December 13, 1993.

Following his historic second spaceflight, Nicollier was appointed professor at the Swiss Federal Institute of Technology in Lausanne.

He again flew in space during STS-75, aboard the shuttle Columbia in 1996. Launched on February 22, STS-75 featured the second test of the Tethered Satellite System, TSS-1R

After a day's delay because of difficulties with the experiment's on-board computer, the deployment of the TSS satellite began well, returning useful data until the satellite was almost fully extended to its planned distance of 12.8 miles. Unfortunately, the tether broke just prior to the completion of the deployment, ending the experiment, but the effort still yielded important data, including confirmation of the fact that tethers can produce electricity while in linked flight in the Earth's ionosphere.

The STS-75 mission also featured the third flight of the United States Microgravity Payload (USMP-3), and the crew worked around the clock to complete their scientific research. Columbia landed on March 9, 1996.

After his third flight, Nicollier was assigned to head the Robotics Branch of the Astronaut Office at the Johnson Space Center.

On December 19, 1999, Nicollier lifted off on a remarkable fourth spaceflight as a mission specialist aboard the shuttle Discovery for the third Hubble Space Telescope maintenance and repair mission. Curtis Brown commanded the flight, and the crew included pilot Scott Kelly, NASA astronauts Michael Foale, John Grunsfeld, and Steve Smith, and ESA astronauts Nicollier and Jean-Francois Clervoy of France.

Clervoy used the shuttle's RMS to capture the HST on December 21. Then, working in teams—Nicollier paired with Foale, and Grunsfeld with Smith—the spacewalking astronauts made long EVAs that totaled 24 hours and 33 minutes over the course of three days while replacing key parts of the space telescope, which had been launched in April 1990 during STS-31 and previously serviced in December 1993 (STS-61) and February 1997 (STS-82).

Grunsfeld and Smith made the first and third EVAs, on December 22 and 24, and Nicollier and Foale spent 8 hours and 10 minutes working on the HST on December 23, 1999 while installing a new computer and a 550-pound fine guidance sensor. With their repair and maintenance duties successfully completed, the STS-103 crew returned to Earth on December 27, 1999, after 120 orbits.

Claude Nicollier has spent more than 1,000 hours in space, including 8 hours and 10 minutes in EVA, during his remarkable career as an astronaut.

In the years since his fourth spaceflight, he has continued to serve NASA in the EVA Branch of the Astronaut Office, while also serving as the agency's Lead ESA astronaut.

NIE, HAISHENG (NIE HAISHENG)

(1964–)

Chinese Taikonaut

A member of the first two-person crew launched by the space program of the People's Republic of China, Haisheng Nie was born in Zaoyang, Hubei Province on September 8, 1964. He joined the Chinese People's Liberation Army (PLA) Air Force and received his initial training as a pilot in 1984.

Nie attended the Air Force Number 7 Flying School, from which he graduated in 1987, and then served as a squadron commander, deputy group commander, and master navigator. He first flew fighter aircraft in 1989, and in January 1998 he was chosen as one of the first group of taikonauts (Chinese cosmonauts). He served as a backup crew member for the historic Shenzhou V flight in 2003, the first manned mission of the Chinese space program.

On October 12, 2005, Nie lifted off from the Jiuquan Satellite Launching Center in the Gobi Desert aboard Shenzhou VI with Junlong Fei. Nie served as flight engineer for the mission (Fei was commander), which was the first Chinese spaceflight to feature a crew of two. The taikonauts made 76 orbits during the flight and returned to Earth on October 16, 2005 with a landing in Siziwang Banner in Inner Mongolia.

Haisheng Nie spent 4 days, 19 hours, and 32 minutes in space during his landmark first spaceflight. A lieutenant colonel in the Chinese Air Force, he has continued to serve as a taikonaut since his Shenzhou VI mission.

NIKOLAYEV, ANDRIAN G.

(1929–2004)

Soviet Cosmonaut

A pioneering cosmonaut and veteran of two spaceflights, Andrian Nikolayev was born in the village of Shersbely, in the autonomous Chuvash republic in the Mariinski Posad district of Russia, on September 5, 1929. He completed his secondary education in 1944 and then pursued a career in the timber industry. He received vocational training at the Mariinski Posad Timber industry training school, and worked for more than two years at the Dereviansk state timber industry enterprise.

Nikolayev then joined the Soviet Air Force and attended the Kirovobad Higher Air Force Academy, where he graduated in 1950 with a specialization in aircraft gunnery. He also attended the Tchernigov Higher Air Force Academy and subsequently served in the Khmelnitskaya region as an airborne gunner.

He went on to attend the Frunze Higher Air Force Academy, from which he graduated in 1954. He served as a pilot and senior pilot, and in February 1958 he was assigned to Air Defense Engineering Aviation Regiment 401 as a squadron adjutant and senior pilot. During his exceptional military career, he rose to the rank of major general in the Soviet Air Force.

Andrian Nikolayev (right) with NASA astronaut Thomas Stafford during training for the Apollo-Soyuz Test Project (ASTP). Nikolayev first flew in space during the August, 1962 flight of Vostok 3—the Soviet Union's third manned space launch. [NASA/cour tesy of nasaimages.org]

He was chosen for training as a cosmonaut as a member of the first group of cosmonaut candidates, Air Force Group One, in March 1960. After a period of intensive training, he became a cosmonaut in January 1961. His initial assignments as a cosmonaut included service as the backup to Gherman Titov during the Vostok 2 mission, which took place in August 1961, and he also attended the Zhukovsky Higher Military Engineering Academy from 1961 to 1964.

On August 11, 1962, he lifted off from the Baikonur Cosmodrome on his first spaceflight, as pilot of Vostok 3. His launch was followed the next day by the launch of Pavel Popovich aboard Vostok 4; together, the two cosmonauts achieved the world's first group flight of manned spacecraft.

Neither Vostok was equipped for rendezvous or docking, but Vostok 3 and Vostok 4 were able to achieve simultaneous flight in orbit at a reasonably close distance, a remarkable advance in the history of spaceflight, particularly in light of the fact that the two flights were just the seventh and eighth manned spaceflights ever.

The day before he began his historic flight, Nikolayev requested and received permission to enter Vostok 3 for a brief, final pre-flight check; the engineers responsible for preparing the vehicle were afraid that his presence might disturb the craft's equipment or systems, which had by that time been inspected and approved as ready for launch.

When he lifted off the next day, he felt a mild jolt at the first separation of the launch vehicle's stages; the force increased substantially as the subsequent stages were jettisoned. Vostok 3 entered an orbit of 166 kilometers by 218 kilometers, and from that perspective, Nikolayev awaited the launch of Popovich in Vostok 4.

Popovich launched a little after 8:00 A.M. on August 12, and as his spacecraft was inserted into orbit, the two cosmonauts made the closest approach of their joint mission, passing each other at a distance variously estimated as close as fewer than five kilometers or at six and a half kilometers.

Nikolayev and Popovich then made their first radio contact in orbit, a major accomplishment of the flight, and a reassuring exercise for the cosmonauts. Nikolayev had earlier experienced difficulties with his communications system when Soviet leader Nikita Khrushchev had tried to contact him to offer congratulations on the successful launch.

After achieving the close pass with Popovich, Nikolayev settled in for a longer flight than any yet attempted up to that time. The ultimate length of his mission had been the focus of an intense debate within the Soviet space program, but as everything continued to go well—despite a precipitous drop in temperature within Popovich's Vostok 4—mission managers decided to extend both flights. Nikolayev received word that he was approved to try for a total flight of four days and 64 orbits.

As he continued on in his orbital odyssey, followed throughout by Popovich in Soyuz 4, he completed the program of Earth observations and photography he had been assigned, surprising some with the degree of detail with which he could distinguish landmarks such as cities, roads, and ships at sea.

After nearly four days in orbit, and having successfully completed the 64 orbits to which his mission had been extended, Nikolayev re-entered the Earth's atmosphere in Vostok 3 on August 15, 1962. He ejected from the craft shortly before it landed, at 6:52 A.M., just minutes before Vostok 4 landed. Nikolayev and Popovich each ejected from their respective craft and landed separately via parachute.

Nikolayev accumulated 3 days, 22 hours, and 22 minutes in space during his landmark Vostok 3 flight. He and Popovich were widely celebrated for their remarkable achievement.

Following his first spaceflight, Nikolayev received the rating of first class pilot. He continued to serve as a cosmonaut and trained for several years as commander of the backup crew for the Soyuz 8 mission, which, in October 1969, was part of a group flight that placed three spacecraft (Soyuz 6, 7, and 8) in orbit simultaneously.

Nikolayev began his second flight in space on June 1, 1970 when he launched from the Baikonur Cosmodrome as commander of Soyuz 9 with flight engineer Vitali Sevastyanov. Designed primarily as a means of gathering medical data about longer stays in space in preparation for the launch of the first Salyut space station the following year, Soyuz 9 also featured the first nighttime launch of the manned Soviet space program.

In orbit, Nikolayev and Sevastyanov conducted a demanding program of medical research that involved frequent tests and strenuous exercise designed to combat the losses in blood plasma and bone mass that occur during long stays in space. Working 16 hours each day throughout the 18-day flight, they also carried out a program of additional scientific work, which included psychological tests, navigation exercises, and photography assignments.

Because the spacecraft had to be oriented toward the Sun in order for its solar arrays to operate at peak electrical efficiency, Soyuz 9 was spin-stabilized—kept in a slow, rotating motion—for much of the flight, which resulted in a variety of unpleasant sensations for the crew members, who reportedly suffered fairly severe discomfort when they returned to Earth.

The cosmonauts did receive a break from their intense, demanding routine when they were given a day off from their medical observations and scientific experiments at about the halfway point of their flight. Nikolayev and Sevastyanov spent their brief leisure time in orbit reading and playing chess before returning to the strict routine that had been set for them.

As he had during his first spaceflight, Nikolayev set a new record for the longest stay in space, with Sevastynov, when they returned to Earth on June 19, 1970 after a flight of 17 days, 16 hours, 58 minutes, and 55 seconds. Although neither cosmonaut suffered permanent damage to his health, the long flight left them both with intense fatigue, and they were hospitalized for 10 days after landing before they were able to return to their normal activities.

With his flight aboard Soyuz 9, Nikolayev helped to amass a wealth of important medical data that would prove invaluable to mission planners who were developing plans for future space station missions.

He was given the rating of cosmonaut second class in 1970, and in 1975 he received the degree of Candidate of Technology.

Andrian Nikolayev spent more than 21 days in space during his two historic space missions.

He left the cosmonaut corps in 1982, but continued to serve the Soviet space program as the first deputy chief cosmonaut at the Gagarin Cosmonaut Training Center (GCTC), a capacity in which he served until August 1992, when he reached the age limit for active duty service and transferred to the reserve forces as a major general in the Air Force.

Among the many honors he received for his remarkable cosmonaut career, he was made an honorary citizen of 14 cities in Russia, as well as one each in Algiers and Mongolia, and a crater on the Moon was named in his honor.

Nikolayev was married to fellow cosmonaut Valentina Tereshkova. He died as the result of a heart attack on July 3, 2004.

NOGUCHI, SOICHI

(1965–)

Japanese Astronaut

A veteran of three spacewalks during the STS-114 return-to-flight mission of the U.S. space shuttle Discovery in 2005, Soichi Noguchi was born in Yokohama, Kanagawa, Japan in 1965. In 1984, he graduated from Chigasaki-Hokuryo High School and then attended the University of Tokyo, where he earned a Bachelor of Engineering degree in aeronautical engineering in 1989 and a Master of Engineering degree in aeronautical engineering in 1991.

He began his career in the manufacturing department of Ishikawajima-Harima Heavy Industries Company, Ltd. (IHI), and was subsequently assigned to the research and development department for aero-engine and space operations in the company's aerodynamics group.

The National Space Development Agency of Japan (NASDA) selected Noguchi for training as an astronaut in June 1996. He underwent two years of training at the NASA Johnson Space Center (JSC) in Houston, Texas for flights aboard the U.S. space shuttle as a mission specialist, and then transferred to the Gagarin Cosmonaut Training Center (GCTC) in Star City, Russia, where he received training for flights aboard the Russian Soyuz spacecraft. He has also been trained to operate the systems and equipment of Japan's Kibo scientific module for the International Space Station (ISS).

On July 26, 2005, Noguchi lifted off from the Kennedy Space Center (KSC) in Florida on his first flight in space, as a mission specialist aboard the space shuttle Discovery during STS-114, the first shuttle flight after the loss of the space shuttle Columbia in February 2003.

Commanded by Eileen Collins, the STS-114 crew also included pilot James Kelly and Noguchi's fellow mission specialists Charles Camarda, Wendy Lawrence, Stephen Robinson, and Andrew Thomas.

Despite an incident at launch in which a piece of foam insulation nearly hit the shuttle, an occurrence frighteningly similar to the case of the Columbia accident two years earlier, in which an insulation strike at launch had caused the shuttle to break up during its later return to Earth, the STS-114 Discovery crew successfully delivered supplies to the International Space Station (ISS), helped with repair and maintenance tasks at the station, and conducted a series of spacewalks designed to test potential repairs of the kind of damage that had caused the loss of the shuttle Columbia.

NASA officials evaluating the launch data conceded that the failure to properly secure the foam insulation was cause enough to postpone planned future shuttle missions, and on July 27, 2005—while STS-114 was in progress—the agency announced that it was grounding the space shuttle fleet until a better solution to the foam insulation problem could be found.

Noguchi and fellow mission specialist Stephen Robinson made the first of their three STS-114 spacewalks on July 30. In an exercise designed to test potential in-flight responses to the sort of damage that had caused the loss of the shuttle Columbia, they tested a variety of repair techniques on pre-damaged samples of thermal tiles and panels. They used a paint-like substance on the tiles and pushed a dark colored filler material into the cracks and crevices on the larger samples.

After several years of training to prepare for the flight, Noguchi and Robinson were comfortable working together and performed their unique tasks with precision and good humor. At one point, they reported the consistency of the paste as being similar to pizza dough.

They also repaired one of the four ISS Control Moment Gyroscopes during their busy first spacewalk, and replaced the station's Global Positioning System (GPS) antenna. The work took a total of 6 hours and 50 minutes.

On August 1 they added another 7 hours and 14 minutes to their STS-114 EVA total, while replacing a second gyroscope at the station.

Then, on August 3, they made a remarkable third spacewalk in which Robinson made the first-ever in-flight repair to the underside of a space shuttle. Noguchi monitored his partner's progress from outside the ISS while Robinson was positioned beneath the shuttle atop Discovery's Remote Manipulator System (RMS) robotic arm. Working deliberately and taking his time, Robinson removed two strips of ceramic-fiber filler that had come loose between several of the shuttle's protective tiles.

They also attached a work platform and the fifth Material International Space Station Experiment (MISSE) to the outside of the ISS during the EVA, which took a total of 6 hours and 1 minute.

After nine days of docked operations and the completion of the complex spacewalking activities, Discovery undocked from the ISS and began its return trip to Earth. Although there had been no compelling evidence that the shuttle had suffered damage at launch, the incident involving the loose insulation and subsequent grounding of the shuttle fleet gave rise to intense public concern about the safety of the crew as the landing approached.

Poor weather delayed the scheduled landing by one day, and caused the planned landing to be diverted from KSC in Florida to Edwards Air Force Base in California, which caused more consternation. NASA officials altered the shuttle's flight path so that Discovery would not pass directly over the city of Los Angeles, in response to public fears about the possibility of another disaster like the Columbia accident, which had showered debris over several southern U.S. states.

Despite the concern, Discovery touched down safely and without incident in a nighttime landing in the early morning hours of August 9, 2005.

Soichi Noguchi spent 13 days, 21 hours, 32 minutes, and 48 seconds in space, including 20 hours and 5 minutes in EVA, during the STS-114 mission.

Since his first spaceflight, Noguchi has continued his astronaut career as a representative of the Japan Aerospace Exploration Agency (JAXA).

NOZOMI

(July 3, 1998–December 9, 2003)
Japanese Spacecraft Designed to Study Mars

A project of the Institute of Space and Aeronautical Science (ISAS) at the University of Japan, the Nozomi spacecraft was designed to orbit Mars and study the planet's upper atmosphere and to test equipment and technologies for use in future planetary explorations.

Known prior to launch as Planet-B, Nozomi lifted off on July 3, 1998. The spacecraft's trip to Mars was delayed several times by frustrating malfunctions, beginning with a faulty valve that caused it to lose fuel during a flyby of Earth that had been planned to set the craft on a trajectory for Mars.

Mission controllers adapted admirably to the crisis, settling on a new plan that would have Nozomi in heliocentric orbit for four more years, resulting in an encounter with Mars in December 2003 at a slower relative velocity, but that plan was waylaid on April 21, 2002 when solar flares wreaked havoc with the craft's communications and power systems and damaged the heating system that normally prevented its hydrazine fuel from freezing. The fuel thawed during flybys of the Earth and Sun, but on December 9, 2003 preparations for Nozomi's insertion into orbit around Mars failed and the mission had to be abandoned.

OBERTH, HERMANN J.

(1894–1989)
Pioneering Scientist and Spaceflight Theorist

A pioneering scientist who made significant contributions to the theoretical framework that led to the first spaceflights and influenced other space visionaries, Hermann Oberth was born in Nagyszeban, Transylvania on June 25, 1894. The city of his birth was also known as Hermannstadt in German (Oberth was of German descent), and subsequently became Sibiu, Romania in the aftermath of World War I.

Oberth pursued an interest in rocketry and spaceflight from an early age. He began his collegiate studies at the University of Klausenburg in Germany in 1912, where he initially trained for a career in medicine. He served as a medic during World War I, and then continued his education after the war's end, switching his field of study from medicine to physics.

When Oberth completed his studies and presented his doctoral thesis, a treatise on the use of rocket technology as a means of launching a vehicle into space, his work was rejected as too fantastic to be worthy of serious study and was deemed irrelevant to any recognized academic discipline.

Despite the criticism, he published his work in a limited edition in 1923, as *Rocket Into Interplanetary Space,* and he later expanded the original document into the more comprehensive *Ways to Space Flight* (1929).

Oberth was ultimately awarded a doctorate degree for his original work by the Babes-Bolyai University in Romania. Non-academics also recognized him as an important scientific innovator. He was sought out by the seminal German film

Pioneering space scientist Hermann Oberth. [NASA/courtesy of nasaimages.org]

director Fritz Lang to act as a consultant for the film "The Woman in the Moon" (1929), which featured one of the earliest depictions of spaceflight on film.

After the controversy involving his doctoral work, Oberth continued his academic career as a professor at the Berlin Technical University, where he counted a young Wernher von Braun among his students. Working together with the bright young scientists on whom he was to have a lasting influence, Oberth launched his first liquid fuel rocket in 1929.

In 1938, he accepted a new position as a professor at the Technische Hochschule in Vienna, and he subsequently worked at the Technische Hochschule in Dresden before the outbreak of World War II.

During the war, Oberth was briefly involved in the development of the German V-2 vengeance weapon at Peenemunde, along with his former student von Braun, who was heading the project. Oberth also worked on solid fuel anti-aircraft rockets for the German Army during the war. He remained in Germany in the years following the end of the war before moving to Switzerland in 1948.

The post-war Italian Navy called on Oberth for help in the development of anti-aircraft systems similar to those he had worked on in Germany, and he lived and worked in Italy for three years while employed by the Navy as a consultant.

In 1953, he published another groundbreaking compendium of his forward-looking ideas about space travel. Among the concepts he outlined in *Man in Space,* he envisioned a space station, spacesuits, and a space-based telescope.

In the mid-1950s, Oberth's former student Wernher von Braun, who by that time was leading the U.S. Army engineers in developing the Redstone launch vehicle that would ultimately be used in the first manned American space program, recruited him again, this time for Project Mercury. Working alongside his former student at the Redstone Arsenal in Huntsville Alabama, Oberth contributed a wealth of innovative ideas to the early days of the United States' space effort. He also served as a consultant to the Convair Corporation during the development of the Atlas rocket, from 1960 to 1962.

After a remarkable—and often difficult—career of four decades as a physicist and visionary spaceflight theoretician, Oberth retired in 1962. He enjoyed a long and fruitful retirement and wrote several more books, including *A Primer For Those Who Would Govern,* before his death on December 28, 1989. He was survived by his wife Tilli and two of their children; a son and a daughter preceded him in death, during World War II.

From the earliest years of the space age, when the engineers of the Soviet and U.S. space programs proved that his theories about spaceflight were sound, Hermann Oberth increasingly received the respect he deserved for his forward-looking fascination with space. A museum in Feucht, Germany celebrates his career and his work, and the Hermann Oberth Society carries on his scientific legacy. He has also been accorded the honor of having a crater on the Moon named for him.

OCHOA, ELLEN

(1958–)

U.S. Astronaut

The first Hispanic woman to fly in space, Ellen Ochoa was born in Los Angeles, California in 1958. In 1975, she graduated from Grossmont High School in La Mesa, California and then attended San Diego State University, where she received a Bachelor of Science degree in physics in 1980. She then attended Stanford University, from which she received a Master of Science degree in electrical engineering in 1981 and a Ph.D. in electrical engineering in 1985.

Her doctoral research, which involved the use of optical systems in information processing, led her to work as a researcher at Sandia National Laboratories and at NASA's Ames Research Center in California. She has also collaborated on three inventions for which she has shared patents as a co-inventor, developing an optical inspection system, a method for optical object recognition, and a novel method for removing noise in images. She is a member of the Optical Society of America (OSA).

During her time at the Ames Research Center, she supervised the work of 35 engineers and scientists in the development of computational systems for aerospace missions while serving as chief of the center's Intelligent Systems Technology Branch.

Ochoa became an astronaut in 1991. Her technical assignments at NASA included service as the crew representative for flight software, computer hardware, and robotics, assistant to the chief of the Astronaut Office for space station operations, and acting deputy chief of the Astronaut Office.

On April 8, 1993, Ochoa became the first Hispanic woman to fly in space when she launched aboard the shuttle Discovery for the STS-56 flight of the Atmospheric Laboratory for Applications and Science (ATLAS-2).

Designed to study the interaction between the Sun and the Earth's middle atmosphere and their impact on the Earth's ozone layer, the ATLAS-2 laboratory consisted of seven instruments, six of which were mounted on a Spacelab pallet in the shuttle's cargo bay. The flight also featured a study of the solar corona, which was carried out by a free-flying SPARTAN satellite. Ochoa played a key role in the SPARTAN deployment and retrieval, using the shuttle's Remote Manipulator System (RMS) robotic arm to release and then capture the satellite at the start and end of its flight.

Discovery's landing was postponed by one day because of poor weather at the Kennedy Space Center (KSC) in Florida, which extended the total mission duration to 9 days, 6 hours, 8 minutes, and 24 seconds, with the shuttle touching down at KSC on April 17, 1993.

Ochoa served as payload commander on her second spaceflight, the STS-66 ATLAS-3 mission in November 1994. Launched on November 3 aboard the shuttle Atlantis, the STS-66 crew worked around the clock to operate the Atmospheric Laboratory for Applications and Science (ATLAS-3), which made its third shuttle flight, and the Cryogenic Infrared Spectrometers and Telescopes for the Atmosphere-Shuttle Pallet Satellite (CRISTA-SPAS), which had been jointly developed by NASA and the German space agency Deutsches Zentrum für Luft- und Raumfahrt e.V. (DARA).

Using the instruments of the ATLAS-3 and the CRISTA-SPAS equipment, the crew gathered data about the Earth's atmosphere and the Sun's energy output and how they affect the Earth's ozone layer. The overall program of study, including the earlier flights of the instruments, was designed to study the Earth's energy balance and atmospheric change during an 11-year solar cycle.

At the end of STS-66, Atlantis returned to Earth at Edwards Air Force Base in California on November 14, 1994, after 175 orbits.

Ochoa's third space mission was STS-96, which launched aboard the space shuttle Discovery on May 27, 1999. The historic first docking of a shuttle and the International Space Station (ISS), STS-96 brought replacement parts and equipment to the ISS in preparation for its eventual habitation by the first long-duration crew. Ochoa coordinated the transfer of supplies during six days of linked flight, as the STS-96 crew transferred 3,000 pounds of equipment and supplies to the station.

She also operated Discovery's Remote Manipulator System (RMS) robotic arm to support Daniel Barry and Tamara Jernigan while they made a 7 hour and 55-minute spacewalk on May 30 to install cranes on the ISS for use in future EVAs at the station. Discovery returned to Earth on June 6, 1999.

On April 8, 2002, she lifted off on a remarkable fourth spaceflight as a mission specialist aboard the shuttle Atlantis during STS-110, the 13th shuttle flight to the ISS. The STS-110 crew delivered and installed the space station's S-Zero (S0) Truss, an effort that required four spacewalks, which were all performed for the first time from the ISS Quest Airlock rather than from the shuttle itself.

Jerry Ross and Lee Morin made two of the four EVAs, alternating with Steven Smith and Rex Walheim over the course of six days. Ochoa operated the ISS robotic arm to support the truss installation and to move crew members during three of the four spacewalks. Atlantis landed on April 19, 2002.

Ellen Ochoa spent more than 40 days and 18 hours in space during four space missions.

For her achievements as an astronaut and engineer, she has been honored with the Hispanic Engineer Albert Baez Award for Outstanding Technical Contribution to Humanity and the Hispanic Heritage Leadership Award.

She has continued to serve NASA since her fourth spaceflight, and in 2007 she was named deputy director of the Johnson Space Center (JSC) in Houston, Texas.

OCKELS, WUBBO J.

(1946–)

ESA Astronaut; First Citizen of the Netherlands to Fly in Space

The first citizen of The Netherlands to fly in space, Wubbo Ockels was born in Almelo, The Netherlands on March 28, 1946. He attended the University of Groningen, where he earned a Ph.D. in physics and mathematics in 1973.

Ockels began his career as a scientist at the Nuclear Physics Accelerator Institute in Groningen, where he worked for five years. In 1978, he completed a doctoral thesis based on his advanced research at the institute.

The European Space Agency (ESA) chose Ockels for training as a payload specialist astronaut, as one of three candidates for the first flight of the ESA-designed Spacelab module aboard the U.S. space shuttle. After a year of intensive training at NASA's Johnson Space Center (JSC) in Houston, Texas, he served as a backup payload specialist and spacecraft communicator (Capcom) for the STS-9 Spacelab flight.

On October 30, 1985, Ockels became the first citizen of The Netherlands to fly in space when he lifted off from the Kennedy Space Center (KSC) in Florida during the STS-61A flight of the space shuttle Challenger, which was devoted to the German scientific module Spacelab D-1.

The STS-61A mission was commanded by Henry Hartsfield, and the crew also included pilot Steven Nagel, mission specialists James Buchli, Guion Bluford, Jr., and Bonnie Dunbar, and Ockels's fellow payload specialists Reinhard Furrer and Ernst Messerschmid of West Germany (Germany was at the time still divided into East and West, prior to the country's reunification after the fall of the Soviet Union).

The mission featured a unique control arrangement in which the shuttle's flight was controlled as usual from JSC Mission Control and the Spacelab operations were controlled by German space officials at the German Space Operations Center at Oberpfaffenhofen in West Germany.

During the flight, the STS-61A crew conducted 75 experiments in basic and applied microgravity research in the Spacelab module. The scientific program included investigations in communications, life sciences, materials science, and navigation, and also included an innovative Vestibular Sled experiment featuring a sled on which an astronaut could ride along a set of rails while scientists studied the individual's reaction in the space environment.

Wubbo Ockels spent 7 days, 44 minutes, and 53 seconds in space during the STS-61A mission.

Ockels has continued to serve the ESA in the years since his historic space shuttle flight, working in the agency's European Space Technology Center (ESTEC) in Noordwijk, The Netherlands, where he leads the ESA office for educational projects outreach activities. He is also a member of the aerospace faculty at the Delft University of Technology.

ONIZUKA, ELLISON S.

(1946–1986)

U.S. Astronaut

Ellison Onizuka was born in Kealakekua, Kona, Hawaii on June 24, 1946. In 1964, he graduated from Konawaena High School in Kealakekua and he then attended the University of Colorado, where he received a Bachelor of Science degree in aerospace engineering in June 1969 and a Master of Science degree in aerospace engineering in December 1969.

During his time at the University of Colorado he was a participant in the ROTC program, and he was recognized as a distinguished military graduate. He entered the U.S. Air Force in January 1970 and served at the Sacramento Air Logistics Center at McClellan Air Force Base in California as an aerospaceflight test engineer.

He attended the U.S. Air Force Test Pilot School, where he graduated in 1975, and then served as squadron flight test engineer and subsequently as chief of the engineering support section of the training resources branch of the Air Force Flight Test Center at Edwards Air Force Base in California. In the latter capacity, he taught at the Test Pilot School and managed flight test modifications to the fleet of aircraft used by the school and the flight test center.

During his outstanding military career, Onizuka received a number of U.S. Air Force honors, included the Commendation Medal, the Meritorious Service Medal, the Outstanding Unit Award, and the Organizational Excellence Award. He was also a recipient of the National Defense Service Medal. He accumulated more than 1,700 hours of flying time and was a lieutenant colonel in the U.S. Air Force at the time of his death during the space shuttle Challenger accident in January 1986.

NASA selected Onizuka for training as an astronaut in 1978. His initial technical assignments at the agency included service as a member of the test and checkout teams and launch support crews for the first two shuttle missions, STS-1 and STS-2, in 1981. He also served in the Shuttle Avionics and Integration Laboratory (SAIL).

Onizuka first flew in space as a mission specialist aboard the space shuttle Discovery during STS-51C, the first shuttle mission dedicated to the classified activities of the U.S. Department of Defense, and the 15th mission of the space shuttle program.

Launched on January 24, 1985, STS-51C was originally planned to be flown aboard the shuttle Challenger, but problems with Challenger's thermal tiles made it necessary to shift the flight to Discovery. The flight was also delayed for a day by freezing weather conditions.

The shift from Challenger to Discovery and concerns about the cold weather became significant in retrospect a year later, when freezing weather conditions contributed to the accident that caused the loss of the STS-51L crew in the Challenger explosion, on January 28, 1986.

Apollo 16 and STS-4 veteran Thomas "Ken" Mattingly served as commander of STS-51C, and the crew also included pilot Loren Shriver, mission specialists Onizuka and James Buchli, and payload specialist Gary E. Payton. Except for the veteran Mattingly, the entire crew was making its first flight in space.

Although the shuttle missions devoted to the Department of Defense were considered classified and little information is known about their payloads, NASA did reveal that the STS-51C crew successfully deployed a U.S. Air Force Inertial Upper Stage booster during the flight. Discovery landed at the Kennedy Space Center (KSC) in Florida on January 23, 1985, after completing 48 orbits over 3 days, 1 hour, 33 minutes, and 23 seconds in space.

Onizuka was next assigned as a mission specialist for the STS-51L flight of the space shuttle Challenger, which launched on January 28, 1986 at the KSC in Florida.

Tragically, Onizuka and his fellow Challenger crew mates were killed when a fault in one of the shuttle's solid rocket boosters caused a fuel leak that ignited and caused a massive explosion just 73 seconds after liftoff. Onizuka was 39 at the time of his death.

Ellison Onizuka was survived by his wife, Lorna, and their two children. He was posthumously awarded the Congressional Space Medal of Honor.

ONUFRIENKO, YURI I.

(1961–)

Russian Cosmonaut

Yuri Onufrienko was born on February 6, 1961 in Ryasnoye, in the Zolochevsk district, Kharkov region of the Ukraine.

In 1982, he received a pilot-engineer's diploma at the V. M. Komarov Eisk Higher Military Aviation School for Pilots. He then served as a pilot and senior pilot in the Air Force, and was selected for cosmonaut training in 1989. After completing his general space training, he began preparation as a test cosmonaut in April 1991.

Concurrent with his military career and his training at the Gagarin Cosmonaut Training Center (GCTC), Onufrienko also continued his education, and in 1994, he received a degree in cartography from the Moscow State University.

He has risen to the rank of colonel in the Air Force, and has logged more than 800 hours of flying time as a pilot in a variety of aircraft.

In 1994, Onufrienko served as backup commander to the Mir 18 crew. He made his first flight in space as commander of Mir 21, launching with Yury Usachev aboard Soyuz TM-23 on February 21, 1996.

Onufrienko and Usachev made a remarkable six EVAs during their Mir 21 mission. Each of the first three, on March 15, May 20, and May 24, had the cosmonauts working outside the station for more than five hours at a time. On May 30, they added more than four additional hours to their spacewalking total, as they worked on the station's recently-delivered Priroda science module (designed to add remote sensing capability to Mir and to support joint U.S. and Russian science activities, Priroda was docked to the space station on April 26, 1996). The cosmonauts made two additional spacewalks in June, adding more than nine more hours to their EVA total during the mission.

Another highlight of their six-month stay aboard Mir came on March 24, when the space shuttle Atlantis docked with the station for the third time, during shuttle flight STS-76. More than 1,500 pounds of supplies and two tons of equipment were transferred from the shuttle to the station during five days of linked flight, and U.S. astronaut Shannon Lucid transferred from the shuttle to Mir to begin her record-breaking long-duration mission aboard the station.

At the end of their Mir 21 mission, Onufrienko and Usachev returned to Earth on September 2, 1996 accompanied by European Space Agency (ESA) astronaut Claudie Andre-Deshays (later known as Claudie Haigneré, following her marriage to fellow French astronaut Jean-Pierre Haigneré). Andre-Deshays' flight, which began on August 17 when she launched with the Mir 22 crew aboard Soyuz TM-24, was an historic milestone, as she became the first French woman to fly in space.

Onufrienko and Usachev spent a total of 193 days, 19 hours, and 7 minutes in space during the flight, and accumulated 30 hours and 21 minutes in EVA during their six spacewalks.

Onufrienko next flew in space as commander of International Space Station (ISS) Expedition 4. Launched aboard space shuttle Endeavour on December 5, 2001 with fellow Expedition 4 crew mates Daniel Bursch and Carl Walz of the United States and a full shuttle crew, Onufrienko served on the ISS for more than six months.

At 196 days, the long mission constituted a new record for spaceflight endurance for U.S. astronauts Bursch and Walz; for Onufrienko, the trip was only about two days longer than his 1996 Mir 22 flight.

The crew performed scientific experiments during the mission and tested the station's systems and equipment. Onufrienko, wearing a Russian Orlan spacesuit, participated in two of the three EVAs conducted during the flight. In nearly identical spacewalks with each of his fellow crew members, he worked with Walz to install a ham radio antenna on the station's Zvezda module during a six-hour-long EVA on January 14, and then he and Bursch installed a second antenna on the module during another six-hour excursion on January 25.

Having successfully completed their activities aboard the ISS, Onufrienko, Bursch, and Walz returned to Earth aboard the space shuttle Endeavour at the end of shuttle flight STS-111, landing on June 19, 2002 at Edwards Air Force Base in California.

During his career as a cosmonaut, Yuri Onufrienko has spent more than 389 days in space and accumulated a total of 42 hours and 23 minutes in EVA during eight spacewalks.

OSWALD, STEPHEN S.

(1951–)

U.S. Astronaut

Stephen Oswald was born in Seattle, Washington on June 30, 1951. In 1969, he graduated from Bellingham High School in Bellingham, Washington, and then attended the United States Naval Academy, where he received a Bachelor of Science degree in aerospace engineering in 1973.

Oswald was designated a naval aviator in 1974 and was assigned to the USS *Midway* from 1975 to 1977. He then attended the U.S. Naval Test Pilot School at Patuxent River, Maryland, and remained at the Naval Air Test Center after graduation to conduct flight tests in the A-7 and F/A-18 aircraft. He also served aboard the USS *Coral Sea* as an F/A-18 flight instructor and catapult officer before resigning from active duty to join the Westinghouse Electric Corporation, where he served as a civilian test pilot. He remained a captain in the U.S. Naval Reserve.

During his exceptional career as a pilot, Oswald has accumulated more than 7,000 hours of flying time in more than 40 different types of aircraft.

He began his association with NASA in 1984 as an aerospace engineer and instructor pilot. The agency selected him for astronaut training in 1985. Among his varied technical assignments, he served as crew representative to the Marshall Space Flight Center (MSFC) in Huntsville, Alabama during the redesign of the space shuttle's solid rocket boosters (SRBs), and also served as chief of the Astronaut Office Operations Development Branch and as assistant director of engineering at the Johnson Space Center (JSC) in Houston.

Oswald first flew in space as pilot of the space shuttle Discovery during the STS-42 International Microgravity Laboratory (IML) mission. A milestone in international scientific cooperation in space, the flight featured 55 major IML experiments designed by scientists from six international organizations representing 11 countries.

Flying into space with Oswald were STS-42 mission commander Ronald Grabe, mission specialist William Readdy, payload commander Norman Thagard, and payload specialists Roberta Bondar of the Canadian Space Agency (CSA) and astronaut Ulf Merbold of Germany, representing the European Space Agency (ESA).

The crew worked around the clock in alternating shifts to conduct the IML experiments, observing the impact of microgravity on a variety of life forms and materials. The mission was extended by one day to allow time for the crew to complete their work, and Discovery made a total of 129 orbits in a little more than eight days, landing at Edwards Air Force Base on January 30, 1992.

Oswald also served as pilot of Discovery during his second spaceflight, STS-56. Launched on April 8, 1993, STS-56 featured the second flight of the Atmospheric Laboratory for Applications and Science (ATLAS-2).

Designed to study the interaction between the Sun and the Earth's middle atmosphere and their impact on the Earth's ozone layer, the ATLAS-2 laboratory consisted of seven instruments, six of which were mounted on a Spacelab pallet in the shuttle's cargo bay. Discovery's landing was postponed by one day because of poor weather at the Kennedy Space Center (KSC) in Florida, which extended the total mission duration to 9 days, 6 hours, 8 minutes, and 24 seconds, with the shuttle touching down at KSC on April 17, 1993.

In the early morning hours of March 2, 1995, Oswald lifted off on a remarkable third flight in space as commander of STS-67, aboard the space shuttle Endeavour. His crew for the mission included pilot William Gregory, payload commander Tamara Jernigan, mission specialists John Grunsfeld and Wendy Lawrence, and payload specialists Ronald Parise and Samuel Durrance.

STS-67 was the second flight of the Astro Observatory (ASTRO-2). Designed to study ultraviolet light, the ASTRO-2 facility featured three of the four telescopes that flew on the first ASTRO mission, STS-35 in 1990, including the Hopkins

Ultraviolet Telescope (HUT), the Wisconsin Ultraviolet Photo-Polarimeter Experiment (WUPPE), and the Ultraviolet Imaging Telescope (UIT). Researchers at The Johns Hopkins University designed the HUT instrument, and the WUPPE was a project by scientists at the University of Wisconsin; the UIT was built at NASA's Goddard Space Flight Center.

The crew activated the ASTRO-2 instruments on the first day of the flight. Data collected during the hundreds of ultraviolet observations supported 23 different science investigations, including an exercise aimed at finding evidence of intergalactic helium, which was thought to be a product of the Big Bang that scientists theorize occurred when the universe formed. Other studies included ultraviolet measurements of the aurora of Jupiter and the atmospheres of Venus and Mars.

Other data collected by the ASTRO-2 instruments included images of spiral and elliptical galaxies, studies of hot stars, examinations of dust clouds in the Milky Way and the Large Magellanic Cloud galaxies, the Wolf-Rayet and Be stars, and three exploding novae.

In addition to its impressive scientific activities, the STS-67 flight was the first shuttle mission to be featured in real-time on the Internet. The Astro-2 home page at the Marshall Space Flight Center in Huntsville, Alabama received nearly 2.5 million inquiries about the mission and the activities of the crew, and the astronauts responded to as many questions as they could during the flight.

Endeavour returned to Earth with a landing at Edwards Air Force Base in California on March 18, 1995, after a flight of more than 16 days, the longest shuttle mission up to that time.

Stephen Oswald spent more than 33 days in space during his astronaut career.

After his third flight, Oswald served as deputy associate administrator for Space Operations at NASA headquarters in Washington, D.C. for more than two years, and then returned to the Astronaut Office at the Johnson Space Center (JSC) in Houston in 1998. He retired from NASA in January 2000.

OVERMYER, ROBERT F.

(1936–1996)

U.S. Astronaut

One of the pioneering astronauts selected for training in the U.S. Air Force Manned Orbiting Laboratory (MOL) program, Robert Overmyer was also a veteran of two space shuttle flights.

Overmyer was born in Lorain, Ohio on July 14, 1936. In 1954, he graduated from Westlake High School in Westlake, Ohio and then attended Baldwin Wallace College, where he earned a Bachelor of Science degree in physics in 1958.

As a member of the U.S. Marine Corps, he received Navy flight training in Kingsville, Texas, and was then assigned to Marine Attack Squadron 214 in 1959.

Continuing his education concurrently with his military career, Overmyer attended the U.S. Naval Postgraduate School, where he earned a Master of Science degree in aeronautics with a major in aeronautical engineering in 1964.

Next, Overmyer served in Iwakuni, Japan with Marine Maintenance Squadron 17 for one year before being selected to attend the U.S. Air Force Test Pilot School at Edwards Air Force Base in California.

In 1966, the Air Force chose Overmyer to become a member of its second group of MOL astronauts. The MOL program was conceived as a military surveillance program that would send two-man crews of specially trained pilots into space aboard modified Gemini spacecraft. Once in orbit, the crew members were to have conducted photo and radar reconnaissance, spying on the activities of nations unfriendly to the United States and its allies.

Over the course of its decade-long development, the MOL program suffered from financial and technical difficulties, and was eventually canceled in 1969. Only one MOL test flight was flown, a 33-minute unmanned sub-orbital test on November 3, 1966, but some of the technology developed for the program was later adapted for use in spy satellites. When the MOL program was canceled, seven of its astronauts, including Overmyer, were transferred as a group to NASA.

During his outstanding military career, Overmyer accumulated more than 7,500 hours of flying time, including more than 6,000 hours in jet aircraft. He rose to the rank of colonel in the U.S. Marine Corps by the time of his retirement from the service in May 1986.

His initial NASA duties included engineering support for the Skylab program and service as a member of the astronaut support crew for the Apollo 17 lunar landing mission and the Apollo-Soyuz Test Project (ASTP) flight. During ASTP, he served as NASA's spacecraft communicator (Capcom) at Soviet Mission Control in Moscow.

In 1976, he served as the primary chase pilot for several space shuttle tests, flying a T-38 jet aircraft while observing the flight of the prototype space shuttle Enterprise during the Shuttle Approach and Landing Test (ALT) program. He also played a key role in preparing the space shuttle Columbia for its maiden voyage, as he oversaw the application of the shuttle's special heat-resistant tiles.

Overmyer first flew in space during STS-5, as pilot of Columbia. Lifting off with mission commander Vance Brand and mission specialists Joseph Allen and William Lenoir, Overmyer piloted Columbia during the first fully operational flight of the space shuttle program. The STS-5 crew deployed two commercial satellites and conducted a program of experiments that included medical tests and three scientific investigations designed by students. Columbia returned to Earth on November 16, 1982, landing at Edwards Air Force Base after a flight of 5 days, 2 hours, 14 minutes, and 26 seconds.

On April 29, 1985, Overmyer lifted off on his second flight into space, as commander of STS-51B, the second Spacelab mission. His crew aboard the shuttle Challenger included pilot Frederick Gregory, mission specialists Don Lind, Norman Thagard and William Thornton, and payload specialists Lodewijk van den Berg and Taylor Wang.

Engineers and scientists at the European Space Agency (ESA) had designed the Spacelab module and the experimental program being carried out in the facility in cooperation with NASA.

The STS-51B crew successfully carried out 14 of the 15 primary Spacelab experiments assigned to the flight, conducting investigations in atmospheric physics, astronomy, fluid mechanics, life sciences, and materials sciences. The flight also marked the first time since the early 1960s that monkeys were carried into space aboard an American spacecraft, and the first time that astronauts and monkeys had ever flown together in space, as part of an investigation into the effects of weightlessness on animals.

After a flight of 7 days, 8 minutes, and 46 seconds, STS-51B ended with a safe landing at Edwards Air Force Base on May 6, 1985.

Robert Overmyer spent more than 12 days in space during his career as an astronaut. He retired from NASA in May 1986.

On March 22, 1996, Robert Overmyer was killed during a test flight of a light aircraft. He was survived by his wife Katherine and their three children.

PADALKA, GENNADY I.

(1958–)

Russian Cosmonaut

Gennady Padalka was born in Krasnodar, Russia on June 21, 1958. He graduated from the Eisk Military Aviation College in 1979 and served in the Russian Air Force as a pilot and as a senior pilot. He was selected for cosmonaut training in 1989, and after completing two years at the Gagarin Cosmonaut Training Center (GCTC), he was qualified as a test cosmonaut in 1991.

He also worked as an engineer-ecologist at the UNESCO International Center of Instruction Systems until 1994.

As a pilot, he has logged 1,500 hours of flying time in various types of aircraft, and he has also served as a parachute training instructor, making over 300 parachute jumps. He has risen to the rank of colonel in the Russian Air Force.

Padalka served as backup commander for the Mir 24 mission while preparing for his first flight in space, as commander of Mir 26. Launched aboard Soyuz TM-28 with Sergei Avdeyev and Yuri Baturin on August 13, 1998, Padalka served aboard the space station for more than six months. He and Avdeyev made two spacewalks, the first a 30-minute excursion to make repairs to Mir's Spektr module on September 15, and the second, on November 10, a nearly six hour EVA to deploy a satellite and experiments outside the station. Padalka returned to Earth on February 28, 1999, sharing his return flight with Ivan Bella of Slovakia, who had arrived at the station with the Mir 27 crew aboard Soyuz TM-29 a few days earlier. Bella was the first citizen of Slovakia to fly in space.

Gennady Padalka has spent more than 386 days in space during two long-duration missions. [NASA/courtesy of nasaimages.org]

Padalka next served as backup commander for the International Space Station (ISS) Expedition 4 mission while preparing for his next mission, as commander of ISS Expedition 9.

With fellow Expedition 9 crew mate Michael Fincke of the United States and European Space Agency (ESA) astronaut André Kuipers of The Netherlands, Padalka launched aboard Soyuz TMA-4 from Baikonur Cosmodrome in Kazakhstan on April 18, 2004. Kuipers returned to Earth with the Expedition 8 crew on April 30, while Padalka and Fincke lived and worked on the ISS for the next six months.

Their first spacewalk, on June 24, featured a scary moment for Padalka and Fincke when Fincke's oxygen supply tank malfunctioned and suddenly lost pressure. The EVA was cut short at 13 minutes. A second EVA, on June 30, was far more successful, as the pair was able to service the space station's gyroscope in under six hours. During their two final EVAs, on August 3 and September 3, they accumulated just under 10 hours of spacewalking while installing antennas and replacing laser reflectors on the station.

Following the successful completion of their long-duration mission aboard the ISS, Padalka and Fincke returned to Earth aboard Soyuz TMA-4 on October 23, 2004. Accompanying them on the return flight was cosmonaut Yuri Shargin, who had flown to the station a week earlier with the Expedition 10 crew.

Gennady Padalka has spent more than 386 days in space during his two long-duration missions. He has also logged more than 22 hours in EVA.

PARAZYNSKI, SCOTT E.

(1961–)

U.S. Astronaut

Scott Parazynski was born in Little Rock, Arkansas on July 28, 1961. He attended school in Dakar, Senegal, Beirut, Lebanon, and at the Tehran American School in Tehran, Iran prior to his 1979 high school graduation from the American Community School in Athens, Greece.

In 1983 he received a Bachelor of Science degree in biology from Stanford University, and he was awarded the National Institutes of Health Pre-doctoral Training Award in Cancer Biology.

Parazynski then entered Stanford Medical School, where he was awarded a NASA graduate student fellowship. He conducted research at NASA's Ames

Research Center, and received the agency's Graduate Student Researchers Award in 1988. He graduated from Stanford Medical School in 1989.

He interned at the Harvard Medical School's Brigham and Women's Hospital, and was working on his residency in emergency medicine in Denver, Colorado at the time of his selection as an astronaut candidate in 1992.

Parazynski is also a commercial pilot and a world-class mountaineer. He has accumulated more than 2,000 hours of flying time, and has climbed to the summit of Cerro Aconcagua in Argentina (22,841 feet above sea level) and to the top of dozens of mountains in Colorado whose peaks rise more than 14,000 feet.

He first flew in space in November 1994, during STS-66. The STS-66 crew worked around the clock in shifts to carry out an intensive program of scientific research, and Parazynski led the crew in exercise on the Interlimb Resistance Device, which he had developed as a means of fighting the atrophy that affects the musculoskeletal system during spaceflight.

During his second space mission, STS-86 in 1997, Parazynski participated in an historic spacewalk as he and Vladimir Titov conducted the first joint EVA by an American astronaut and a Russian cosmonaut.

Accumulating 5 hours and 1 minute in EVA, Parazynski and Titov tested the Simplified Aid for EVA Rescue (SAFER) backpacks and other spacewalking tools, retrieved four experiments from Mir's exterior, and deployed a solar array cap that would be installed during a later EVA to seal Mir's damaged Spektr module.

Parazynski's third spaceflight also featured a history-making moment when Mercury program pioneer John Glenn returned to space with the STS-95 launch of the shuttle Discovery on October 29, 1998.

During STS-95, Parazynski also played a key role in the deployment of the SPARTAN 201 free flyer, which was deployed for two days of independent flight while gathering data about the solar wind. He served as navigator during the shuttle's rendezvous with the satellite, which was successfully retrieved. Parazynski also used Discovery's remote manipulator system (RMS) during a test of space vision systems and monitored several of the life science investigations that examined the relationship between spaceflight and the aging process, for which John Glenn served as a subject.

Parazynski next flew in space in April 2001, during STS-100, the sixth shuttle flight devoted to the assembly of the International Space Station (ISS). The STS-100 crew delivered the Canadarm2 robotic arm to the station, and Parazynski and Canadian Space Agency astronaut Chris Hadfield made two spacewalks to install the device.

During their first spacewalk, on April 22, Parazynski and Hadfield spent over 7 hours in EVA while they installed the Canadarm2 and a UHF antenna. Then, in a second EVA two days later, they tested their work and made sure the Canadarm2 device was properly installed.

Parazynski also used Endeavour's RMS to install and remove the Raffaello Multi-Purpose Logistics Module (MPLM) at the start and end of the STS-100 delivery of equipment and supplies to the ISS.

Parazynski's fifth spaceflight, STS-120, launched aboard the shuttle Discovery on October 23, 2007. The STS-120 crew delivered the Harmony Node to the ISS, and Parazynski participated in four spacewalks during the mission, including an unscheduled excursion to fix a damaged solar array. During STS-120, he accumulated more than 27 hours in EVA.

Scott Parazynski has spent more than 58 days in space, including over 41 minutes in EVA, during five spaceflights.

PATSAYEV, VIKTOR I.

(1933–1971)

Soviet Cosmonaut

A member of the first crew to successfully live and work aboard an orbiting space station, Viktor Patsayev was born in Aktyubinsk, Kazakhstan on June 19, 1933. He attended the Penza Industry Institute, from which he graduated in 1955, and helped to develop the Salyut series of space stations as an engineer at the Korolev Design Bureau.

Patsayev was chosen for training as a cosmonaut as a member of the civilian specialist group three selection of cosmonaut candidates in May 1968.

On June 6, 1971, he lifted off from the Baikonur Cosmodrome in Kazakhstan as research engineer for the Soyuz 11/Salyut 1 long-duration mission, with commander Georgy Dobrovolsky and flight engineer Vladislav Volkov. The Salyut 1 space station had been launched unmanned on April 19, 1971, with the intention of the crew of Soyuz 10 being the first to occupy the station in orbit. A malfunction had forced an early end to the Soyuz 10 flight, however, and the Soyuz 11 crew took up the task of being the first crew to live and work aboard the world's first space station.

Patsayev, Dobrovolsky, and Volkov had originally been the backup crew for the flight; they replaced original crew members Aleksei Leonov, Valeri Kubasov, and Pyotr Kolodin when Kubasov was diagnosed with a lung condition.

Once in orbit, Patsayev and his crew mates experienced none of the difficulty that had cut the Soyuz 10 flight short, and they were able to dock their Soyuz 11 craft with Salyut 1 and transfer into the station as planned. They tested the station's systems and equipment, carried out navigational activities, and conducted a series of science investigations that included medical tests, cosmic ray investigations, studies of the Earth's atmosphere, and observations of astronomical objects and cloud formations.

Their milestone stay in space quickly became a national sensation in the Soviet Union, as Patsayev, Dobrovolsky, and Volkov made daily television broadcasts from their home in orbit. They showed off the inside of Salyut 1 and described their daily routines aboard the station with good humor and in some detail. On June 19, Dobrovolsky and Volkov helped Patsayev celebrate his 38th birthday during a broadcast viewed by millions of Russian citizens who ultimately embraced the three cosmonauts as national heroes even before the crew had completed its stay aboard the station.

In the years since the milestone flight, details that were closely held at the time by the totalitarian Soviet state have revealed the heroics of the Soyuz 11 crew in greater detail. Unknown to their large audience during the regular television broadcasts, Patsayev, Dobrovolsky, and Volkov faced a crisis on June 17 when an electrical fire broke out on Salyut 1. They were able to contain and extinguish the fire quickly, but no mention of the incident was made in the Soviet press. Although it was not related to the Soyuz malfunction that later cost the cosmonauts their lives, the fire may have caused them to leave the station earlier than originally planned.

They boarded Soyuz 11 on June 29, 1971 and prepared for their return trip to Earth. The initial stages of their return went as planned, until they reached the point at which a series of explosive bolts in the frame holding the spacecraft's Orbital Module and Descent Module were to have fired in sequence. The bolts were designed to separate the two modules, with the cosmonauts continuing their descent in the Descent Module, and the Orbital Module being jettisoned to burn up in the Earth's atmosphere.

When the bolts fired to separate the modules, they fired all at once, rather than in the carefully prescribed sequence necessary for their proper performance. The error caused the two modules to separate with a harsh jolt, which in turn caused a vent in the Descent Module to open prematurely. The life-sustaining atmosphere within the spacecraft quickly escaped through the open vent, into space.

Immediately recognizing the gravity of the situation, Patsayev was able to close the vent about halfway before the Descent Module became completely depressurized. At that point, the crew lost consciousness, and the three cosmonauts died less than a minute later.

The loss of the beloved cosmonauts at the end of their long journey was a shocking national tragedy for the Russian people. After three weeks in space during which they had displayed courage in performing their otherworldly tasks and joy in performing them well, Patsayev, Dobrovolsky, and Volkov were deeply mourned by their fellow citizens and by all those around the world who had followed their flight.

Viktor Patsayev spent 23 days, 18 hours, and 22 minutes in space during the Soyuz 11/Salyut 1 mission. In honor of his having given the ultimate measure of dedication in living out the ideals of exploring space, Asteroid 1791 was named for him.

PERRIN, PHILIPPE

(1963–)

French Astronaut

A veteran of three spacewalks during the STS-111 mission of the space shuttle Endeavour in June 2002, Philippe Perrin was born in Meknes, Morocco on January 6, 1963. In 1985, he earned an engineering degree from the Ecole Polytechnique in Paris.

Perrin began his military career with the French Navy while he was still a student at the Ecole Polytechnique. He was deployed on a six-month tour of duty in the Indian Ocean during his Navy service.

After graduation, he entered the French Air Force. He received flight training and was recognized as first in his class when he received his pilot's wings in 1986. He then served at Strasbourg Air Force Base for four years, and flew the Mirage F1 CR aircraft during deployments in Africa and Saudi Arabia while attached to the 33rd Reconnaissance Wing. In 1991, he was awarded the French Overseas medal for service during the Gulf War.

Perrin began his association with France's space program in 1992, when he was temporarily assigned to the French space agency Centre National d'Études Spatiales (CNES). He trained for two months at the Gagarin Cosmonaut Training Center (GCTC) in Russia, and attended the French Test Pilot School Ecole du Personnel Navigant d'Essais et de Reception (EPNER) at Istres Air Force Base, where he was licensed as a test pilot in 1993. He returned to military service that same year, as senior operations officer during Operation Southern Watch, attached to the 2nd Air Defense Wing at Dijon Air Force Base.

In 1995, Perrin became chief pilot deputy responsible for the development of the Mirage 2000–5 aircraft at the Bretigny Test Center. He was also certified as an airline pilot in 1995.

During his outstanding military career, Perrin has flown 26 combat missions and has accumulated more than 3,000 hours of flying time in more than 30 types of aircraft. He is a colonel in the French Air Force.

CNES chose him for training as an astronaut in July 1996, and he underwent two years of training at NASA's Johnson Space Center (JSC) in Houston, Texas for assignment to flights aboard the U.S. space shuttle as a mission specialist.

Perrin joined the European Space Agency (ESA) in December 2002. He first flew in space as a mission specialist during STS-111 in June 2002, aboard the space shuttle Endeavour.

Commanded by Kenneth Cockrell, the STS-111 crew also included pilot Paul Lockhard, Perrin's fellow mission specialist Franklin Chang-Diaz, and the ISS Expedition 5 crew—commander Valery Korzun, flight engineer Sergei Treschev, and flight engineer Peggy Whitson.

Using the Leonardo Multi-Purpose Logistics Module to transfer equipment and supplies to the ISS, the shuttle and space station crews moved 8,000 pounds of material onto the station and removed 4,500 pounds of items no longer needed at the ISS. Perrin operated Endeavour's Remote Manipulator System (RMS) robotic arm to return the Leonardo module to the shuttle's payload bay at the end of the transfer work.

Perrin participated in three spacewalks during STS-111, starting on June 9, when he and Chang-Diaz attached a fixture to the space station's P6 Truss, transferred micro-meteoroid shields from the shuttle to the station, and inspected and photographed a gyroscope on the ISS Z1 truss segment.

During their second EVA, two days later, Perrin and Chang-Diaz installed communications, electric and video cables and completed the installation of the ISS Mobile Base System (MBS) work platform.

They made their final STS-111 spacewalk on June 13 when they replaced the wrist joint in the ISS Canadarm2 robotic arm.

In addition to successfully achieving its assigned cargo transfer and station maintenance goals, the STS-111 flight also delivered Korzun, Treschev, and Whitson to begin their stay as the station as the ISS Expedition 5 crew and collected ISS Expedition 4 crew members Yuri Onufrienko, Daniel Bursch, and Carl Walz for their return trip to Earth. Onufrienko, Bursch, and Walz had been in space for 196 days.

Endeavour landed at Edwards Air Force Base in California on June 19, 2002.

Philippe Perrin spent more than 13 days in space, and more than 19 hours in EVA, during STS-111.

PETERSON, DONALD H.

(1933–)

U.S. Astronaut

One of the pioneering astronauts chosen for training in the U.S. Air Force's Manned Orbiting Laboratory (MOL) program, Donald Peterson flew in space during the first flight of the space shuttle Challenger.

Peterson was born in Winona, Mississippi on October 22, 1933. He graduated from Winona City High School and then attended the United States Military Academy at West Point, where he earned a Bachelor of Science degree in 1955.

He served with the Air Training Command as a flight instructor and training officer and also served with the Tactical Air Command for one year as a fighter pilot.

Continuing his education concurrently with his military career, Peterson attended the Air Force Institute of Technology at Wright-Patterson Air Force Base in Ohio, where he received a Master's degree in nuclear engineering in 1962. He served three years as a nuclear systems analyst with the Air Force Systems Command.

Peterson also graduated from the Aerospace Research Pilot School at Edwards Air Force Base in California. During his exceptional military career he accumulated more than 5,300 hours of flying time, including 5,000 hours in jet aircraft. He had risen to the rank of colonel in the United States Air Force by the time of his retirement from the service.

In June 1967 the Air Force selected Peterson as a member of its third group of MOL astronauts. The MOL program was designed as a series of small space stations that would enable U.S. military personnel to conduct manned surveillance from space. Two man crews were to have been placed in orbit in a modified Gemini spacecraft and, during missions planned to last about 30 days each, would conduct photographic and radar reconnaissance of the activities of nations unfriendly to the United States.

The MOL program was simultaneously but separately developed alongside NASA's civilian space effort throughout the 1960s. The cost of the Air Force program escalated steadily as the years passed, and only one unmanned, sub-orbital MOL test flight was flown before the program was canceled in June 1969.

Technology originally designed for use in the MOL project was later adapted for use in military spy satellites. When the MOL program was canceled, those MOL astronauts who were under the age of 35 were invited by NASA to put their training to use at the civilian agency, and Peterson was among the seven eligible Air Force astronauts who transferred to NASA in August 1969.

His initial assignments at the space agency included service as a member of the astronaut support crew for the Apollo 16 lunar landing mission.

Peterson began his first flight in space on April 4, 1983 as a mission specialist aboard the first flight of the space shuttle Challenger, STS-6. Challenger was commanded by Paul Weitz and the crew also included Peterson's fellow mission specialist Story Musgrave and pilot Karol Bobko who, like Peterson, had been selected as an MOL astronaut and then transferred to NASA when the Air Force program was canceled.

The STS-6 crew deployed the first Tracking and Data Relay Satellite (TDRS-1), and on April 7, 1983, Peterson and Musgrave made the first spacewalk of the shuttle program when they ventured outside of Challenger for 4 hours and 10 minutes to test EVA equipment and tools, including the emergency maneuvering unit (EMU).

At the end of the flight, Challenger returned to Earth at Edwards Air Force Base on April 9, 1983.

Donald Peterson spent more than five days in space during the STS-6 mission. He left NASA's astronaut corps in 1984.

PETTIT, DONALD R.

(1955–)

U.S. Astronaut

A distinguished scientist and veteran of a long-duration stay aboard the International Space Station (ISS), Donald Pettit was born in Silverton, Oregon on April 20, 1955. He graduated from Silverton Union High School in 1973 and then attended Oregon State University, where he earned a Bachelor of Science degree in chemical engineering in 1978. He went on to attend the University of Arizona, where he earned a Doctorate degree in chemical engineering in 1983.

Pettit began his career as a scientist at the Los Alamos National Laboratory in New Mexico in 1984. He began his association with NASA during that period, when he worked on experiments involving the agency's KC-135 aircraft, and he subsequently worked on teams responsible for developing technologies to be used in NASA's plans to explore the Moon and Mars. In addition, he served as a member of the group that redesigned the space station Freedom in the early 1990s, which led to the creation of the ISS.

NASA chose Pettit for training as an astronaut in 1996. He underwent an intensive period of training and was qualified in 1998 for assignment as a mission specialist.

Pettit began his first flight in space with the STS-113 launch of the space shuttle Endeavour, which lifted off from the Kennedy Space Center (KSC) in Florida on November 23, 2002. Commanded by James Wetherbee, the STS-113 crew included pilot Paul Lockhart, mission specialists John Herrington and Michael Lopez-Alegria, and the ISS Expedition 6 crew, commander Kenneth Bowersox and flight engineers Pettit and Nikolai Budarin of Rosaviakosmos.

In addition to bringing the Expedition 6 crew members to the station to begin their long-duration mission, the STS-113 crew also delivered the ISS P1 Truss segment, which Herrington and Lopez-Alegria installed on the station during three long spacewalks.

Pettit, Bowersox, and Budarin remained aboard the ISS when Endeavour undocked on December 2, replacing Expedition 5 crew members Valery Korzun, Sergei Treschev, and Peggy Whitson, who returned to Earth aboard the shuttle.

During his long stay in orbit with Bowersox and Budarin, Pettit served as NASA's science officer, responsible for overseeing an extensive program of scientific experiments that included an emphasis on health and medical observations, including studies of the effects of EVA on pulmonary function, the impact that radiation has on astronauts during EVA, and the reactivation of Epstein-Barr virus brought on by spaceflight. The Expedition 6 science work also included materials processing experiments and protein crystal growth exercises.

Pettit participated in two spacewalks during his long stay at the ISS. On January 15, 2003, he and Bowersox made their first trip outside the ISS to perform maintenance work on the P1 Truss. They made a second EVA on April 8, 2003, to reorganize the cables linking the station's various truss segments.

The Expedition 6 crew members welcomed their replacements on April 26, 2003, when ISS Expedition 7 commander Yuri Malenchenko of Rosaviakosmos and flight engineer Edward Lu of NASA arrived in Soyuz TMA-2. Pettit, Bowersox, and Budarin completed their long-duration mission on May 3, 2003 when they returned to Earth in Soyuz TMA-1. Their landing proved a harrowing ordeal; a computer error forced them to make a ballistic re-entry, and they missed their intended landing site by about 300 miles. They were fortunate to survive the difficult experience without injury.

During his ISS Expedition 6 mission, Donald Pettit spent more than 161 days in space, including more than 13 hours in EVA during two spacewalks.

PHILLIPS, JOHN L.

(1951–)

U.S. Astronaut

John Phillips was born in Fort Belvoir, Virginia on April 15, 1951. In 1966, he graduated from Scottsdale High School in Scottsdale, Arizona, and then attended the

United States Naval Academy, after having been named a National Merit Scholar. He graduated second in his class of 906 students while earning a Bachelor of Science degree in mathematics in 1972.

After completing his initial flight training, he was designated a Naval Aviator in 1974 and subsequently served overseas aboard the USS *Oriskany* and the USS *Roosevelt*. He later served as a recruiter in Albany, New York, and also served at Naval Air Station North Island in California, where he flew the CT-39 Sabreliner aircraft.

Continuing his education concurrently with his military career, Phillips attended the University of West Florida, where he earned a Master of Science degree in aeronautical systems in 1974.

After completing his military service, Phillips left active duty Navy service in 1982 to attend the University of California at Los Angeles (UCLA), where he studied geophysics and space physics, earning a Master of Science in 1984 and a Doctorate degree in 1987. He continued to serve in the U.S. Navy reserve from 1982 to 2002.

He began his association with NASA during his graduate studies at UCLA, when he worked on research based on data gathered by the agency's Pioneer Venus space probe. He was subsequently awarded an Oppenheimer Fellowship to conduct post-doctoral research at the Los Alamos National Laboratory in New Mexico, where he became a full staff member in 1989.

During his career at Los Alamos, he served as principal investigator for the solar wind plasma experiment flown aboard the Ulysses spacecraft, which was launched from the space shuttle Discovery during STS-41 on October 6, 1990. Among the many honors he has received for his work as a scientist, he was honored with Los Alamos's Distinguished Performance Award.

Phillips has published more than 150 scientific papers based on his research. During his military career, he accumulated more than 4,400 hours of flying time and achieved 250 landings on aircraft carriers. He retired from military service in 2002 with the rank of captain in the U.S. Naval Reserve.

NASA chose Phillips for training as an astronaut in 1996. He began his first flight in space as a mission specialist aboard the space shuttle Endeavour during STS-100, which lifted off from the Kennedy Space Center (KSC) in Florida on April 19, 2001.

Endeavour docked with the International Space Station (ISS) on April 21, and the STS-100 and ISS Expedition 2 crews opened the hatches between the two vehicles for the first time two days later. Over the course of the next few days, the two crews moved 6,000 pounds of equipment and supplies onto the space station, including the first commercial payloads delivered to the ISS on behalf of U.S. companies.

The STS-100 crew returned to Earth on May 1, 2001. During his first space mission, Phillips spent 11 days, 21 hours, and 30 minutes in space.

After his first flight in space, Phillips served as a member of the backup crew for the long-duration ISS Expedition 7 mission.

On April 15, 2005, he began his second space mission, as Flight Engineer and NASA Science Officer for the ISS Expedition 11 long-duration flight. He launched from the Baikonur Cosmodrome in Kazakhstan in Soyuz-TMA 6 with Expedition 11 commander Sergei Krikalyov of the Russian Federal Space Agency and European Space Agency (ESA) astronaut Roberto Vittori of Italy, who was traveling to the ISS for a short visit to conduct the ESA Eneide program of scientific research.

After successfully completing his scientific work, Vittori returned to Earth with the Expedition 10 crew on April 24, while Phillips and Krikalyov remained at the ISS and settled into a productive routine of scientific work.

On July 28, 2005, Phillips and Krikalyov welcomed the STS-114 crew of the space shuttle Discovery, the first space shuttle crew to fly in the aftermath of the loss of the space shuttle Columbia in February 2003. As part of the procedures being tested during STS-114, the crew worked to detect potentially troubling areas of damage on the exterior of the shuttle, and Phillips and Krikalyov carefully photographed the underside of Discovery while STS-114 commander Eileen Collins piloted the shuttle through a slow end-over-end flight known as a Rendezvous Pitch Maneuver. Their photos were then transmitted to Mission Control, where teams of engineers scoured the vehicle for any sign of serious damage and, fortunately, found no major problems.

After a brief visit with the STS-114 crew and the shuttle's safe return to Earth, Phillips and Krikalyov prepared themselves for another highlight of their long stay: a spacewalk of nearly 5 hours. Beginning on August 18, 2005 and stretching into the early hours of the following day, their EVA work included the installation of a television camera on the station's exterior and the retrieval of several experiment packages.

At the end of their long stay, Philips and Krikalyov returned to Earth in Soyuz TMA-6 on October 11, 2005, with space tourist Gregory Olsen, who had arrived at the ISS a short while earlier with the Expedition 12 crew. During his second spaceflight, Phillips spent more than 179 days in space.

PIONEER SATELLITE PROGRAMS

(1958–1960; 1972–1973; 1978)

U.S. Satellite Programs Designed to Explore the Moon (1958–1960), the Outer Solar System (1972–1973), and the Planet Venus (1978)

The prefix "Pioneer" has been given to several unmanned projects of the U.S. space program, beginning with the nation's first satellites aimed at exploring the Moon in the late 1950s, the early 1970s Pioneer 10 and 11 missions to explore the outer solar system, including Jupiter and Saturn, and the 1978 Pioneer Venus orbiter and probe.

Pioneer Lunar Exploration Program (1958–1960)

Representing the first U.S. attempts to explore the Moon with unmanned satellites, the original Pioneer program was begun by engineers and scientists in the U.S. Air Force and the U.S. Army in the late 1950s, shortly before the formation of

The planet Saturn as photographed by Pioneer 11, August 31, 1979. Saturn's moon Titan is at upper left. [NASA/courtesy of nasa images.org]

the National Aeronautics and Space Administration (NASA), and then continued under the direction of the civilian space program.

The goal of launching a spacecraft that could reach the Moon required the development of a powerful rocket that could be used as a launch vehicle. In the case of the 1950s Pioneer program, several such vehicles were developed.

U.S. Air Force engineers designed a three-stage launcher that incorporated the Thor intermediate range ballistic missile (IRBM) as its first stage, with two other rockets (Able and Altair) attached above as the second and third stages of the launch vehicle. The Thor-Able-Altair combination created enough thrust to carry a maximum payload of 88 pounds (40 kilograms).

Equipped with an infrared camera, a microphone, and an instrument designed to measure magnetic fields, the first Pioneer satellites weighed in just below the limit for the Air Force launcher, at 84 pounds (38 kilograms). They were intended to travel to the Moon and enter into lunar orbit.

A second launch vehicle, designed by U.S. Army engineers to carry a much smaller payload, also consisted of three stages but had as its first stage the Army's Jupiter IRBM. Because the Jupiter launcher was capable of carrying a payload of no more than 15.5 pounds (seven kilograms), the Pioneer satellites launched atop the Army's rocket were, by necessity, much smaller than the first Pioneers.

In its fledgling attempts at lunar exploration, NASA also developed a launcher for Pioneer satellites, incorporating the U.S. Air Force's more powerful Atlas intercontinental ballistic missile (ICBM) into its approach. NASA engineers used Thor and Able upper stages to complete their launcher. The inclusion of the powerful Atlas ICBM meant that NASA's Pioneer probes could be far heavier than previous versions of the satellite. As a result, the three spacecraft the agency attempted to send to the Moon in late 1959 and in 1960 were packed with instruments that were both greater in number and in sophistication than the rudimentary equipment carried during the earlier attempts.

Unfortunately, the three NASA Pioneer launch attempts all ended in failure.

Pioneer 0 (August 17, 1958)

Launched atop a U.S. Air Force multi-stage Thor launch vehicle from Cape Canaveral on August 17, 1958, the first Pioneer satellite was lost a little more than a minute after lifting off when a fault in one stage of the Thor launch vehicle caused the rocket to explode. The attempt was the first by any nation to launch a spacecraft intended to travel beyond Earth orbit; it had traveled 16 kilometers in 77 seconds when the fault—later determined to most likely have been a rupture in a first stage fuel or oxygen line—caused the vehicle to explode. The U.S. Air Force and the U.S. Department of Defense oversaw the first Pioneer launch.

Pioneer 1 (October 11, 1958)

The first space mission flown under the aegis of the newly formed National Aeronautics and Space Administration (NASA), Pioneer 1 was launched atop a U.S. Air Force multi-stage Thor launch vehicle on October 11, 1958 and was intended to travel to and then enter into orbit around the Moon. Pioneer 1 achieved a flight that reached nearly halfway from the Earth to the Moon before a malfunction caused the second stage of its launch vehicle to shut down prematurely. The stranded space-craft was then pulled back toward Earth by the force of the planet's gravitational pull and was destroyed during re-entry through the Earth's atmosphere two days after launch, on October 13, 1958, after a total flight of about 43 hours. Although it did not reach the Moon, Pioneer 1 did gather useful data about the Van Allen radiation belts.

Pioneer 2 (November 8, 1958)

On November 8, 1958, Pioneer 2, the third and final Pioneer spacecraft to be launched atop a U.S. Air Force multi-stage Thor launch vehicle, lifted off from Cape Canaveral with the intent of its traveling to the Moon and entering lunar orbit. The flight ended prematurely, however, when the third stage of the launch vehicle failed. Pioneer 2 reached a maximum altitude of 963 miles before it plunged back into the Earth's atmosphere and was destroyed. Although it did not complete its intended mission, Pioneer 2 did return useful data, including the first evidence that the Earth's equatorial region has a higher concentration of radiation than other areas of the planet.

Pioneer 3 (December 6, 1958)

The first of the 1950s Pioneer series of spacecraft to be launched by the Jupiter launch vehicle developed by the U.S. Army, Pioneer 3 was designed to fly past the Moon. Because the Jupiter launcher was capable of launching a payload of no more than 15.5 pounds (seven kilograms), the Pioneer 3 satellite was, by necessity, much smaller than the first three Pioneers. The spacecraft's weight at launch was 5.87 kilograms.

Pioneer 3 was launched on December 6, 1958. The spacecraft achieved a maximum altitude of 63,155 miles (102,360 kilometers), but a malfunction in the first stage of its launch vehicle prevented it from traveling past the Moon as it had been intended. The Pioneer 3 spacecraft was destroyed during re-entry into the Earth's atmosphere on December 7, 1958, after a total flight of 38 hours and 6 minutes. Pioneer 3 was the first satellite to indicate that the radiation surrounding the Earth is in the form of two separate radiation belts.

Pioneer 4 (March 3, 1959)

Launched atop a multi-stage U.S. Army launch vehicle that incorporated the Jupiter IRBM, Pioneer 4 lifted off on March 3, 1959 with the intent of its flying past the Moon and entering into orbit around the Sun. The satellite performed well and became the first U.S. spacecraft to escape the Earth's gravitational pull. Pioneer 4 successfully flew by the Moon on March 4, at a distance as close as 37,300 miles from the lunar surface. The distance was farther than mission planners had hoped, and a great deal farther than the flyby achieved by the Soviet Union with Luna 1, which had launched on January 2, 1959 and passed the Moon at a distance as close as 3,675 miles.

Pioneer 4 was nonetheless the first successful lunar flyby mission of the U.S. space program, and after its relatively close encounter with the Moon, the satellite continued on into space and subsequently entered into orbit around the Sun, as planned. Mission controllers continued to receive signals from Pioneer 4 for 82 hours, and tracked its flight to a distance of 655,000 kilometers.

Pioneer P-3 (November 26, 1959)

After the first five Pioneer satellite launches, three of which used the U.S. Air Force Thor launcher and two which utilized the U.S. Army Jupiter launch vehicle, NASA made its first attempt to launch a Pioneer lunar orbiter satellite with a multi-stage launcher based on the Atlas ICBM. The Pioneer P-3 spacecraft was launched on November 26, 1959, and shortly after lift-off lost the shroud intended to protect the satellite during its trip into space. The force of the launch caused the carefully constructed rocket stack to topple after the loss of the shroud, and within a minute and a half after lift-off, the mission ended with the first and second stages (the Atlas and Thor rockets) continuing upward while the upper stage Able rocket and the Pioneer P-3 fell into the Atlantic Ocean.

Pioneer P-30 (September 25, 1960)

Intended to travel to the Moon and enter into lunar orbit, the Pioneer P-30 spacecraft was launched atop an Atlas ICBM launch vehicle on September 25, 1960. A malfunction in the second stage of the launch vehicle caused its progress to stop at a maximum altitude of about 370 kilometers, and the spacecraft subsequently returned to Earth, with a portion of the vehicle surviving re-entry and falling into the Indian Ocean.

Pioneer P-31 (December 15, 1960)

NASA's third attempt to send a satellite into orbit around the Moon by using a three-stage Atlas-Thor-Able rocket launch vehicle, Pioneer P-31 was launched from Cape Canaveral on December 15, 1960. One minute and eight seconds after launch, a malfunction in the first stage of the launch vehicle (the Atlas ICBM) caused the vehicle to explode, sending Pioneer P-31 into the Atlantic Ocean. The failed attempt was the eighth and final launch of the first U.S. Pioneer satellite program, which had had as its goal the exploration of Moon.

Pioneer Outer Solar System Exploration Program (1972–1973)

Long after the Pioneer prefix had been given to the agency's frustrating initial attempts to explore the Moon via unmanned satellites, NASA revived the Pioneer name for the Pioneer 10 and Pioneer 11 satellite flights of 1972 and 1973, which were designed to send the first spacecraft to the outer solar system to explore Jupiter and Saturn.

Pioneer 10 (March 3, 1972)

The first spacecraft designed to explore the outer solar system and the first to explore the planet Jupiter, Pioneer 10 was launched on March 3, 1972.

Equipped with 15 instruments to collect data about Jupiter and its moons, Pioneer 10 included a charged particle detector, an imaging photopolarimeter, an ionizing detector, an infrared radiometer, magnetometer, a plasma analyzer, several telescopes, and an ultraviolet photometer.

The spacecraft was powered by four radioisotope thermonuclear generators (RTGs) that initially generated 155 watts, which decreased to 140 watts by the time the vehicle arrived at Jupiter nearly two years after launch.

Because the mission profile called for Pioneer 10 to enter a trajectory that would lead it to escape the galaxy after its exploration of Jupiter, the spacecraft also carried a plaque with drawings of a man and a woman and a diagram indicating the location of the Sun and the Earth in our galaxy.

Pioneer 10 passed within 200,000 kilometers of Jupiter on December 3, 1973. After its closest approach to the planet, it entered a trajectory designed to cause it to escape from the solar system. Mission managers at the NASA Ames Research Center in California tracked the spacecraft's progress until March 31, 1997, and its signal was last detected on January 23, 2003. Over the course of its 25-year mission, Pioneer 10 cost a total of $350 million.

Pioneer 11 (April 6, 1973)

The second spacecraft designed to explore the outer solar system and the planet Jupiter, and the first to explore the planet Saturn, Pioneer 11 was launched on April 6, 1973.

Equipped similarly to Pioneer 10, Pioneer 11 carried a charged particle detector, an imaging photopolarimeter, an ionizing detector, an infrared radiometer, magnetometer, a plasma analyzer, several telescopes, and an ultraviolet photometer.

The spacecraft was powered by two radioisotope thermonuclear generators (RTGs) that output 144 watts by the time the craft arrived at Jupiter, nearly two years after launch, and which decreased to 100 watts by the time the vehicle passed by Saturn.

Because its mission profile called for Pioneer 11 to enter a trajectory that would lead it to eventually escape the galaxy, the spacecraft also carried a plaque with drawings of a man and a woman and a diagram indicating the location of the Sun and the Earth in our galaxy.

Pioneer 11 passed within 34,000 kilometers of Jupiter on December 4, 1974, and passed within 21,000 kilometers of Saturn on September 1, 1979.

After its explorations of Jupiter and Saturn, the spacecraft entered a trajectory designed to cause it to escape from the solar system. Mission managers at the NASA Ames Research Center in California tracked Pioneer 11 until September 30, 1995, when its electrical output fell below the minimal power necessary to operate any of the spacecraft's scientific instruments.

Pioneer Venus Project (1978)

In addition to using the Pioneer prefix for its early lunar exploration efforts and the deep space probes of the early 1970s, NASA also gave the Pioneer name to its 1978 exploration of the planet Venus. Consisting of an orbiter and a multi-probe that were launched separately over a period of several months in mid-1978, the Pioneer Venus Project successfully completed a wide-ranging study of the atmosphere of Venus.

Pioneer Venus 1 (Venus Orbiter, May 20, 1978)

The Pioneer Venus Orbiter was launched atop a multi-stage Atlas-Centaur launch vehicle on May 20, 1978, and it was successfully inserted into an elliptical orbit around Venus on December 4, 1978. The spacecraft carried 17 instruments to study various aspects of the atmosphere, including its composition, clouds, infrared emissions, UV light emissions, thermal properties, ionosphere, and interaction with the solar wind. Additional objectives of the orbiter included studies of the planet's topography and surface characteristics, its magnetic field and gravity field, and the gathering of data about gamma ray burst events.

The successful operation of the spacecraft's instruments and the orbiter's continued operation for the next 15 years marked the Pioneer Venus Orbiter mission as a major step in the advance of interplanetary science. The spacecraft successfully conducted its initial observations, including initial radar mapping and ionospheric measurements, during its first several years in orbit. In 1991, its radar mapper was reactivated to continue its observations as the orbiter passed over previously inaccessible southern portions of Venus.

Continuing its mission into the 1990s, the Pioneer Venus Orbiter returned a wealth of valuable scientific data about Venus until its fuel supply was finally exhausted in August 1993, when the spacecraft entered Venus's atmosphere and was destroyed.

Pioneer Venus 2 (Venus Probe, August 8, 1978)

The second element of the Pioneer Venus Project was the Pioneer Venus Multiprobe, which was launched on August 8, 1978. Also designed to study various aspects of the planet's atmosphere, the Multiprobe spacecraft consisted of a bus, a cylinder about two and a half meters in diameter, and four probes: one large probe containing seven instruments, and three smaller probes equipped with a variety of sensors and a net flux radiometer to study the distribution of radiative energy in the atmosphere in three distinct areas of the planet.

The bus was equipped with a neutral mass spectrometer and an ion mass spectrometer to study the composition of the planet's upper atmosphere.

Released on November 16, 1978, the large probe contained a neutral mass spectrometer and a gas chromatograph to gauge the composition of the atmosphere, a cloud particle size spectrometer and nephelometer to study the composition and distribution of clouds, radiometers designed to measure solar flux penetration and the distribution of infrared radiation, and temperature, pressure, and acceleration sensors.

The three small probes were released on November 20. On December 9, all four probes and the bus entered the atmosphere and began to collect data. Designed only to observe the upper atmosphere, the bus was not equipped with a heat shield and was destroyed during descent after successfully completing its observations. The parachute on the large probe was successfully deployed, and the probe performed as expected. The three small probes also functioned well, with the probe aimed at the "day side" of the planet continuing to transmit radio signals to mission controllers on Earth for more than an hour after it impacted the surface.

PLUTO: MAJOR FLIGHTS.

See New Horizons

POGUE, WILLIAM R.

(1930–)

U.S. Astronaut

William Pogue was born in Okemah, Oklahoma on January 23, 1930. He attended Oklahoma Baptist University, where he received a Bachelor of Science degree in education in 1951.

After graduation, Pogue enlisted in the U.S. Air Force. He served with the Fifth Air Force during the Korean War, flying fighter bombers in combat. After the end of the war, in 1955, he became a member of the U.S. Air Force Thunderbirds.

Continuing his education concurrently with his military career, Pogue attended Oklahoma State University, where he earned a Master of Science degree in mathematics in 1960.

He then joined the faculty of the Air Force Academy in Colorado Springs, Colorado as an assistant professor in the mathematics department, a position he held until 1963. In 1963, he began a two-year tour with the British Ministry of Aviation as a test pilot in the U.S. Air Force exchange program with the British Royal Air Force (RAF). He graduated from the Empire Test Pilot's School in Farnborough, England in 1965.

Later that same year he became an instructor at the Air Force Aerospace Research Pilot School at Edwards Air Force Base in California, where he was serving at the time of his selection by NASA as an astronaut.

During his outstanding career as a pilot, combat pilot, flight instructor and test pilot, Pogue accumulated nearly 5,200 hours of flying time in more than 50 types of aircraft, including 4,200 hours in jet aircraft. In addition to his extensive military experience, he is also qualified as a civilian flight instructor. He retired from the U.S. Air Force in 1975 with the rank of colonel.

NASA selected Pogue for training as an astronaut in April 1966, as one of its Group Five selection of pilot astronauts. He served as a member of the astronaut support crew for the Apollo 7 mission, the Apollo 11 first lunar landing, and the Apollo 14 lunar landing.

On November 16, 1973, Pogue lifted off on a remarkable long-duration mission as pilot of the Skylab 4 Apollo command module, a spacecraft identical to the vehicle that previous crews had used to fly to the Moon. With Skylab 4 commander Gerald Carr and scientist astronaut Edward Gibson, Pogue traveled to the first U.S. space station, Skylab, for a long-duration stay that was longer than any previous flight up to that time. Their record-setting mission kept them in space for 84 days, 1 hour, and 16 minutes, a record that stood for five years, until the 96-day Soviet Soyuz 26/Salyut 6 mission of cosmonauts Yuri Romanenko and Georgi Grechko in 1977–1978.

Skylab 4 was the third manned Skylab flight; the space station itself had been designated Skylab 1 when it was launched unmanned on May 14, 1973.

Traveling 34.5 million miles during 1,214 orbits in just under three months aboard the station, the Skylab 4 crew conducted observations of comets and stellar objects, Earth resources studies, solar physics observations, and space medicine and physiology experiments.

On November 22, 1973, Pogue and Gibson ventured outside Skylab for the first spacewalk of the flight. They accumulated 6 hours and 33 minutes in EVA while repairing an antenna and taking photographs.

Pogue made his second EVA of the Skylab 4 flight on December 25, 1973, with Gerald Carr. During a spacewalk of 7 hours and 1 minute, Pogue and Carr were able to capture stunning images of the comet Kohoutek, which had only been discovered in March. Carr and Gibson took more photos of the comet during another EVA, on December 29.

Working long hours during the productive flight, Pogue, Carr, and Gibson conducted 56 scientific experiments, 26 science demonstrations, and 13 student science investigations. They returned to Earth on February 8, 1974.

The Skylab 4 crew was the last to occupy the space station. Skylab remained in orbit for another five years; it re-entered Earth's atmosphere and fragmented into pieces on July 11, 1979.

For his participation in the superb Skylab 4 flight, Pogue was awarded the Robert J. Collier Trophy, the Dr. Robert H. Goddard Memorial Trophy, and the General Thomas D. White U.S. Air Force Space Trophy.

William Pogue accumulated 84 days, 1 hour, and 15 minutes in space, including 13 hours and 34 minutes in EVA, during the Skylab 4 mission. After retiring from NASA, he became a consultant to the aerospace industry and a producer of videos about spaceflight for the general public. In 1975, he became a fellow of the Oklahoma State University Academy of Arts and Sciences.

POLAND: FIRST CITIZEN IN SPACE.

See Intercosmos Program

POLESCHUK, ALEXANDER F.

(1953–)

Russian Cosmonaut

Alexander Poleschuk was born in Cheremkhovo, in the Irkutsk region of Russia, on October 30, 1953. In 1977, he graduated from the Moscow Aviation Institute with a diploma in mechanical engineering.

Following graduation, he began his career as a test engineer at RSC Energia, the Rocket/Space Corporation. His engineering work has included extensive testing of repair and assembly techniques for use in space, and he has accumulated a large amount of experience in working in simulated weightlessness.

He was selected as a civilian specialist candidate for cosmonaut training in 1989, and underwent two years of intensive training to earn his qualification as a test cosmonaut in 1991. His additional training included advanced preparation for Soyuz flight and Mir space station operations, and he served as a member of the backup crew for the flight of Soyuz TM-15.

On January 24, 1993, he began his first space mission as a member of the Mir 13 long-duration crew, launched aboard Soyuz TM-16 with Gennadi Manakov. The two cosmonauts spent the next six months aboard Mir. During their stay, they conducted two spacewalks. The first, on April 19, was a five and a half hour effort in which they installed solar array drives on the station. They completed the job during a second EVA on June 18.

Another highlight of the Mir 13 mission was the first testing of the androgynous peripheral docking subassembly of the station's Kristall module.

Poleschuk and Manakov returned to Earth on July 22, 1993.

Alexander Poleschuk spent more than 179 days in space during the Mir 13 mission, and accumulated just under 10 hours in EVA during two spacewalks.

POLYAKOV, VALERI V.

(1942–)

Russian Cosmonaut; Set Record for Longest Spaceflight

A veteran of two remarkable long-duration missions aboard the Russian space station Mir, including the longest flight in the history of space exploration, Valeri Polyakov was born in Tula, Russia on April 27, 1942.

He changed his birth name, Valeri Ivanovich Korshunov, to Valeri Polyakov at the age of 15 when his stepfather adopted him. In 1959, he completed his initial education when he graduated from Secondary School Number Four, in Tula. He then attended the Moscow Medical Institute, where he earned a medical degree in 1965.

He was chosen for training as a cosmonaut as a member of the medical group three selection of cosmonaut candidates in March 1972.

Continuing his education concurrently with his cosmonaut career, Polyakov earned a Candidate of Medical Sciences degree in 1976 and specialized in astronautic medicine at the Ministry of Public Health's Institute of Medical and Biological Problems (IMBP) in Moscow. He completed his basic cosmonaut training in 1979.

On August 19, 1988, he began his first space mission aboard Soyuz TM-6 with Vladimir Lyakhov and Abdul Ahad Mohmand, the first citizen of Afghanistan to fly in space. They traveled to the space station Mir, where Lyakhov and Mohmand made a short visit as part of the Soviet Intercosmos program, which provided spaceflight opportunities for citizens of nations that were aligned with or friendly to the Soviet Union. Lyakhov and Mohmand then returned to Earth on September 7 aboard Soyuz TM-5, while Polyakov remained aboard Mir to begin his long-duration stay with the Mir 3 crew, Vladimir Titov, and Musa Manarov.

Conducting medical research—and serving himself as the primary subject of his work—Polyakov remained aboard Mir when Titov and Manarov left the station in December 1987. He continued his work with Mir 4 crew members Alexander Volkov and Sergei Krikalyov, conducting research and serving as on-board physician until he left the station with Volkov and Krikalyov in April 1989, when the three cosmonauts returned to Earth aboard Soyuz TM-7.

At the end of his exceptional first spaceflight, Polyakov had logged 240 days, 22 hours, and 35 minutes in space.

He became deputy director of IMBP after his first space mission, a capacity in which he continued to serve until 1997.

Polyakov began his second space mission, which by its end would be the single longest flight in the history of space exploration, on January 8, 1994, when he lifted off with Viktor Afanasyev and Yury Usachev in Soyuz TM-18. Again traveling

Valeri Polyakov (seen here aboard the Mir space station in 1995) has lived in space for more than 678 days. [NASA/courtesy of nasaimages.org]

to Mir, he settled into a routine of medical research and regularly performed the physical exercise necessary to combat losses in bone density and blood plasma that regularly occur during long stays in space. He also supported fellow crew members in their work and was aboard the station during several historic milestones, including the first rendezvous flight with Mir by an American space shuttle, which took place during the STS-63 flight of the shuttle Discovery in February 1995, and the first joint Russian-European Space Agency (ESA) mission, EuroMir '94, which brought ESA astronaut Ulf Merbold of Germany to Mir in October 1994 for a one-month stay.

Polyakov served with the Mir 15, Mir 16, and Mir 17 crews during his epic second space mission. He returned to Earth aboard Soyuz TM-20 on March 22, 1995, with Mir 17 crew members Aleksandr Viktorenko and Yelena Kondakova.

During his record-breaking 1994–1995 stay aboard Mir, Polyakov spent 437 days, 17 hours, and 59 minutes in space.

Valeri Polyakov spent a career total of 678 days, 16 hours, and 34 minutes in space during his two long-duration visits to the Mir space station.

He retired from the cosmonaut corps in June 1995. In 1997, he earned a Doctorate of Medical Sciences degree, and subsequently joined the faculty of the International Academy of Astronautics.

PONTES, MARCOS C.

(1963–)

Brazilian Astronaut; First Citizen of Brazil to Fly in Space

The first citizen of Brazil to fly in space, Marcos Pontes was born in Bauru, Sao Paulo, Brazil on March 11, 1963. In 1980, he graduated from Liceu Noroeste High School in Bauru, Sao Paulo, and he then attended the Brazil Air Force Academy, where he received a Bachelor of Science degree in aeronautical technology in 1984.

He received training in jet aircraft while attached to the 2/5 Instruction Aviation Group in Natal, Rio Grande do Norte and then joined the 3/10 Strike Aviation Group in Santa Maria, Rio Grande do Sul.

Pontes specialized in the investigation of aircraft accidents as a flight safety officer for 14 years, and in 1989 he received additional training in aeronautical engineering. He was subsequently trained as a test pilot and in that capacity he helped to develop weapons systems and missiles. For his performance in the test pilot training course, he was honored with both the Space and Aeronautics Institute Award and the Empres Brasileira de Aeronautica Award.

Continuing his education concurrently with his military career, Pontes attended the Instituto Technologico de Aeronautica, where he graduated with distinction and received a Bachelor of Science degree in aeronautical engineering in 1993.

He was then chosen to attend the United States Naval Postgraduate School in Monterey, California in 1996, where he graduated with distinction with a Master of Science degree in systems engineering in 1998. Upon his graduation, the Brazilian space agency chose him for training as an astronaut.

During his outstanding military career, Pontes has accumulated more than 1,900 flying hours in over 20 different types of aircraft, including the F-15, F-16, F-18, and MIG-29. He is qualified as a ground attack flight instructor and as an advanced air controller for attack missions, and has risen to the rank of lieutenant colonel in the Brazil Air Force.

After an intensive period of training at NASA's Johnson Space Center (JSC) in Houston, Texas, Pontes was qualified to fly on future missions aboard the U.S. space shuttle as a mission specialist. His initial assignments at NASA included service in the space station operations branch of the agency's Astronaut Office.

On March 29, 2006, Pontes became the first citizen of Brazil to fly in space when he lifted off from the Baikonur Cosmodrome in Kazakhstan aboard Soyuz TMA-8 with International Space Station (ISS) Expedition 13 crew members Pavel Vinogradov of the Russian Federal Space Agency Roscosmos and flight engineer Jeffrey Williams of NASA.

A great national event in Brazil, Pontes's flight to the ISS was also known as the Centennial mission, in honor of Brazil's pioneering pilot Alberto Santos Dumont. ISS Expedition 12 crew members William McArthur, Jr. of NASA and flight engineer Valery Tokarev of Roscosmos greeted Pontes, Vinogradov, and Williams. After a stay of just under 10 days, Pontes returned to Earth with McArthur and Tokarev aboard Soyuz TMA-7 on April 8, 2006.

Marcos Pontes spent 9 days, 21 hours, and 18 minutes in space during his historic first space mission.

POPOV, LEONID I.

(1945–)

Soviet Cosmonaut

A veteran of three space missions, including a long-duration flight in which he spent six months aboard the Salyut 6 space station, Leonid Popov was born in Aleksandriya, Ukraine on August 31, 1945. He attended the Chernigov Higher Air Force School, where he graduated in 1968 with a degree in electrical engineering, and then attended the Gagarin Military Academy in Monino, where he graduated in 1976. During his military career, Popov rose to the rank of major general in the Soviet Air Force.

He was chosen for training as a cosmonaut as a member of the Air Force Group Five selection of cosmonaut candidates in April 1970. His initial duties as a cosmonaut included service as a member of the backup crew for the Soyuz 32/Salyut 6 long-duration mission, which launched in February 1979.

Popov first flew in space as commander of the Soyuz 35/Salyut 6 long-duration mission, which launched on April 9, 1980. Traveling to Soyuz 6 with flight engineer Valeri Ryumin, Popov spent six months in space, setting a new record for the longest stay in space up to that time while serving with Ryumin as the fourth long-duration crew of Salyut 6.

The original crew assignments for the long-duration flight had designated Popov as commander with Valentin Lebedev as flight engineer, but Lebedev was injured during training and Ryumin, who had served 175 days aboard Salyut 6 as a member of the station's third long-duration crew, was chosen to replace him. Boris Andreyev had been Lebedev's original backup, but Ryumin was chosen for the mission because of his previous spaceflight experience.

Despite the confusion surrounding the crew assignments prior to launch, Popov and Ryumin were able to settle into a productive routine in the early days of their stay at Salyut 6.

During their long stay, they welcomed four visiting crews, including three that featured an international guest cosmonaut who had been provided an opportunity to fly in space as part of the Soviet Intercosmos Program. Evolved out of the need to replace a Soyuz spacecraft docked at a space station before it reached its 90-day safe operations limit, the Intercosmos flights were used both to switch a fresh Soyuz with the craft docked at the station and to provide opportunities for citizens from nations aligned with or friendly to the Soviet Union to fly in space.

The first visitors to arrive during Popov's Salyut 6 long-duration mission were Soviet cosmonaut Valeri Kubasov and guest cosmonaut Bertalan Farkas, the first citizen of Hungary to fly in space. They arrived on May 26 in Soyuz 36, and left about a week later in Soyuz 35, leaving their Soyuz 36 craft docked at the station.

The two other Intercosmos flights to visit Popov and Ryumin featured the first spaceflight of a citizen of Vietnam, Pham Tuan, and a citizen of Cuba, Arnaldo Tamayo-Mendez. The first visit, carried out against the backdrop of the American boycott of the 1980 Moscow Olympics and featuring a pilot of the Vietnamese Air Force who had fought United States forces during the Vietnam War, featured particularly harsh anti-American propaganda and engendered unprecedented bitterness in the West. The practical intent of the flight, the switching of Soyuz 37 for Soyuz 36, was successfully carried out without any difficulty.

In addition to the flights highlighted by the presence of their three international visitors, Popov and Ryumin were also visited by Russian cosmonauts Yuri Malyshev and Vladimir Aksyonov, who arrived on June 5 during a test of the Soyuz T-2 spacecraft. The flight was the first manned test of the new, second-generation Soyuz T-series of vehicles, which were gradually phased into the Soviet program and replaced the original Soyuz after the flight of Soyuz 40 in May 1981.

When not welcoming other crews to the station, Popov and Ryumin carried out their assigned scientific work, performed space station engineering and maintenance chores, and kept up a routine of regular exercise necessary to combat the losses in blood plasma volume and bone density that regularly occur during long stays in space. At the end of their long stay, they returned to Earth on October 11, 1980, after a then-record 184 days, 20 hours, and 12 minutes in space.

Popov made his second spaceflight as commander of Soyuz 40, the final flight of a Soyuz craft of the original design, which was subsequently replaced by the Soyuz T-series of spacecraft. Launched on May 14, 1981, Popov and guest cosmonaut Dumitru Prunariu traveled to Salyut 6 for an Intercosmos visit, with Prunariu becoming the first citizen of Romania to fly in space.

They visited with the station's long-duration resident crew members Vladimir Kovalyonok and Viktor Savinykh for about a week, and, following the usual routine for an Intercosmos flight, they participated in televised ceremonies and speeches celebrating the cooperative nature of the flight.

At the end of their visit to Salyut 6, Popov and Prunariu returned to Earth in Soyuz 40, as the usual switching of Soyuz spacecraft was not necessary for Kovalyonok and Savinykh, whose mission would be concluded within the safe operational window of their Soyuz craft (they landed just four days after Popov and Prunariu).

On August 19, 1982, Popov lifted off as commander of his third flight in space, when he launched aboard Soyuz T-7 with flight engineer Aleksandr Serebrov and research engineer Svetlana Savitskaya, the second woman to fly in space (Valentina Tereshkova had been first, during Vostok 6 in June 1963). They traveled to the Salyut 7 space station, where they visited for about a week before returning to Earth aboard Soyuz T-5 on August 27, after a flight of just under eight days.

Popov also served as a member of the backup crew for the Soyuz T-13/Salyut 7 repair mission, which launched in June 1985.

Leonid Popov spent more than 200 days in space during his three space missions.

Popov left the cosmonaut corps in 1987 and attended the Military General Staff Academy in Moscow, from which he graduated in 1989. He subsequently worked at the Russian Ministry of Defense before concluding his active duty military career in 1995.

POPOVICH, PAVEL R.

(1930–)

Soviet Cosmonaut

A pioneering cosmonaut and veteran of two spaceflights, Pavel Popovich was born in the village of Uzin, in the Kiev region of the Ukraine, on October 5, 1930. In 1947, he completed his secondary education, and he then attended the Belaya Tserkov vocational school, where he trained for a career as a carpenter.

He continued his education at the industrial technical school in Magnitogorsk, where he graduated in 1951 after being qualified as a building technician and industrial training master. During that same period, he also pursued an interest in flying and was trained at the Magnitogorsk Aeronautics Club.

He began his association with the Soviet military in 1952 as a cadet at the Stalingrad Military Pilot School in Novosibirsk, and subsequently attended Military Pilot School Number 52 in the village of Vozzhayevka and the Air Force officer's training school in Grosny, where he completed his training in December 1954.

Popovich was initially assigned to FAR 265 as a pilot, and in June 1957 he became a squadron adjutant and senior pilot in the Baltic Air district.

He was chosen for training as a cosmonaut as a member of the first selection of cosmonaut candidates, Air Force Group One, in March 1960. After a period of intensive training, he became a cosmonaut in January 1961.

On August 12, 1962, Popovich began his first space mission, just the eighth launch in the history of manned spaceflight, aboard Vostok 4. He lifted off from the Baikonur Cosmodrome one day after Andrian Nikolayev's launch in Vostok 3 as part of the world's first group flight of manned spacecraft.

Although the Vostok vehicles were not equipped for rendezvous or docking activities, Popovich and Nikolayev were able to travel in orbit in relatively close proximity to each other, which represented a remarkable achievement for such an early point in the history of manned spaceflight.

Launched in Vostok 4 atop a multi-stage SS-6 Sapwood launch vehicle shortly after 8:00 A.M. on August 12, Popovich made his closest approach to Nikolayev, who was already orbiting in Vostok 3, as he entered his prescribed orbit of 159 kilometers by 211 kilometers. Varying accounts of their close encounter described the nearest pass at less than five kilometers or at a distance of six and a half kilometers.

Shortly after entering orbit, Popovich established radio contact with Nikolayev, therein achieving another important milestone in the mission profile for the group flight. Then, after celebrating the achievement of their primary goal of simultaneously flying in orbit, each cosmonaut turned his attention to his individual program of scientific work, which included Earth observations and photography assignments for both cosmonauts. The two pilots had carefully trained for the science work, as it required each of them to manually orient his spacecraft to carry out the assignments.

Popovich and Nikolayev successfully returned the first color video images of the Earth and of their cramped quarters, and excerpts were broadcast on Soviet television.

Although he reported that he was well and that his spacecraft appeared to be operating properly, engineers monitoring his flight became alarmed when a malfunction caused the temperature within Vostok 4 to drop precipitously, and medical personnel misunderstood one of his transmissions and became convinced that he was trying to signal that he was experiencing nausea and flight sickness.

At the time, mission planners were debating the possibility of extending one or both of the flights, in order to test the effects that a longer stay in space might have on the cosmonauts and their vehicles, but worries about the possible ill effects of an extended stay were exacerbated by the concerns surrounding Popovich's health and the status of his Vostok 4 spacecraft. Regardless, it was gradually deduced that he was in no imminent danger, and Soviet space officials ultimately decided to extend both flights—which would result in a new record of four days in space for Nikolayev, who had launched a day earlier than Popovich—and then to bring the cosmonauts back to Earth at virtually the same time.

Popovich tried to communicate his willingness to continue for additional orbits beyond the extended period, but his perspective on the issue was not fully understood until after the end of the mission.

Despite the miscommunications, Popovich successfully completed 48 orbits during his landmark Vostok 4 flight. He re-entered the Earth's atmosphere on August 15, 1962, and Vostok 4 landed just minutes after Vostok 3. Both cosmonauts ejected from their vehicles shortly before the two spacecraft landed; Popovich touched down beneath his parachute after a flight of 2 days, 22 hours, and 57 minutes. Neither man suffered any adverse effects to his health, despite the concerns of medical personnel during the mission.

With Nikolayev and Vostok 1 and 2 veterans Yuri Gagarin and Gherman Titov, Popovich was widely celebrated for his remarkable first flight and was embraced by the Russian people as a national hero.

Upon completion of his first spaceflight, he received the rating of first class military pilot.

Following the Vostok and Voskhod programs, as the attention of Soviet space officials turned to the development of the Soyuz spacecraft and plans for manned lunar landings, Popovich was trained for a planned flight around the Moon that had a hoped-for launch tentatively set for July 1969. This was not to be a lunar landing mission; instead, the proposed flight was to have been the third manned Soviet flight around the Moon, following four earlier, unmanned, test flights. As things worked out, the unmanned tests did not go well enough to approve even the first manned flight, and a series of difficulties in developing the huge N-1 launch vehicle that was roughly analogous to the U.S. Saturn V further hampered plans for manned Soviet lunar missions.

Continuing his education concurrently with his career as a cosmonaut, Popovich earned a pilot-engineer-cosmonaut certificate from the Zhukovsky Air Force Engineering Academy in 1968.

He was also trained for flights in the classified Almaz military space station program, which would subsequently place cosmonauts in space for extended missions during which they would conduct photo surveillance of the activities of nations deemed to be unfriendly to the Soviet Union.

Popovich returned to space as commander of Soyuz 14. He lifted off from the Baikonur Cosmodrome with flight engineer Yuri Artyukhin on July 3, 1974 to travel to the Salyut 3 space station, which had been launched unmanned on June 24. Their mission turned out to be the first completely successful Soviet space station mission.

During a stay of just over two weeks aboard Salyut 3, Popovich and Artyukhin carried out a program that included medical investigations, scientific experiments, and engineering and maintenance tasks aboard the station, all of which were covered in the Soviet media and thereby revealed to observers of the Soviet space program in the West.

In addition to their higher profile duties, however, the cosmonauts also conducted a carefully concealed series of military photo reconnaissance assignments. The classified spying activities were known within the Soviet hierarchy as the Almaz program, and had first been intended for the Salyut 2 space station, which had failed in April 1973 before it could be occupied.

The Soviet Union's Almaz program of placing trained military observers in space to spy on the nation's enemies was similar to the United States' Manned Orbiting Laboratory (MOL) program developed by the U.S. Air Force in the 1960s. The MOL project was abandoned by the decade's end, however, without flying a single manned flight.

During his Soyuz 14/Salyut 3 mission, Popovich spent more than 15 days in space. After his second space mission, he received the rating of cosmonaut second class.

Pavel Popovich spent more than 18 days in space during his two space missions.

Popovich received the degree of candidate of technical sciences in 1977. He left the cosmonaut corps in 1982 to become deputy chief of the scientific research program at the Gagarin Cosmonaut Training Center (GCTC) in Star City. He retired from the military in 1993 with the rank of major general.

Among his many awards and honors, he received special commendation for his contributions to the effort to preserve Soviet space history for future generations as president of the Cosmonautics Museums Association in the mid-1960s. He also received two Gold Star Hero of the Soviet Union medals, two Orders of Lenin, and the Order of the Red Star.

In retirement, Pavel Popovich and his wife Marina, a retired test pilot and colonel in the Soviet Air Force, have supported research into the investigation of unidentified flying objects (UFOs).

PRECOURT, CHARLES J.

(1955–)

U.S. Astronaut

Charles Precourt was born in Waltham, Massachusetts on June 29, 1955. In 1973, he graduated from Hudson High School in Hudson, Massachusetts and then attended the United States Air Force Academy, where he was recognized as a distinguished graduate in 1977, receiving a Bachelor of Science degree in aeronautical engineering. He was also part of an exchange student program during his time at the Air Force Academy, which enabled him to attend the French Air Force Academy in 1976.

He received his initial training as a pilot at Reese Air Force Base in Texas, where in 1978 he was awarded the Air Training Command Trophy as the outstanding graduate of his undergraduate pilot training class.

In 1982, he was assigned to Bitburg Air Base in Germany.

Precourt attended the U.S. Air Force Test Pilot School at Edwards Air Force Base in California in 1985, and then served at Edwards after graduation as a test pilot. In 1989, he was awarded the David B. Barnes Award as the school's outstanding instructor pilot.

Continuing his education concurrently with his military career, he attended Golden Gate University, where he received a Master of Science degree in engineering management in 1988.

Precourt went on to attend the United States Naval War College in Newport, Rhode Island, where he was recognized as a distinguished graduate in 1990, when he received a Master of Arts degree in national security affairs and strategic studies.

During his outstanding military career, Precourt has accumulated more than 7,500 hours of flying time in more than 60 types of military and civilian aircraft. He had risen to the rank of colonel in the U.S. Air Force by the time of his retirement from the service in March 2000.

He also holds commercial pilot, multi-engine instrument, glider, and certified flight instructor ratings, and is a member of the Experimental Aircraft Association. He has built and flown a Varieze experimental aircraft.

NASA selected Precourt for training as an astronaut in 1990, and he first flew in space as a mission specialist during STS-55, aboard the space shuttle Columbia, in 1993.

STS-55 was the second Spacelab flight dedicated to German scientific research. The launch was preceded by a pad abort on March 22, when, just three seconds before the shuttle was scheduled to lift off, Columbia's computers detected an incomplete ignition in the number three Space Shuttle Main Engine (SSME). The fault was later traced to a liquid oxygen leak; at the time, however, the T-3 abort was a nerve-wracking ordeal for crew members and mission controllers.

Fortunately, the launch on April 26 proceeded without incident. Once in orbit, the crew worked around the clock in alternating shifts to complete 88 experiments, and the flight was extended by one day to give the astronauts more time to finish their work. The experiment program included investigations in a wide range of disciplines, including astronomy, the Earth's atmosphere, life sciences, materials science, physics, and robotics. The crew also operated the Shuttle Amateur Radio Experiment (SAREX), which brought them into radio contact with students at 14 schools across the globe. Columbia landed on May 6, 1993, after 160 orbits.

Precourt served as pilot of the space shuttle Atlantis during his second spaceflight, STS-71, the historic first docking of a space shuttle and the Russian space station Mir, and also the milestone 100th human spaceflight launched from Cape Canaveral in Florida.

Launched on June 27, 1995, the STS-71 crew included Russian cosmonauts Anatoly Solovyov and Nikolai Budarin, who would remain on the station as the Mir 19 crew. On the return trip to Earth, the Atlantis crew included Mir 18 cosmonauts Vladimir Dezhurov and Gennady Strekalov and NASA astronaut Norman Thagard, who had lived and worked at Mir since March 1995.

The docking went exceptionally well and was especially remarkable in light of the long history of political and technological differences that had separated the U.S. and Soviet/Russian space programs throughout the Cold War.

The STS-71 crew visited the space station shortly after they and the Mir 18 crew opened the hatches at each end of the docking mechanism. Thagard, Strekalov, and Dezhurov welcomed the shuttle crew during a brief ceremony, and then officially relinquished command of Mir to the Mir 19 crew.

In their five days of linked flight, Atlantis and Mir together constituted the largest spacecraft yet flown in orbit, with a combined weight of about 500,000 pounds. The flight also marked the beginning of a period of cooperation between American and Russian crews that would lead to the commonplace international cooperation of the International Space Station (ISS) program.

While their spacecraft were docked, the astronauts and cosmonauts transferred equipment and supplies, with the shuttle picking up a large number of medical samples that the Mir 18 crew had collected during their stay and the space station receiving EVA tools, water, oxygen, and nitrogen from Atlantis's environmental control system to raise the air pressure within Mir.

Atlantis also carried a Spacelab module in its payload bay, and the crew used the pressurized module to perform life science experiments. Using Thagard, Strekalov, and Dezhurov as test subjects, the STS-71 shuttle mission specialists carried out a total of 15 investigations in a variety of disciplines that included behavioral performance and biology, cardiovascular and pulmonary functions, fundamental biology, human metabolism, hygiene, sanitation and radiation, microgravity research, and neuroscience.

With its crew of eight for the return trip (the STS-71 astronauts and the three returning Mir 18 crew members), Atlantis equaled the previous record for the most crew members on a single shuttle at one time (previously set during the STS-61A flight of the space shuttle Challenger, in 1985). The landmark mission ended on July 7, 1995 with the shuttle's landing at the Kennedy Space Center (KSC) in Florida.

In October 1995 Precourt became NASA's director of operations at the Gagarin Cosmonaut Training Center (GCTC) in Russia. He returned to the United States in May 1996 to become acting assistant director (technical) of the Johnson Space Center (JSC) in Houston, Texas. He served in that capacity until September 1998.

On his third space mission, Precourt commanded the crew of STS-84, aboard the shuttle Atlantis in May 1997. Launched on May 15, STS-84 was the sixth shuttle-Mir docking flight. Michael Foale traveled to the station with the crew to begin his long-duration stay on Mir as part of the ongoing American presence aboard the Russian space station. Foale replaced Jerry Linenger, who returned to Earth on Atlantis after having lived and worked aboard Mir for 123 days. The crew also transferred about four tons of equipment and supplies between the shuttle and the space station during five days of docked flight. Atlantis landed at KSC in Florida on May 24, 1997.

On June 2, 1998, Precourt launched on a fourth spaceflight, as commander of STS-91, aboard the space shuttle Discovery. The ninth and final docking of a shuttle and the Russian space station Mir, STS-91 marked the successful completion of the first phase of the joint American-Russian cooperative space program. The Mir docking was the first for Discovery.

During four days of docked flight, the STS-91 astronauts worked with the Mir 25 crew to transfer supplies and equipment to the station, and to move long-term U.S. experiments and equipment into the shuttle for the return trip to Earth.

The Discovery crew also collected NASA astronaut Andrew Thomas, who had lived aboard Mir for 130 days. Thomas was the last of seven U.S. astronauts to make a long-duration visit to Mir during 907 consecutive days of continual American presence aboard the Russian station. The shuttle landed June 12, 1998, after 154 orbits.

Charles Precourt spent more than 38 days in space during his career as an astronaut.

After his fourth spaceflight, he worked as chief of the Astronaut Corps from 1998 to 2002, and then served as deputy manager for the ISS. He left NASA in 2005 to work for the Thiokol Corporation in Brigham City, Utah.

PROBA

(2001–)

ESA Technology Test Satellite

A micro-satellite designed by the European Space Agency (ESA) and built by an industrial consortium led by Verhaert Space Systems of Belgium, the Project for On-Board Autonomy (Proba) was launched from Sriharikota, India on October 22, 2001.

ESA's first micro-satellite, Proba's mission profile has as its main objective the testing of construction technologies and procedures for future ESA spacecraft. In a mission extended well beyond its initial one-year plan, the spacecraft has successfully captured about 300 high quality images per month utilizing the Compact High Resolution Imaging Spectrometer (CHRIS), which was developed by the British National Space Center (BNSC) and SIRA Space of the United Kingdom.

Images from the instrument have been used in practical Earth-based applications such as the study of flood-prone areas in China and the study of the long-term effects of forest fires in Germany. More than 60 principal investigators at scientific institutions in Europe have made use of the spacecraft in their research.

RAMON, ILAN

(1954–2003)

Israeli Astronaut; First Citizen of Israel to Fly in Space

Ilan Ramon was born in Tel Aviv, Israel on June 20, 1954. He graduated from high school in 1972, and then attended the Israeli Air Force Flight School, from which he graduated as a fighter pilot in 1974. He then received A-4 Basic training and in 1976 he was assigned to Mirage III-C training and operations.

In 1980, he became a member of Israel's first F-16 Squadron and traveled to the United States to receive F-16 training at Hill Air Force Base in Utah. He was subsequently assigned as the F-16 Squadron Deputy Squadron Commander B.

Continuing his education concurrently with his military service, Ramon attended the University of Tel Aviv, where he earned a Bachelor of Science degree in electronics and computer engineering in 1987.

Following his graduation, he became Deputy Squadron Commander A of the Israeli F-4 Phantom Squadron, and in 1990 he attended Squadron Commanders Course. He was F-16 Squadron Commander from 1990 to 1992, and he then served as Head of the Aircraft Branch of the Israeli Air Force Operations Requirement Department. In 1994, he became Head of the Operational Requirements Department for Weapon Development and Acquisition, and was promoted to the rank of colonel.

During his outstanding military career, Ramon accumulated more than 3,000 hours of flying time in the A-4, F-4, and Mirage III-C aircraft, and more than 1,000 hours of flying time in the F-16 aircraft.

NASA selected Ramon as a payload specialist in 1997, and his subsequent training focused on the operation of a multispectral camera capable of recording images of aerosol in the desert.

On January 16, 2003, Ramon lifted off aboard the space shuttle Columbia as a payload specialist during STS-107, becoming the first citizen of Israel to fly in space. The STS-107 crew worked around the clock in two alternating shifts to complete 80 experiments during their remarkable flight.

Tragically, the shuttle Columbia had been seriously damaged at launch (more seriously than NASA officials realized during the mission) and was destroyed by the stresses of re-entry into the Earth's atmosphere on February 1, 2003. Ramon and his fellow crew members were killed when Columbia broke apart over the southern United States, just 16 minutes before the shuttle was scheduled to land.

Ilan Ramon spent 15 days, 22 hours, and 20 minutes in space during the STS-107 mission. He was posthumously awarded the U.S. Congressional Space Medal of Honor, the NASA Space Flight Medal, and the Distinguished Public Service Medal. He was survived by his wife Rona and their four children.

RANGER PROGRAM

(August 1961–March 1965)

U.S. Lunar Exploration Satellite Program

First proposed by the scientists at the Jet Propulsion Laboratory of the California Institute of Technology, the Ranger series of lunar exploration missions was the second unmanned U.S. Moon program after the Pioneer launches of the late 1950s.

The Ranger spacecraft. [NASA/courtesy of nasa images.org]

The Ranger Program was conducted in two separate series of launches. The initial five attempts were designed primarily as scientific exploration satellites intended to photograph the lunar surface and to conduct tests of surface samples via a small (25 inches, or 63 centimeters, in diameter) landing capsule equipped with a seismometer. The experiment capsule included a small piece of balsa wood that was affixed to its bottom to cushion the impact of the device's landing on the Moon.

As NASA's preparations for the manned Apollo lunar landings solidified in the mid-1960s, additional Ranger launches were planned with a new mission profile that called for higher resolution photography. The landing capsule was eliminated in favor of more sophisticated television equipment capable of capturing images of the surface at a distance as close as six and a half feet (two meters) before the spacecraft impacted the surface. Apollo mission planners used these photos as they began their search for landing sites for the later lunar explorers.

Ultimately successful in impacting the Moon with five spacecraft and carrying out three missions that resulted in the return of thousands of images of the lunar surface, the Ranger program was followed by the Lunar Orbiter and Surveyor programs, which also prepared the way for the Apollo lunar landing missions.

The total cost of the nine Ranger missions was about $170 million.

The first five spacecraft in the Ranger series were of similar design and had similar mission profiles. Featuring an aluminum tower atop a hexagonal base, with the craft's solar panels extended from the base, and the first Rangers were equipped with several specialized telescopes, electrostatic analyzers, a magnetometer, particle detectors, instruments to detect and measure cosmic rays and cosmic dust, and counters for measuring solar X-ray scintillations.

Launched atop a multi-stage launch vehicle that featured an Agena B rocket mounted on an Atlas launcher, the Ranger vehicles were propelled into Earth orbit and then placed on the proper trajectory for the trip to the Moon. At the end of its flight, the Ranger spacecraft was designed to take photos of the lunar surface for a period of 10 to 40 minutes before impact.

Ranger 1 (Launched August 23, 1961)

Designed as a test of the spacecraft and its systems, the first Ranger satellite was launched on August 23, 1961. Because it was intended to fly only in Earth orbit, Ranger 1 did not carry a television system or a lunar experiment capsule. NASA engineers were able to direct the craft into its intended Earth "parking" orbit, but the spacecraft's engine failed, ending the mission. Ranger 1 was destroyed during re-entry into the Earth's atmosphere on August 30, 1961.

Ranger 2 (Launched November 18, 1961)

Similarly equipped and with a mission profile virtually identical to Ranger 1, Ranger 2 was launched on November 18, 1961. The spacecraft suffered a fate virtually identical to that of its predecessor, as it attained its planned Earth orbit but was unable to enter the proper trajectory to fly to the Moon because of a fault in its Agena rocket stage. Ranger 2 was destroyed during re-entry on November 20, 1961.

Ranger 3 (Launched January 26, 1962)

The first operational Ranger equipped with a lunar experiment capsule and a television system designed to return photographs of the lunar surface prior to the spacecraft's impact, Ranger 3 was launched on January 26, 1962.

The guidance system designed to control the flight of the spacecraft's Agena rocket stage suffered a malfunction before the planned course correction that would have placed Ranger 3 on the right path to the Moon, and a failure in the craft's communications system prevented the acquisition of television images. Ranger 3 flew past the lunar surface at a distance of 36,800 kilometers on January 28, 1962 and entered orbit around the Sun.

Ranger 4 (Launched April 23, 1962)

Ranger 4, the first U.S. spacecraft to impact the lunar surface, was launched on April 23, 1962. A timer in the craft's computer system suffered a malfunction during the trip to the Moon, which prevented the spacecraft's solar panels from unfolding and caused the vehicle's navigation system to fail. As a result, Ranger 4 crash landed on the Moon's far side on April 26, 1962, without returning photographs or other data.

Ranger 5 (Launched October 18, 1962)

Launched on October 18, 1962, Ranger 5 was placed in Earth orbit as planned and was successfully propelled into the proper trajectory to travel to the Moon after a mid-course correction burn of its Agena rocket stage. At that point in the mission, however, a failure in one of the spacecraft's on-board systems caused the vehicle's batteries to run down, and NASA mission managers lost contact with the satellite. Ranger 5 ultimately flew past the Moon at a distance of 725 kilometers.

Ranger 6 (Launched January 30, 1964)

The first of the Ranger missions designed to gather data to support the planned flights of the Apollo manned lunar program, Ranger 6 was equipped with improved navigation, attitude control, and television equipment and a next-generation hydrazine-fueled engine that was responsible for the mid-course correction that would place the spacecraft on the proper trajectory for the trip to the Moon.

For the first time during a Ranger flight, all the spacecraft's systems appeared to function as expected following launch on January 30, 1964, including the ascent to the Earth parking orbit and the mid-course burn of the Agena stage of the launch vehicle. Ranger 6 traveled to the Moon and successfully impacted close to its intended target site, on the eastern edge of the Sea of Tranquility. Unfortunately, the television system aboard the spacecraft had briefly, and unintentionally, turned on early in the flight due to an electrical fault and was damaged as a result. By the time Ranger 6 landed on the Moon, the camera system was inoperable and no photographs were returned.

Ranger 7 (Launched July 28, 1964)

Launched on July 28, 1964, Ranger 7 was the first entirely successful mission of the Ranger series. The spacecraft entered Earth orbit as planned and was placed on the appropriate course to the Moon. The solar panels deployed properly, the communications system worked properly, and the vehicle's attitude control system functioned successfully throughout the flight.

The craft's cameras were activated when Ranger 7 was 2,110 kilometers from the surface of the Moon, and 4,308 high-resolution images were returned before the spacecraft impacted the lunar surface in the Mare Cognitum region on July 31, 1964.

Ranger 8 (Launched February 17, 1965)

Ranger 8 transmitted 7,137 images of the lunar surface—achieving the most successful flight of the entire Ranger series—before it impacted in the Sea of Tranquility on February 20, 1965.

Launched on February 17, Ranger 8 began transmitting photographs 23 minutes before impact, which took place after a flight of just under 65 hours at a speed of about 2.68 kilometers per second.

Ranger 9 (Launched March 21, 1965)

Equipped with six television cameras to maximize its photographic coverage of the lunar surface, Ranger 9 was launched on March 21, 1965. The flight turned out to be the final launch of the program, and was the third successful Ranger mission.

Arriving at the Moon on March 24, Ranger 9 began taking photographs at a distance of 2,363 kilometers from the surface and transmitted 5,814 images of the Alphonsus Crater before crash landing. The final portion of the Ranger 9 flight, including the images the spacecraft transmitted in real-time, was televised by the major U.S. television networks.

Although additional flights were planned at the time, NASA officials ultimately decided to end the Ranger program with the Ranger 9 flight. U.S. lunar exploration continued with the unmanned Lunar Orbiter and Surveyor programs, which, like Ranger, were carried out in support of the later, manned Apollo lunar program.

READDY, WILLIAM F.

(1952–)

U.S. Astronaut

William Readdy was born in Quonset Point, Rhode Island on January 24, 1952. In 1970, he graduated from McLean High School in McLean, Virginia and then attended the United States Naval Academy, where he graduated with honors in 1974, earning a Bachelor of Science degree in aerospace engineering.

He was trained in the A-6 Intruder aircraft at Naval Air Station Oceana, Virginia and was then deployed to the North Atlantic Ocean and the Mediterranean Sea aboard the USS *Forrestal*.

In 1980, he graduated from the U.S. Naval Test Pilot School at Patuxent River, Maryland, where he was recognized as a distinguished graduate. He was assigned to the Strike Aircraft Test Directorate as a project test pilot and test pilot instructor, and in 1984 he was named the U.S. Naval Test Pilot School Instructor of the Year. He was subsequently deployed to the Caribbean and the Mediterranean aboard the USS *Coral Sea*.

Readdy began his association with NASA in October 1986 when he joined the agency as a research pilot and program manager for the modified Boeing 747 Shuttle Carrier Aircraft. With his new assignment, he transferred from active duty to the U.S. Naval Reserve.

During his outstanding military career, Readdy accumulated 7,000 hours of flying time in more than 60 different types of fixed wing aircraft and helicopters and made more than 550 landings on aircraft carriers. He retired from the U.S. Navy in August 2000.

NASA selected him for training as an astronaut in 1987. His initial technical assignments at the agency included service as chief of the Operations Development Branch, director of NASA's operations at Star City in Russia, and as the first manager of Space Shuttle Program Development responsible for upgrading the space shuttle. He also served as associate administrator of the Space Operations Mission Directorate at NASA headquarters in Washington, D.C., and chaired the Space Flight Leadership Council, which was responsible for overseeing the STS-114 return to flight mission in July 2005.

Readdy first flew in space in January 1992, as a mission specialist aboard the space shuttle Discovery during the STS-42 International Microgravity Laboratory (IML) mission. A milestone in international scientific cooperation in space, the flight featured 55 major IML experiments designed by scientists from six international organizations representing 11 countries.

The STS-42 crew worked around the clock in alternating shifts to conduct the IML experiments, observing the impact of microgravity on a variety of life forms and materials. The mission was extended by one day to allow time for the crew to complete their work, and Discovery made a total of 129 orbits in a little more than eight days before landing at Edwards Air Force Base on January 30, 1992.

On his second spaceflight, STS-51 in September 1993, Readdy served as pilot of the space shuttle Discovery. Launched on September 12, the STS-51 crew deployed the Advanced Communications Technology Satellite (ACTS) on the first day of the flight and then deployed the Orbiting And Retrievable Far And Extreme Ultraviolet Spectrograph-Shuttle Pallet Satellite (ORFEUS-SPAS) the following day.

The ACTS spacecraft was propelled into geosynchronous orbit by the Transfer Orbit Stage (TOS) booster rocket, which was used for the first time on STS-51.

A joint project of the U.S. and German space programs, ORFEUS-SPAS was the first of a series of ASTRO-SPAS missions designed to make astronomical observations from space. Managers at the SPAS Payload Operations Control Center at KSC controlled the instrument, thus marking the first time that a shuttle payload was managed from the Center. The ORFEUS-SPAS craft was retrieved after six days of free flight and returned to Earth aboard the shuttle at the end of the flight.

The STS-51 crew returned to Earth on September 22, 1993, after 157 orbits and more than nine days in space. Discovery touched down on KSC Runway 15 at 3:56 A.M., in the first nighttime landing of a shuttle at KSC.

On September 16, 1996, Readdy lifted off on his third spaceflight as commander of STS-79 aboard the space shuttle Atlantis.

His crew for the flight included pilot Terrence Wilcutt and mission specialists Thomas Akers, John Blaha, Jerome Apt, and Carl Walz.

STS-79 was the fourth docking mission of a space shuttle and the Russian space station Mir. The flight also featured the first exchange of U.S. astronauts aboard the station as part of a continuous American presence aboard Mir that constituted the first phase of cooperation in the planned International Space Station (ISS) program. John Blaha replaced Shannon Lucid aboard Mir, and Lucid returned to Earth on Atlantis on September 26 after having spent 188 days in space.

During five days of docked flight, the STS-79 crew transferred 4,000 pounds of supplies from the shuttle to the station. Atlantis landed at KSC on September 26, 1996.

William Readdy spent more than 28 days in space during his career as an astronaut. He retired from NASA in 2005 to found the aerospace consulting firm Discovery Partners International in Arlington, Virginia.

REAGAN, RONALD WILSON

(1911–2004)

President of the United States, 1981–1989

Born in Tampico, Illinois on February 6, 1911, Ronald Reagan culminated his long political career with his election as the 40th President of the United States on November 4, 1980. He went on to reelection four years later, and served two full terms as President.

His influence on the future course of the U.S. space program was considerable, and several space milestones were achieved during the course of his presidency.

After six years during which no manned U.S. spaceflights were conducted, the launch of the first space shuttle mission, STS-1, took place on April 12, 1981. Although the flight was the culmination of many years of development during the 1970s, it was celebrated as a moment of great pride that seemed particularly appropriate for the upbeat tenor of the new president's successful campaign and his initial months in office, and helped to establish President Reagan as a staunch supporter of the U.S. space program.

Fewer than three years into the space shuttle program, Reagan announced a major new space initiative during his January 25, 1984 State of the Union address, when he directed NASA to begin the development of a project that would initially be known as Space Station Freedom, and which was ultimately transformed into the International Space Station (ISS).

President Ronald Reagan and First Lady Nancy Reagan greet the STS-4 crew, Thomas Mattingly and Henry Hartsfield, at Edwards Air Force Base, July 4, 1982. [NASA/courtesy of nasa images.org]

The announcement of the ambitious space station project represented a high point in the development of U.S. space activities during Reagan's presidency. Two years later, the president delayed his State of the Union address in the aftermath of the loss of the space shuttle Challenger, instead devoting his remarks on that day to a poignant reflection on the tragedy and a strong statement of his desire to see the U.S. space program move forward.

To that end, Reagan appointed a commission led by former U.S. Secretary of State William Rogers that was given the task of investigating the Challenger accident. Among its members, the commission included Neil Armstrong, veteran test pilot Charles "Chuck" Yeager, and Sally Ride, the first American woman to fly in space.

The Rogers Commission presented its final report to the president on June 9, 1986. The investigators found that the cause of the shuttle explosion was the failure of an "O-ring" gasket on one of the vehicle's solid rocket boosters (SRBs). Weakened by icy weather prior to launch, the O-ring became brittle and failed, allowing fuel to leak onto the shuttle's massive external fuel tank, where it ignited the fatal explosion.

In addition to identifying the specific cause of the accident, the commission members also blamed the circumstances leading to the accident on a climate of poor management at NASA and a lack of communication within the organization and between the agency and its suppliers.

Reagan's continued support of NASA throughout the rest of his presidency helped prod the space program forward as the agency struggled to rebound from the Challenger disaster. The first shuttle flight after the accident, STS-26, launched on September 29, 1988, as Reagan neared the end of his second term.

Another aspect of Reagan's involvement with space research and technology was reflected in his championing the Strategic Defense Initiative (SDI), a system of anti-missile defense that initially envisioned the deployment of X-ray lasers capable of intercepting and destroying intercontinental ballistic missiles that might be used against the United States in an attack. The concept of deploying portions of the SDI system in space led to the program becoming popularly known as the "Star Wars" program, and Reagan made use of the reference to the popular science fiction film series when he characterized the Soviet Union as an "evil empire" in a major speech.

Although the SDI project was ultimately not deployed as originally envisioned, the U.S. Congress appropriated a total of $44 billion to develop a variety of experimental systems as part of the research carried out by the SDI Organization within the U.S. Department of Defense. The president authorized the creation of the SDI Organization in 1984.

Although it was widely criticized by the president's opponents as too costly, impractical, and ultimately vulnerable to competing technologies, Reagan's supporters pointed to his use of SDI as a bargaining chip in his dealings with Soviet leader Mikhail Gorbachev during their October 1986 summit meeting in Reykjavik, Iceland as being instrumental in the two nations' signing the Intermediate-Range Nuclear Forces Treaty.

Reagan stunned both his detractors and supporters during the Reykjavik summit when he offered to share the SDI research with the Soviet Union, in an attempt to ensure that neither nation would have an incentive for adventurous testing of the other's capabilities. Although he was greatly concerned by the U.S. SDI development, Gorbachev did not consider Reagan's offer to share the research to be serious or sincere.

Research into SDI and its related technologies was later used in an array of other projects, including subsequent anti-missile systems.

Ronald Reagan died on June 5, 2004 after a long battle with Alzheimer's disease. A measure of his enduring influence on the U.S. space program was evident in the continuous presence of U.S. astronauts and Russian cosmonauts aboard the ISS beginning in November 2000 as part of a cooperative spaceflight program that would have seemed at best unlikely in the days when he first proposed the Space Station Freedom project and publicly predicted the fall of the Soviet Union.

REILLY, JAMES F., II

(1954–)

U.S. Astronaut

James Reilly was born at Mountain Home Air Force Base in Idaho on March 18, 1954. In 1972, he graduated from Lake Highlands High School in Dallas, Texas and then attended the University of Texas, where he earned a Bachelor of Science degree in geosciences in 1977, a Master of Science degree in geosciences in 1987, and a Doctorate degree in geosciences in 1995.

During his exceptional career as a geologist, Reilly has participated in a research expedition to Antarctica and worked in deep submergence vehicles operated by the Harbor Branch Oceanographic Institution and by the U.S. Navy. He worked as an exploration geologist for Santa Fe Minerals Inc. in Dallas, Texas in 1979, and, from 1980 to 1994, he was employed by Enserch Exploration Inc. of Dallas, where he began as an oil and gas exploration geologist and became chief geologist of the company's offshore region division. NASA selected him for training as an astronaut in December 1994.

Reilly first flew in space as a mission specialist aboard the STS-89 flight of the space shuttle Endeavour, which launched on January 22, 1998. The flight was the eighth docking mission between a space shuttle and the Russian space station Mir. Endeavour docked with Mir on January 24 for five days of operations during which the shuttle crew transferred more than 8,000 pounds of equipment and supplies to the station and exchanged STS-89 crew member Andrew Thomas for fellow NASA astronaut David Wolf, who returned to Earth with the STS-89 crew after completing a long-duration stay at Mir.

At the conclusion of the STS-89 mission, Endeavour landed at the Kennedy Space Center (KSC) in Florida after a flight of slightly fewer than nine days in space.

During his second spaceflight, STS-104 in July 2001, Reilly participated in three spacewalks while installing the Quest airlock module at the International Space Station (ISS). Launched on July 12, 2001 aboard the space shuttle Atlantis, STS-104 was the ninth shuttle mission devoted to the assembly of the ISS.

Reilly and fellow mission specialist Michael Gernhardt made their first STS-104 spacewalk on July 13, 2001, working outside the space shuttle for about six hours while beginning the installation of the Quest airlock module. They then spent about six and a half hours in EVA on July 18 while installing oxygen tanks on the airlock, and on July 21, they finished the installation while adding slightly more than five hours to their EVA total for the mission.

The STS-104 crew returned to Earth on July 24, 2001 after 12 days, 18 hours, and 36 minutes in space. During their three STS-104 spacewalks, Reilly and Gernhardt spent a total of 16 hours and 30 minutes in EVA.

Reilly launched on his third flight into space, STS-117, on June 8, 2007, aboard the space shuttle Atlantis. He again traveled to the ISS, where he participated in two of the four spacewalks the STS-117 crew made while installing the ISS S3/S4 Truss Segment. With fellow mission specialist John Olivas, Reilly first ventured outside the shuttle on June 11 to begin attaching the truss segment. He and Olivas spent 6 hours and 15 minutes in EVA during their first excursion.

Their crew mates Patrick Forrester and Steven Swanson made the second spacewalk of the flight, and Reilly and Olivas then made their second EVA, in which they worked to retract recalcitrant ISS solar panels and performed other maintenance and repair tasks during a spacewalk of just under 8 hours. Forrester and Swanson then finished up the STS-117 spacewalking chores, bringing the total EVA time for the mission to a remarkable 27 hours and 58 minutes.

STS-117 concluded with Atlantis landing at Edwards Air Force Base in California on June 22, 2007 after a flight of just under 14 days. Reilly spent 14 hours and 40 minutes in EVA during his two STS-117 spacewalks.

James Reilly accumulated more than 35 days in space, including more than 31 hours in EVA during five spacewalks, during his career as an astronaut.

He retired from NASA in May 2008 to pursue a private sector career and subsequently became vice president of research and development at Photo Stencil Corporation in Colorado Springs, Colorado.

REISMAN, GARRETT E.

(1968–)

U.S. Astronaut

Garrett Reisman was born in Morristown, New Jersey on February 10, 1968. In 1986, he graduated from Parsippany High School in Parsippany, New Jersey and then attended the University of Pennsylvania, where in 1991 he earned Bachelor of Science degrees in economics and mechanical engineering and applied mechanics. He went on to attend the California Institute of Technology, where he earned a Master of Science degree in mechanical engineering in 1992 and a Ph.D. in mechanical engineering in 1997.

Reisman began his engineering career in 1996 as an employee of the space and technology division of TRW Inc. in Redondo Beach, California. NASA selected him for training as an astronaut in 1998. His initial assignments at the agency included a two-week stay at the Aquarius underwater laboratory as part of the fifth NASA Extreme Environment Mission Operations (NEEMO) crew.

He began his first space mission, a three-month stay at the International Space Station (ISS), on March 11, 2007 when he launched with the STS-123 crew of the space shuttle Endeavour. At the ISS, he became a flight engineer for the station's Expedition 16 crew, with Peggy Whitson of NASA, the first female commander of an ISS mission, and Yuri Malenchenko of Russia's Federal Space Agency.

During the period of docked operations shared by the STS-123 and ISS Expedition 16 crews, Reisman participated in his first spacewalk. With STS-123 mission specialist Richard Linnehan, he spent more than seven hours in EVA while delivering the logistics module of the Japanese Kibo scientific laboratory to the space station.

On April 10, 2008, ISS Expedition 17 crew members Sergei Volkov and Oleg Kononenko arrived at the station aboard Soyuz TMA-12 to replace Whitson and Malenchenko. Reisman remained aboard the ISS to become a member of the Expedition 17 crew for two months, until the arrival of the STS-124 flight of the space shuttle Discovery in June 2008. With the docking of Discovery, Gregory Chamitoff took over Reisman's duties as an Expedition 17 flight engineer, and Reisman returned to Earth on June 14, 2008 with the STS-124 crew.

During his time at the ISS as a member of Expeditions 16 and 17, Garrett Reisman spent more than 95 days in space, including 7 hours and 1 minute in EVA.

REITER, THOMAS

(1958–)

ESA Astronaut

As a representative of the European Space Agency (ESA), Thomas Reiter spent a remarkable 179 days aboard the Russian Mir space station during the EuroMir '95 long-duration mission.

Reiter was born in Frankfurt, Germany on May 23, 1958. In 1977, he graduated from Goethe High School in Neu-Isenburg, and he then attended the German Armed Forces University in Neubiberg, where he earned a Master's degree in aerospace technology in 1982.

He subsequently moved to the United States to pursue flight training in military jet aircraft at Sheppard Air Force Base in Texas, and he then returned to Germany to serve in a fighter-bomber squadron, where he flew the Alpha Jet aircraft. Reiter served as a flight operations officer and a deputy squadron commander and then received training at the German Flight Test Center in Manching, where he was certified as a Class Two test pilot in 1990.

Thomas Reiter with STS-121 crew mates (from left): Stephanie Wilson, Michael Fossum, Steven Lindsey, Piers Sellers, Mark Kelly, Reiter, and Lisa Nowak. [NASA/courtesy of nasaimages.org]

In 1992, he received further test pilot training at the Empire Test Pilot School in Boscombe Down, England, where he graduated in December of that year, earning the rating of Class One test pilot.

During his outstanding military career, Reiter accumulated over 2,300 hours of flying time in more than 15 different types of military aircraft.

Reiter began his association with the ESA as a contributor to preliminary studies of the Hermes spacecraft and also assisted in the design of the ESA Columbus module of the International Space Station (ISS). The Hermes project was later abandoned because of financial pressures and an ESA change in plans that emphasized greater participation in the manned programs of the United States and Russia.

Reiter was selected by the ESA for training as an astronaut in 1992. He was among six candidates chosen from more than 22,000 applicants. He completed his initial astronaut training at the European Astronaut Center (EAC) in Cologne, Germany, and then moved to Star City, Russia in August 1993 to begin training at the Gagarin Cosmonaut Training Center (GCTC) for the long-duration EuroMir '95 mission.

He served as a backup to fellow German ESA cosmonaut Ulf Merbold for the Soyuz TM-20 flight to Mir in 1994.

On September 3, 1995, Reiter lifted off aboard Soyuz TM-22 with Russian cosmonauts Sergei Avdeyev and Yuri Gidzenko and traveled to Mir to begin the long-duration EuroMir '95 mission.

At Mir, Reiter conducted the expansive program of scientific experiments that had been developed for the flight at the ESA's Astronaut Center. He conducted a total of 40 European-designed experiments during the flight.

In addition to his long-duration accomplishment, Reiter also achieved another historic first for the ESA when he and Avdeyev ventured outside the station on October 20, 1995 to perform some external maintenance and to install the European Space Exposure Facility (ESEF). With this 5 hour and 11-minute excursion, Reiter made the first spacewalk by an ESA astronaut (Jean-Loup Chrétien of France had made the first spacewalk by a European in 1998, but did so as a representative of the French national space agency).

Reiter added another three hours and six minutes to his EVA total on February 8, 1996, when he and Gidzenko made a second spacewalk to retrieve the ESEF and to repair the antenna on the Mir Kvant 2 module.

Thomas Reiter spent more than 179 days in space, including more than 8 hours in EVA, during his first flight in space.

After his historic flight, Reiter was trained at GCTC to operate the Soyuz TM-series of spacecraft. He also served as a member of ESA's European Robotic Arm development team.

In 1997, he was assigned to the German Air Force to serve as an operational group commander of a fighter bomber wing, flying the Tornado aircraft.

On July 4, 2006, Reiter launched aboard the STS-121 flight of the shuttle Discovery to travel to the ISS, where he served as a member of the ISS Expedition 13 and Expedition 14 crews while conducting the ESA Astrolab program of scientific research. On August 3, he participated in a spacewalk of just under six hours to install EVA hardware on the station's exterior. He returned to Earth on December 22, 2006, after a flight of 171 days.

REPUBLIC OF KAZAKHSTAN—SPACE PROGRAM

(1991–)

National Space Program of the Republic of Kazakhstan

After hosting one of the two most important launch facilities in the world for the first five decades of the space exploration era, Kazakhstan launched its first satellite as an independent nation on June 18, 2006.

The Soviet Union opened the Baikonur Cosmodrome in south central Kazakhstan on June 2, 1955. The remote location was chosen to avoid the scrutiny of the West and particularly that of the United States, with which the Soviets were intensely involved in a Cold War competition to prove the superiority of their missile technology by achieving ever more impressive milestones in spaceflight.

During the first decade of the Soviet space program and throughout the successful unmanned Sputnik program and the manned Vostok and Voskhod flights, a community was built around the launch facility to provide housing and other support structures for the individuals employed at the cosmodrome. The city was named Leninsk in 1966; the actual name of the town nearest the site is Tyuratam. In 1995, after the fall of the Soviet Union, Leninsk was renamed Baikonur to more accurately reflect its purpose and history.

The Baikonur Space Center itself was named for a mining town that is actually located some 300 kilometers north of the launch facility site.

Six of the cosmonauts who flew in space during the Soviet and Russian space programs, into the International Space Station (ISS) era, were born in Kazakhstan. Vladimir Shatalov was the first native of Kazakhstan to fly in space, during Soyuz 4 in January 1969; he flew twice more during his remarkable career and was followed by fellow Kazakhstan-born cosmonauts Viktor Patsayev, Aleksandr Viktorenko (who made four flights), Toktar Aubakirov, Talgat Musabayev (three flights), and Yuri Lonchakov (who has flown in space twice).

When the Soviet Union dissolved in 1991, the leaders of the Republic of Kazakhstan extended an option to the Russian Republic to continue to operate the Russian space program from Baikonur under a lease arrangement in which Russia pays Kazakhstan $115 million annually. The term of the lease extends to 2050.

The two nations also agreed in 2004 to cooperate on the building of a new launch complex at Baikonur known as the Baiterek Launch Site.

Western observers were given their first extensive view of Baikonur in the 1970s, when the U.S. crew members of the Apollo-Soyuz Test Project (ASTP) and their support staff visited the facility during the joint U.S.-Soviet training program for the cooperative flight. In the modern era, the site has become an important nexus for international space activities, with many international visitors and several ISS crews launching from Baikonur, and its new status as a symbol of international cooperation is representative of Kazakhstan's emergence as a strong republic in the post-Soviet era.

Using the facility for the first time for the launch of one of its own space vehicles, the Republic of Kazakhstan launched its first commercial satellite, KazSat 1, on June 18, 2006, atop a Russian launch vehicle.

RESNIK, JUDITH A.

(1949–1986)

U.S. Astronaut

Judith Resnik was born in Akron, Ohio on April 5, 1949. In 1966, she graduated from Firestone High School in Akron and then attended Carnegie-Mellon University, where she received a Bachelor of Science degree in electrical engineering in 1970.

She began her professional career at the RCA Corporation in Moorestown, New Jersey in 1970, and the following year relocated to the company's Springfield, Virginia office. She was honored with the company's graduate study program award in 1971.

As a design engineer for RCA, she helped to develop custom integrated circuits for phased-array radar control systems and control system equipment, and she also worked on sounding rocket and telemetry systems that RCA was developing for NASA.

In 1974, Resnik joined the National Institutes of Health in Bethesda, Maryland, where she served in the Laboratory of Neurophysiology as a biomedical engineer and staff fellow, conducting research in the physiology of visual systems.

Continuing her education concurrently with her engineering career, Resnik attended the University of Maryland and earned a Doctorate degree in electrical engineering in 1977. Her exceptional academic abilities earned her the recognition of the American Association of University Women, which awarded her a fellowship for the year 1975–1976.

Resnik went on to become a senior systems engineer at Xerox Corporation in El Segundo, California in 1978, and was working for Xerox when NASA selected her for astronaut training. She became an astronaut in 1979.

Resnik was a member of the Institute of Electrical and Electronic Engineers (IEEE) and the IEEE Committee on Professional Opportunities for Women, and a senior member of the Society of Women Engineers. She also belonged to the American Association for the Advancement of Science and the American Institute of Aeronautics and Astronautics (AIAA).

In addition to her exceptional technical career, Resnik was also a classical pianist.

On August 30, 1984, she became the second American woman to travel in space when she launched as a mission specialist during STS-41D, the first flight of the space shuttle Discovery.

The maiden flight of Discovery was fraught with long delays and frustrating postponements, and the crew experienced the first pad abort of the space shuttle program. Originally scheduled for launch on June 25, STS-41D was first postponed when one of the shuttle's computers failed. Another near-launch the following day (at T-9 seconds, in the shuttle's first pad abort) had to be called off because of an apparent fault in one of the Space Shuttle Main Engines (SSMEs), which resulted in Discovery being removed from the launch pad entirely so the engine could be replaced.

Changes in the flight schedule necessitated by the long delay caused NASA administrators to cancel the two missions scheduled to follow 41D and redistribute the payloads involved. An August 29 attempt to launch Discovery was delayed by a software problem, and the eventual launch on August 30, 1984 was briefly delayed when a plane strayed into the air space near the launch site.

Henry Hartsfield, Jr. was commander of STS-41D; his crew included pilot Michael Coats, mission specialists Resnik, Steven Hawley, and Richard Mullane, and payload specialist Charles Walker.

Once it finally got under way, the first flight of the shuttle Discovery was a resounding success. During the mission the crew launched three commercial satellites: Satellite Business System SBS-D, SYNCOM IV-2 (also known as LEASAT 2), and TELSTAR 3-C. The crew also tested a solar wing designed by the Office of Application and Space Technology. Equipped with a variety of solar arrays, the large (102 feet long by 13 feet wide) solar wing was designed to prove that large solar arrays could be used to provide power for future space facilities, including the International Space Station (ISS).

Other payloads included the first use of the Continuous Flow Electrophoresis System (CFES) III, which was operated by Charles Walker of McDonnell Douglas, an IMAX camera, and a crystal growth experiment that had been designed as part of the Shuttle Student Involvement Program (SSIP).

In a unique exercise, the STS-41D crew used Discovery's Remote Manipulator System (RMS) robotic arm to remove ice particles from the outer skin of the shuttle, leading mission controllers to dub the astronauts "icebusters."

The busy, productive flight came to a close after 96 orbits when Discovery landed at Edwards Air Force Base in California on September 8, 1984, after a flight of 6 days, 56 minutes, and 4 seconds.

Resnik was next assigned to fly as a mission specialist aboard the space shuttle Challenger during STS-51L, which launched from the Kennedy Space Center (KSC) in Florida on January 28, 1986.

Tragically, Resnik and her STS-51L crew mates were killed when a fault in one of the shuttle's huge solid rocket boosters caused a fuel leak that ignited and caused a massive explosion just 73 seconds after liftoff. She was 36 at the time of her death.

Judith Resnik spent more than six days in space during her remarkable career as an astronaut. As an outstanding engineer and the second American woman to fly in space, she serves as a role model for many. She was posthumously awarded the Congressional Space Medal of Honor.

RICHARDS, RICHARD N.

(1946–)

U.S. Astronaut

Richard "Dick" Richards was born in Key West, Florida on August 24, 1946. In 1964, he graduated from Riverview Gardens High School in St. Louis, Missouri, and then attended the University of Missouri, where he received a Bachelor of Science degree in chemical engineering in 1969.

Richards was commissioned in the U.S. Navy and was designated a naval aviator in August 1970. Continuing his education concurrently with his military career, he attended the University of West Florida, where he received a Master of Science degree in aeronautical systems in 1970.

While assigned to Tactical Electronic Warfare Squadron 33 at Norfolk Naval Air Station in Virginia from 1970 to 1973, he flew support missions in A-4 Skyhawk and F-4 Phantom aircraft. He then flew the F-4 while deployed aboard the USS *America* (CV-66) in the North Atlantic and the USS *Saratoga* (CV-61) in the Mediterranean Sea.

In 1976, he was selected for test pilot training at the U.S. Naval Test Pilot School at Patuxent River, Maryland, where he was honored as a distinguished graduate and subsequently served in the Naval Air Test Center's Carrier Systems Branch and the F/A-18A Program Office of the Strike Aircraft Test Directorate.

As a project test pilot helping to develop systems for automatic landings on aircraft carriers, Richards conducted approach, landing, and catapult flying tests, and as a participant in the Initial Sea Trials of the F/A-18A aircraft aboard the USS *America* in 1979, he made the aircraft's first shipboard catapults and arrested landings.

He was honored for his outstanding work by being named the Naval Air Test Center Pilot of the Year for 1980.

He was assigned to Fighter Squadron 33 when NASA selected him for astronaut training in 1980.

During his military career as a pilot and test pilot, Richards accumulated more than 5,300 hours of flying time, made more than 400 landings on aircraft carriers, and rose to the rank of captain in the U.S. Navy.

Richards first flew in space during STS-28, aboard the space shuttle Columbia. Launched on August 8, 1989, STS-28 was the fourth shuttle mission devoted to the classified activities of the U.S. Department of Defense. Richards served as pilot of Columbia for the mission; he flew into space with STS-28 commander Brewster Shaw and mission specialists James Adamson, David Leestma and Mark Brown. The flight ended after 5 days, 1 hour and 8 seconds, with Columbia's landing at Edwards Air Force Base in California on August 13, 1989.

During his second spaceflight, Richards served as commander of STS-41, which launched aboard the shuttle Discovery on October 6, 1990.

STS-41 was devoted to the deployment of the Ulysses solar polar orbiter, which had been developed by the European Space Agency (ESA).

Ulysses had evolved out of a plan that originally envisioned NASA and ESA collaborating on two spacecraft designed to study the Sun. NASA withdrew from the project in 1981, but ESA continued to develop its portion of the mission and the re-named Ulysses was launched with a mission profile calling for intense study of the Sun's polar areas.

The deployment of the Ulysses spacecraft featured the first combination of an Inertial Upper Stage (IUS) and a Payload Assist Module-S (PAM-S) motor, which together placed the probe on the appropriate trajectory for its journey to the Sun.

After 66 orbits and a flight of just over four days, Discovery landed at Edwards Air Force Base in California on October 10, 1990.

On his third trip into space, Richards was commander of STS-50, which launched aboard the shuttle Columbia on June 25, 1992. The flight set a new record for the longest shuttle mission up to that time, and advanced the study of microgravity through the first flight of the U.S. Microgravity Laboratory (USML-1).

Columbia had undergone an extensive series of modifications prior to STS-50. Engineers at Rockwell International Corporation outfitted the shuttle with more than 50 new or altered pieces of equipment, many of which were designed to make the shuttle more well suited for longer-duration flights. With the installation of the Extended Duration Orbiter (EDO) hardware, Columbia became the first shuttle equipped for longer missions.

Utilizing the USML-1 and its pressurized Spacelab module, the STS-50 crew conducted experiments in a variety of areas. USML-1 was equipped with a crystal growth furnace, a drop physics module, the Surface Tension Driven Convection Experiments, experiments in Zeolite crystal growth and protein crystal growth, a Glovebox facility, the Space Acceleration Measurement System (SAMS), a Generic Bioprocessing Apparatus (GBA), Astroculture-1, the Extended Duration Orbiter Medical Project, and a solid surface combustion experiment.

The flight was extended by one day because of poor weather at the originally scheduled landing site, at Edwards Air Force Base in California, and the landing was subsequently diverted to the Kennedy Space Center (KSC) in Florida. With Columbia's landing on July 9, 1992, STS-50 set a new record for the longest shuttle mission up to that time, as the crew made 221 orbits in 13 days, 19 hours, 30 minutes, and 4 seconds.

On September 9, 1994, Richards launched on a fourth spaceflight—his third space mission as commander—aboard Discovery during STS-64. The flight featured the first untethered spacewalk by American astronauts in 10 years and also successfully tested an innovative optical radar system.

Richards' crew mates for STS-64 were pilot Blaine Hammond, Jr. and mission specialists Jerry Linenger, Susan Helms, Carl Meade, and Mark Lee.

The radar exercise was part of NASA's "Mission to Planet Earth" initiative. An experimental system that utilized laser pulses instead of radio waves, the Lidar In-space Technology Experiment (LITE) instrument was trained on a variety of targets, including cloud structures, dust clouds, and storm systems, among others. Groups in 20 countries around the globe collected data using ground- and aircraft-based radar instruments to help verify the data collected during the LITE experiment.

In a test of NASA's Simplified Aid for EVA Rescue (SAFER) backpacks, STS-64 mission specialists Mark Lee and Carl Meade made a spacewalk of 6 hours and 51 minutes on September 16, in the first untethered American spacewalk in a decade. The SAFER devices were designed as a backup for spacewalking astronauts who might become untethered while conducting extravehicular activities.

The crew also deployed and retrieved the Shuttle Pointed Autonomous Research Tool for Astronomy (SPARTAN-201), which collected data about the solar wind and the Sun's corona during two days of free flight, and the Shuttle Plume Impingement Flight Experiment (SPIFEX), which was carried on the end of the shuttle's robotic arm while collecting data about the potential impact of the shuttle's Reaction Control System thrusters on space structures like the Mir space station or the future International Space Station (ISS). Discovery landed at Edwards Air Force Base on September 20, 1994, after a flight of 10 days, 22 hours, and 51 minutes.

Dick Richards spent more than 33 days in space during his career as an astronaut.

He left NASA's Astronaut Office in 1995 and subsequently served in the agency's Space Shuttle Program Office.

RIDE, SALLY K.

(1951–)

U.S. Astronaut; First American Woman to Fly in Space

The first American woman to fly in space, Sally Ride was born in Los Angeles, California on May 26, 1951. She graduated from Westlake High School in Los Angeles in 1968 and then attended Stanford University, where she earned a Bachelor of Science degree in physics and a Bachelor of Arts degree in English in 1973, a Master of Science degree in physics in 1975, and a Doctorate degree in physics in 1978.

She was selected for training as an astronaut as a member of NASA's eighth group of astronaut candidates, in January 1978. After an intensive period of training, she became an astronaut in 1979. Her initial duties at NASA included service as a spacecraft communicator ("Capcom") during the STS-2 and STS-3 shuttle flights in 1981 and 1982.

On June 18, 1983, Ride became the first female American astronaut to fly in space when she lifted off as a mission specialist aboard the space shuttle Challenger during STS-7. Robert Crippen served as commander of STS-7, and the crew also included pilot Frederick Hauck and Ride's fellow mission specialists John Fabian and Norman Thagard.

During the flight, the STS-7 crew successfully deployed the Canadian Anik C-2 communications satellite and the PALAPA-B1 satellite for Indonesia. They also

Sally Ride, the first American woman to fly in space, aboard the space shuttle Challenger during STS-7. [NASA/courtesy of nasa images.org]

used the shuttle's Remote Manipulator System (RMS) robotic arm to conduct the first deployment and retrieval test of the shuttle program, releasing and then retrieving the Shuttle Pallet Satellite SPAS-01.

Challenger also carried a number of scientific experiments during STS-7, including a payload for the NASA Office of Space and Terrestrial Applications and seven Getaway Special experiments.

Poor weather delayed Challenger's return to Earth at the end of the flight, and the landing was diverted from the Kennedy Space Center (KSC) in Florida to Edwards Air Force Base in California, where the STS-7 crew landed on June 24, 1983 after a flight of 6 days, 2 hours, 23 minutes, and 59 seconds.

Ride also served as a mission specialist aboard the shuttle Challenger during her second space mission, STS-41G—which was again commanded by veteran astronaut Robert Crippen—in October 1984. Lifting off from KSC on October 5, the STS-41G crew also included pilot Jon McBride, mission specialists David Leestma and Kathryn Sullivan, and payload specialists Marc Garneau of the Canada Space Agency (CSA) and Paul Scully-Power.

During the flight, Kathryn Sullivan became the first American woman to conduct a spacewalk when she and David Leestma ventured outside the shuttle for 3 hours and 29 minutes on October 11 to conduct a test of the Orbital Refueling System (ORS), which proved that it is possible to refuel satellites in orbit.

The STS-41G crew also conducted experiments for NASA's Office of Space and Terrestrial Applications and deployed the Earth Radiation Budget Satellite. After a flight of more than eight days, STS-41G came to an end with Challenger landing at KSC on October 13, 1984.

After her second flight in space, Ride was assigned to the planned STS-61M shuttle flight, which was subsequently canceled after the explosion of the shuttle Challenger in January 1986. In the difficult period following the accident, she served as a member of the commission appointed by President Ronald Reagan to investigate the loss of the Challenger and its crew. Led by former U.S. Secretary of State William Rodgers, the commission concluded that the circumstances that led to the accident were, in part, due to a climate of poor management and a lack of communication within NASA and between the space agency and is suppliers. After an exhaustive study, the commission also found the specific cause of the accident, a failed O-ring gasket on one of the shuttle's sold rocket boosters (SRBs).

When the commission completed its work and submitted its report to the President, Ride returned to NASA, where she became special assistant to the Administrator for long range and strategic planning at the agency's headquarters in Washington, D.C. In that capacity she wrote the influential report "Leadership and America's Future in Space" and founded the agency's Office of Exploration.

During her two space shuttle flights, Sally Ride spent more than 14 days in space and made 230 orbits of the Earth.

Ride left NASA in 1987 to join Stanford University's Center for International Security and Arms Control, and in 1989 she joined the faculty of the University of California at San Diego as a professor of physics and was named director of the California Space Institute at the university.

In the wake of the space shuttle Columbia accident in February 2003, Ride was asked to serve on the Columbia Accident Investigation Board (CAIB). Chaired by retired U.S. Navy Admiral Harold W. Gehman, Jr., the CAIB concluded that the Columbia tragedy was due, at least in part, to a "broken safety culture" at NASA. The board found that the shuttle had been damaged more seriously than mission managers realized after Columbia was hit by a piece of foam insulation at launch. In addition to determining the cause of the accident, the CAIB also issued a large number of specific recommendations designed to increase the safety of future shuttle flights.

Since leaving NASA, Ride has written several books for children, and after serving on the CAIB, she took a leave of absence from the University of California to focus on her role as president and CEO of Sally Ride Science, a company devoted to creating science education materials for elementary and middle school students.

Her books include *To Space and Back, Voyager: An Adventure to the Edge of the Solar System, The Mystery of Mars,* and *The Third Planet: Exploring the Earth from Space.*

ROBINSON, STEPHEN K.

(1955–)

U.S. Astronaut

Stephen Robinson was born in Sacramento, California on October 26, 1955. In 1973, he graduated from Campolindo High School in Moraga, California and then attended the University of California at Davis, where he received a Bachelor of Science degree in mechanical and aeronautical engineering in 1978. He began his long association with NASA during his undergraduate years when worked at NASA's Ames Research Center in Mountain View, California as a participant in the University of California's student co-op program.

He began full-time work at Ames in 1979. Continuing his education concurrently with his career as a research scientist, he attended Stanford University, where he earned a Master of Science degree in mechanical engineering in 1985 and a Doctorate degree in mechanical engineering with a minor in aeronautics and astronautics in 1990. His graduate research focused on turbulence physics, and he also studied the dynamics of the human eye. In 1989, he was awarded the Ames Honor Award for his work.

After receiving his Doctorate degree, Robinson moved to Hampton, Virginia to lead a team of 35 researchers in the study of aerodynamics and fluid physics, as chief of the Experimental Flow Physics Branch of NASA's Langley Research Center. In 1992, he co-authored a research study that was honored as the Outstanding Technical Paper in applied aerodynamics by the American Institute of Aeronautics and Astronautics.

He was awarded the NASA/Space Club G. M. Low Memorial Engineering Fellowship in 1993, and while pursuing research at the Massachusetts Institute of Technology (MIT), he studied techniques for capturing satellites during EVA and techniques for performing construction tasks in space. He also conducted neurovestibular research on the crew of STS-58 during his time at MIT.

Robinson also conducted research as a visiting scientist at the U.S. Department of Transportation's Volpe National Transportation Systems Center in Cambridge, Massachusetts. In that capacity he studied environmental modeling for flight simulation, GPS-guided instrument approach procedures, and moving-map displays.

He then returned to NASA's Langley Research Center, where he conducted further studies as a research scientist.

He is also a member of the Experimental Aircraft Association, and has accumulated more than 1,756 hours of flying time in a variety of aircraft.

NASA selected Robinson for astronaut training in December 1994, and he first flew in space as a mission specialist aboard the shuttle Discovery during STS-85 in August 1997. Launched on August 7, STS-85 featured the second flight of the joint NASA-German space agency Deutsches Zentrum für Luft- und Raumfahrt e.V. (DARA) Cryogenic Infrared Spectrometers and Telescopes for the Atmosphere-Shuttle Pallet Satellite (CRISTA-SPAS). The CRISTA-SPAS payload was deployed for eight days of atmospheric study before it was retrieved on August 16. After it was retrieved, the CRISTA-SPAS spacecraft was used in a test of procedures that would later be used during the assembly of the International Space Station (ISS). Robinson operated the shuttle's robotic arm, the Remote Manipulator System (RMS) and the experimental Japanese Manipulator Flight Demonstration (MFD) robotic arm during the flight. Discovery returned to Earth on August 19, 1997.

Robinson's second flight in space was as payload commander during STS-95 in 1998, which featured the return to space of Mercury program pioneer John Glenn.

As the primary operator of Discovery's RMS, Robinson played a key role in the launch and retrieval of the SPARTAN 201 free flyer, which gathered data about the solar wind during two days of independent flight. During STS-95, the shuttle also carried a SPACEHAB pressurized module in its payload bay to facilitate the crew's scientific experiments, and hardware that was being tested for its flight readiness prior to use on a later mission in which it would be used in the servicing of the Hubble Space Telescope (HST).

Robinson next served as the Astronaut Office representative for the ISS robotic arm, and worked as a spacecraft communicator ("Capcom") in Mission Control during several shuttle missions. He also served as a member of the backup crew for the ISS Expedition 4 long-duration mission.

Following the tragic loss of the shuttle Columbia in February 2003 and the long, painful recovery period after the accident, Robinson was a member of the STS-114 crew that led America back into space, aboard the space shuttle Discovery on July 26, 2005. The STS-114 crew spent nearly 14 days in space, including 9 days docked to the ISS during which they delivered equipment and supplies to the space station using the Raffaello Multi-Purpose Logistics Module (MPLM).

Robinson and fellow mission specialist Soichi Noguchi of the Japan Aerospace Exploration Agency (JAXA) made three spacewalks during STS-114 to test methods for repairing the protective heat shield of a shuttle during flight. Designed as a response to the sort of damage that had doomed Columbia, the repair test was one of a number of innovations introduced during the flight.

Another innovation of the flight was the elaborate system of cameras and data acquisition instruments that recorded the launch in great detail, and which plainly showed a chunk of foam insulation flying past the shuttle as it left the launch pad. The chilling video provided clear evidence that more than two years' worth of work on eliminating the possibility that the errant insulation might damage the shuttle, as it had with Columbia, had not solved the problem. In Discovery's case, there did not seem to be a direct hit by the foam, and NASA's engineers and mission controllers concluded that there was no damage. The failure to control the foam was, however, enough for NASA officials to ground all future shuttle flights until a better fix could be found.

Despite the reassurances of the space agency's officials that the crew was not in danger, the flying insulation turned the flight into a national drama, with the American public following the mission's progress with nervous anticipation of the crew's scheduled return to Earth.

For their part, Robinson and Noguchi made the most of their chance to put their long training to use in their innovative spacewalks. During their first spacewalk, on July 30, they spackled a specially prepared repair material onto pre-damaged samples of thermal tiles and panels. Using a paint-like substance for the tiles and squeezing a darkly colored filler material into cracks and crevices on the larger samples, the spacewalkers worked with precision and good humor, carefully carrying out their tasks and at one point reporting the consistency of the paste as being like pizza dough. They also restored power to one of the station's four Control Moment Gyroscopes during their first EVA, and they replaced the ISS Global Positioning System (GPS) antenna.

They replaced a second gyroscope during their second EVA two days later, bringing all four of the devices back to optimal use.

In a remarkable third spacewalk on August 3 that featured an unprecedented under-the-shuttle repair, Robinson rode Discovery's RMS robotic arm to the vehicle's underbelly. Noguchi kept a watchful eye on his spacewalking partner from a perch on the ISS; mission specialist Andrew Thomas coordinated the delicate maneuver, providing a communications link between Robinson and the rest of the crew and mission controllers while mission specialist Wendy Lawrence and pilot James Kelly operated the RMS.

Necessitated by the same sort of fears that arose in the wake of the displaced insulation at launch, the goal of Robinson's daring six-hour EVA was to remove two strips of ceramic-fiber filler that had come loose between a few of the 24,300 protective tiles on the shuttle. The protruding strips had been identified during the extensive in-flight inspections of the shuttle, and NASA officials feared that the dangling filler material could ignite during re-entry. Weighing the dangers posed by the displaced filler versus the danger of sending an astronaut underneath the shuttle during flight, they opted to attempt the on-orbit repair.

Once he was maneuvered into position, Robinson carefully took hold of the displaced filler material and pulled it loose while observers followed his progress via a camera attached to his helmet. Despite the unprecedented nature of his work and the tension accompanying both the EVA and the flight in general, Robinson remained calm and attentive to his work, noting the ease with which he was able to remove the filler material and the grand view he had from his unique position before he returned to the shuttle at the end of the historic spacewalk.

During their three spacewalks, Robinson and Noguchi spent a total of 20 hours and 5 minutes in EVA.

Poor weather in Florida forced the scheduled landing to be postponed for a day, adding to the apprehension surrounding the end of the flight. The landing had to be diverted to Edwards Air Force Base in California, which raised a multitude of new concerns on the part of a wary public. In response to worries about the shuttle's flight path, NASA officials altered their plans so Discovery would not fly directly over Los Angeles; and, where past landings at Edwards had been open to the public, post-9/11 security procedures sharply limited public access to the base.

In spite of all the worries, Discovery landed safely in the early morning hours of August 9, 2005. The nation breathed a sigh of relief as the superb return-to-flight mission demonstrated the astronauts' courage and skill under enormous pressure.

During his career as an astronaut, Stephen Robinson has spent more than 34 days in space, including 20 hours and 5 minutes in EVA.

ROMANENKO, YURI V.

(1944–)

Soviet Cosmonaut

A veteran of three spaceflights, including two long duration missions, Yuri Romanenko was born in the settlement of Koltubanovsky, in the Buzulusk district of

Russia on August 1, 1944. He was educated in Kaliningrad, and then attended the Chernigov Higher Air Force School, where he graduated in 1966 with a certificate in the use of aircraft in combat.

He worked as a concrete and metalworker before attending the Air Force school, and after graduation he was assigned to Training Air Regiment 702 at the school. He served as an instructor pilot for three years and then as a senior instructor pilot until he was selected for cosmonaut training. He received the rating of military instructor pilot second class in March 1969.

Romanenko was chosen for training as a cosmonaut as a member of the Air Force Group Five selection of cosmonaut candidates in April 1970. After a period of intensive training, he became a cosmonaut in 1972. That same year he received the rating of instructor of Air Force Paradrop Training.

His initial assignments as a cosmonaut included service as a member of the backup crew for Soyuz 16, in December 1974, and as a backup crew member during the Soyuz 19 Apollo-Soyuz Test Project mission, which flew in July 1975.

Romanenko was also a member of the backup crew for Soyuz 25 in 1977, which would have been the first long-duration mission to the Salyut 6 space station if a faulty docking mechanism had not prevented the primary crew from docking with the station.

On December 10, 1977, Romanenko lifted off as commander of Soyuz 26, along with Georgi Grechko, who served as flight engineer. Like the Soyuz 25 crew, which had been frustrated in its efforts, Romanenko and Grechko were also given the task of docking with Salyut 6 and occupying the station as its first long-duration crew.

Fortunately, the cosmonauts were able to dock their Soyuz 26 spacecraft at the station's second docking port without any of the difficulties encountered by the Soyuz 25 crew (Salyut 6 was the first Soviet station to have two docking ports; the first five Salyuts were equipped with one).

To help engineers on the ground to figure out whether the previous docking problem indicated a failure on Soyuz 25 or at Salyut 6, Romanenko and Grechko were given the task of making a spacewalk to examine, and if necessary, to repair, the Salyut 6 docking mechanism.

On December 19, 1977, the cosmonauts began their EVA inspection of the station's docking port with Grechko venturing outside of Salyut 6 and Romanenko following his progress from the station's airlock. They spent a total of 1 hour and 28 minutes on the careful examination and found no obvious problems with the docking mechanism.

At one point the attention-consuming inspection resulted in near-calamity for Romanenko. As he floated out of the airlock to gain a better view of Grechko's work at the docking port, he apparently failed to attach the tether that was designed to keep him fastened to the airlock. Fortunately, both cosmonauts realized the seriousness of the situation before Romanenko floated too far from the station, and neither was hurt in the scramble to grab onto the tether and secure it.

Assured of the safety of the flight's commander and the clean bill of health for the Salyut 6 docking port, mission controllers continued to study the prior docking problem while Romanenko and Grechko settled into a productive routine aboard the new space station. They conducted medical and scientific experiments, carried out photography assignments and astronomical observations, and monitored and maintained the station's systems. They also performed the regular exercise necessary to combat the losses of bone density and blood plasma volume that occur during long stays in space.

As the first long-duration crew aboard the first of the next generation Salyut space stations (Salyut 6 being the first station that could be refueled and resupplied, thanks to its second docking port), they largely set the standard for the working routine of future long-duration crews.

They received the first visit from an unmanned Progress cargo spacecraft in January 1978, when Progress 1 arrived with supplies and equipment and achieved the first refueling of an orbiting space station.

Romanenko and Grechko also received human visitors that same month, when cosmonauts Vladimir Dzhanibekov and Oleg Makarov arrived aboard Soyuz 27. Because the Soyuz spacecraft was subject to a limit of 90 days of safe operation in space, after which its onboard systems began to deteriorate, it was necessary to switch one Soyuz craft for another if a crew was to stay in space for a longer period. The Soyuz 27 flight was a test of the switching flights, with Dzhanibekov and Makarov leaving their spacecraft docked at the station and returning to Earth in Soyuz 26 after a brief visit with Romanenko and Grechko.

Another set of visitors arrived in March, when Alexei Gubarev and Vladimir Remek, the first citizen of Czechoslovakia to fly in space, arrived at Salyut 6 for a brief stay as part of the Soviet Intercosmos program, which was designed to provide opportunities for citizens of nations aligned with or friendly to the Soviet Union to fly in space.

Then, with their long-duration mission successfully completed, Romanenko and Grechko returned to Earth on March 16, 1978, landing in Soyuz 27 after a then-record 96 days, 10 hours, and 7 seconds in space. Their remarkable flight broke the previous long-duration mark, which had been set by the American Skylab 4 crew in 1974.

After his superb first flight, Romanenko served as backup commander for the Soyuz 33 Intercosmos flight, which flew in April 1979.

During his second flight in space, Romanenko commanded the Soyuz 38 Intercosmos mission. He lifted off on September 18, 1980 with Arnaldo Tamayo-Mendez, the first citizen of Cuba to fly in space. Romanenko and Tamayo-Mendez visited Salyut 6 long-duration crew members Leonid Popov and Valeri Ryumin, and the four cosmonauts participated in ceremonies and speeches that were aired on Soviet television. After a brief visit, Romanenko and Tamayo-Mendez returned to Earth aboard Soyuz 38 on September 26, 1980 after a flight of just under eight days.

Continuing his education concurrently with his cosmonaut career, Romanenko undertook correspondence studies at the Gagarin Air Force Academy, from which he graduated in 1981 with a specialty in command staff tactical aviation.

Romanenko served as backup commander to Leonid Popov during the Soyuz 40 Intercosmos flight, which took place in May 1981.

On February 5, 1987, Romanenko lifted off on a remarkable third spaceflight, during which he would spend more than 325 days in orbit. As commander of Soyuz TM-2, he traveled to the Mir space station with flight engineer Aleksandr Laveykin to begin the long stay; Romanenko and Laveykin had originally been assigned to the mission as the backup crew, but the primary crew of Vladimir Titov and Aleksandr Serebrov were grounded when a medical exam indicated that Serebrov was experiencing a health problem.

Romanenko and Laveykin settled into a productive routine as they conducted their assigned program of scientific work, carried out astronomical observations and photography, and performed their daily exercises, along with station maintenance and engineering chores.

They also prepared for the arrival of the first Mir expansion module, Kvant 1, which was launched unmanned on March 31 with a large complement of scientific instruments for astronomical observations that had been specially built by the European Space Agency (ESA) and its member nations. Unlike the expansion modules that had in previous years been attached to the earlier Salyut space stations, the Kvant 1 module was equipped with both a docking port and a docking probe, so it could be linked to Mir and still have a docking apparatus available for visits from other vehicles.

On April 5, an attempt to automatically dock Kvant 1 with Mir failed, and a second attempt four days later also failed. Worried mission managers asked Romanenko and Laveykin to make a spacewalk to try to determine the cause of the problem.

On April 11, the cosmonauts ventured outside of Mir for an arduous 3 hour and 35-minute inspection of the docking probe on the Kvant module and the docking port on the space station. As they examined the mechanisms on the two vehicles, they suddenly came across an obvious problem: a piece of plastic was wedged in Mir's rear docking port. Further investigation would later reveal that the plastic had been left behind during the automated docking of the unmanned Progress 28 cargo resupply spacecraft in March. Romanenko and Laveykin were able to tear the plastic away from the docking port, and after their successful EVA, Kvant 1 was docked with Mir on April 12 without further difficulty.

Romanenko and Laveykin made two more spacewalks during their long mission. On June 12, they spent nearly 2 hours in EVA when they began the installation of an additional solar panel on the station's solar array, and on June 16 they finished the job, adding another 3 hours and 15 minutes to their EVA total. They accumulated a total of 8 hours and 48 minutes in EVA during the three spacewalks.

In early July, the cosmonauts faced another challenge when Soviet medical officials monitoring the crew's health detected signs of a heart ailment in Laveykin. Concerned mission controllers decided to arrange a replacement for Laveykin, who was later found to be free of any serious health issues.

Romanenko's new crew mate, Aleksandr Aleksandrov, arrived aboard the Soyuz TM-3 Intercosmos flight on July 22, with Aleksandr Viktorenko and Muhammad Faris, the first citizen of Syria to fly in space.

Romanenko continued his stay aboard Mir for several more months, until Soyuz TM-4 arrived in December with Vladimir Titov, Musa Manarov, and Anatoli Levchenko. Titov and Manarov remained aboard the station as its third long-duration crew, and Romanenko returned to Earth aboard Soyuz TM-3 with Aleksandrov and Levchenko on December 29, 1987. The veteran commander had been in space for a total of 326 days, 11 hours, 37 minutes, and 59 seconds during his remarkable Mir long-duration mission.

On December 30, 1987, Romanenko received the rating of first class cosmonaut.

During his career as a cosmonaut, Yuri Romanenko spent more than 430 days in space, including more than 10.5 hours in EVA during four spacewalks.

After his third flight, he left the cosmonaut corps to serve as chief of division three at the Gagarin Cosmonaut Training Center (GCTC). Having reached the age limit for active duty military service in 1995, he was transferred to the reserve with the rank of colonel.

Among many honors he has received for his contributions to the Soviet space program, Romanenko has been awarded two Gold Star Hero of the Soviet Union medals, three Orders of Lenin, the Order of the Red Star, and the K. E. Tsiolkovsky Gold Medal of the U.S.S.R. Academy of Sciences.

ROMANIA: FIRST CITIZEN IN SPACE.

See Intercosmos Program

ROMINGER, KENT V.

(1956–)

U.S. Astronaut

Kent Rominger was born in Del Norte, Colorado on August 7, 1956. In 1974, he graduated from Del Norte High School and then attended Colorado State University, where he received a Bachelor of Science degree in civil engineering in 1978.

As a participant in the Aviation Reserve Officer Candidate (AVROC) program, he received his commission in the U.S. Navy in 1979. Designated a Naval Aviator the following year, he received F-14 Tomcat training and was subsequently deployed on the USS *Ranger* and the USS *Kitty Hawk* while assigned to Fighter Squadron Two (VF-2) from 1981 to 1985. Rominger also attended the U.S. Navy Fighter Weapons School during his time with Fighter Squadron Two.

He participated in the cooperative program of the U.S. Naval Postgraduate School and the U.S. Naval Test Pilot School, and in 1987 he was recognized as a distinguished graduate of the Naval Test Pilot School and received a Master of

Science degree in aeronautical engineering from the Naval Postgraduate School. He was then assigned to the Carrier Suitability Branch of the Strike Aircraft Test Directorate at Patuxent River, Maryland as an F-14 Project Officer. In that capacity, he flew the initial carrier suitability tests of the F-14B aircraft and made the first catapult and arrested landing trial of the upgraded Tomcat aircraft.

The Naval Air Test Center recognized the superior quality of his work by naming him the center's Test Pilot of the Year for 1988.

In 1990, he was deployed to the Persian Gulf aboard the USS *Nimitz* as part of Operation Desert Storm, where he served as Operations Officer attached to Fighter Squadron 211 (VF-211).

He received the Ray E. Tenhoff Award from the Society of Experimental Test Pilots in 1990, and was recognized as the West Coast Tomcat Fighter Pilot of the Year in 1992.

During his military career, Rominger has accumulated more than 5,000 hours of flying time in more than 35 different types of aircraft and has made 685 landings on aircraft carriers. He is a captain in the U.S. Navy.

NASA selected Rominger for astronaut training in 1992. His technical assignments at the agency included service in the Operations Development Branch of the Astronaut Office, and he has also served as chief of the Shuttle Operations Branch and deputy director of Flight Crew Operations. From 2002 to 2006, he was chief of the Astronaut Office.

Rominger first flew in space as pilot of the space shuttle Columbia during STS-73, the second flight of the United States Microgravity Laboratory (USML-2). STS-73 lifted off on October 20, 1995 after six scrubbed launch attempts.

Once underway, the STS-73 crew worked around the clock in two alternating shifts to conduct the USML experiments, using the pressurized Spacelab module in the shuttle's payload bay. They performed a wide array of experiments, covering biotechnology, combustion science, commercial space processing, fluid physics, and materials science. Highlights included the growth of a record 1,500 protein crystal samples, a comprehensive study of fluid mechanics oscillations, and the growth of five small potatoes in the Astroculture plant growth environment.

One highlight of the flight not related to its science agenda featured the crew, via videotape, throwing out the ceremonial first pitch of the fifth game of the Major League Baseball World Series.

Their exhaustive scientific activities complete, the STS-73 Columbia crew landed on November 5, 1995, after 15 days, 21 hours, 52 minutes, and 28 seconds, the second-longest shuttle flight up to that time.

Rominger again served as pilot of Columbia during his second spaceflight, STS-80, in 1996, which at 17 days, 15 hours, 53 minutes, and 18 seconds, set a new record for the longest space shuttle flight.

The STS-80 crew launched and retrieved two free-flying spacecraft: the Orbiting and Retrievable Far and Extreme Ultraviolet Spectrometer-Shuttle Pallet Satellite II (ORFEUS-SPAS II), which made its second shuttle flight during the mission, and the Wake Shield Facility-3 (WSF-3), which was deployed for the third time. In two weeks of observations, the ORFEUS-SPAS made 422 studies of about 150 astronomical bodies; the Wake Shield Facility successfully cultivated seven thin film semiconductor materials for use in advanced electronics during its three days of free flight. The record-setting STS-80 mission ended with Columbia landing on December 7, 1996, at KSC.

On his third space mission, STS-85, which launched on August 7, 1997, Rominger served as pilot of the shuttle Discovery. STS-85 featured the second flight of the Cryogenic Infrared Spectrometers and Telescopes for the Atmosphere-Shuttle Pallet Satellite (CRISTA-SPAS), which had been jointly developed by NASA and the German space agency Deutsches Zentrum für Luft- und Raumfahrt e.V. (DARA). The CRISTA-SPAS payload was deployed for eight days of atmospheric study before it was retrieved on August 16. After it was retrieved, the CRISTA-SPAS spacecraft was used in a test of procedures that would later be used during the assembly of the International Space Station (ISS). Discovery returned to Earth on August 19, 1997.

During his fourth space mission, in 1999, Rominger was commander of the STS-96 crew for the historic first docking of a space shuttle and the ISS. The STS-96 crew delivered replacement parts and equipment to the space station and prepared the facility for its later habitation by the first long-duration crew.

Mission specialists Tamara Jernigan and Daniel Barry made a 7 hour and 55-minute spacewalk during the flight while installing two cranes and equipment for use by future spacewalkers at the station, and crew members also transferred 3,000 pounds of equipment and supplies during six days of docked operations at the ISS. Discovery returned to Earth in the early hours of June 6, 1999.

On April 19, 2001 Rominger launched on a fifth spaceflight, this time as commander of STS-100 aboard the shuttle Endeavour.

The sixth shuttle flight devoted to the assembly of the ISS, STS-100 delivered the Canadarm2 robotic arm to the station, and Scott Parazynski and Canadian Space Agency (CSA) astronaut Chris Hadfield made two spacewalks to install the 57.7-foot Remote Manipulator System robotic arm.

The STS-100 crew also used the Raffaello Multi-Purpose Logistics Module (MPLM) to deliver equipment and supplies to the ISS. The shuttle spent 8 of the nearly 12 days of its flight docked with the station, and when the two spacecraft undocked, the Endeavour crew filmed footage of the ISS for the IMAX 3-D movie "The International Space Station." Endeavour returned to Earth on May 1, 2001, after 186 orbits.

Kent Rominger spent more than 1,600 hours in space during his career as an astronaut. He left the astronaut corps on September 30, 2006.

ROOSA, STUART A.

(1933–1994)

U.S. Astronaut

A pioneering astronaut who piloted the command module Kitty Hawk in lunar orbit during the Apollo 14 Moon landing, Stuart Roosa was born in Durango,

Colorado on August 16, 1933. He received his initial education at Justice Grade School and Claremore High School in Claremore, Oklahoma, and then attended Oklahoma State University, the University of Arizona, and the University of Colorado, where he graduated with honors and received a Bachelor of Science degree in aeronautical engineering.

Roosa began his 23-year active duty military career in 1953. He flew F-84F and F-100 aircraft as a fighter pilot while assigned to Langley Air Force Base in Virginia, and then served as chief of service engineering at Tachikawa Air Base for two years. He received his U.S. Air Force flight training commission after completing the Aviation Cadet Program at Williams Air Force Base in Arizona, and received gunnery training at Del Rio Air Force Base and Luke Air Force Base.

In July 1962, he was assigned to Olmstead Air Force Base in Pennsylvania as a maintenance flight test pilot. He flew F-101 aircraft in that capacity until 1964 and was then chosen to attend the Aerospace Research Pilots School at Edwards Air Force Base in California. He graduated in September 1965 and remained at Edwards to serve as an experimental test pilot.

During his military career, Roosa accumulated 5,500 hours of flying time, including 5,000 hours in jet aircraft. He had risen to the rank of colonel in the U.S. Air Force by the time of his retirement from the service in 1976.

NASA chose Roosa for training as an astronaut as a member of its Group Five selection of pilot astronauts in April 1966. His initial assignments at the space agency included service as a member of the astronaut support crew for Apollo 9.

On January 31, 1971, he launched as command module pilot aboard Apollo 14, with mission commander Alan Shepard and lunar module pilot Edgar Mitchell.

Roosa's abilities as a pilot played a key role early in the flight, when the first attempt to dock the Apollo 14 command module Kitty Hawk with the lunar module Antares failed to achieve a solid connection between the two vehicles.

After several tries at gently linking the command and lunar modules, mission controllers worried that perhaps some sort of debris that might be preventing the docking. They suggested that the command module pilot try a more severe approach to the docking procedure, and as a result, Roosa's next attempt purposefully charged Kitty Hawk toward the third stage of the Saturn launch vehicle, which still contained the lunar module. With a jarring thud, the two spacecraft finally pulled together and held, and the Apollo 14 flight was cleared to continue.

The rest of the trip to the Moon went well, and Shepard and Mitchell successfully landed on the Moon and carried out two EVAs.

While his crew mates explored the lunar surface, Roosa remained in lunar orbit, alone in the Kitty Hawk command module, for a period of 33 hours. During his lengthy stay in orbit, he conducted an extensive program of scientific experiments, made careful observations of the lunar surface, and carried out a series of photography assignments that included the capturing of images helpful to the planners of future Apollo lunar landing missions, particularly in the choice of future landing sites.

Once Shepard and Mitchell completed their stay on the surface, they returned to orbit to rejoin Roosa in the command module. The two moonwalkers returned with nearly 100 pounds of Moon rocks and soil samples.

At the end of the remarkable flight, Roosa, Shepard, and Mitchell returned to Earth with a splashdown in the Pacific Ocean on February 9, 1971, after a flight of 9 days and 42 minutes.

Roosa subsequently served as backup command pilot for Apollo 16 and Apollo 17 and continued to work for NASA during the early development of the space shuttle program.

Continuing his education concurrently with his career as an astronaut, he attended Harvard Business School, where he completed the advanced management course in 1973.

Roosa retired from the space agency and from the U.S. Air Force in 1976 to pursue a private sector career. He initially served as corporate vice president for international operations for U.S. Industries, Inc. and as president of USI Middle East Development Company, Ltd. In July 1977 he joined Charles Kenneth Campbell Investments as vice president for advanced planning, and in 1981, he became president and owner of Gulf Coast Coors, Inc. in Gulfport, Mississippi.

Stuart Roosa died on December 12, 1994 from complications due to pancreatitis. He was survived by his wife Joan, three sons, and a daughter.

ROSAVIAKOSMOS.

See Russian Federal Space Agency

ROSCOSMOS.

See Russian Federal Space Agency

ROSETTA

(2004–)

ESA Comet Rendezvous Mission

A project of the European Space Agency (ESA), Rosetta is a comet rendezvous mission.

Launched by an Ariane 5 launch vehicle from Kourou, French Guiana on March 2, 2004, the Rosetta spacecraft is scheduled to encounter the comet Churyumov-Gerasimenko in May 2014.

To help scientists gather data about the origin of comets and the relationship between cometary and interstellar material, Rosetta will orbit the comet and map its nucleus, and then release a lander, Philae, onto the surface of the comet in November 2014.

En route to its primary objective, Rosetta successfully encountered the main belt asteroid 2867 Steins on September 5, 2008. The spacecraft achieved a flyby of the asteroid utilizing an on-board navigation camera that determines the vehicle's position in relation to the object being studied.

Discovered in 1969 and named for Soviet astronomer Karlis Steins, asteroid 2867 Steins is a rare type that is difficult to place in any of the accepted systems of asteroid classification.

The first such system was developed in 1975; it categorized individual asteroids as C-types (thought to consist of carbon compounds), S-types (stony asteroids), or the catch-all "U-type."

Classification systems have since been refined in the 14-category Tholen classification, defined by American astronomer David Tholen in 1984, and in the Small Main-Belt Asteroid Spectroscopic Survey of 2002 (which enlarged the Tholen categories to a total of 24 asteroid types).

Rosetta is also scheduled to encounter asteroid 21 Lutetia in July 2010.

ROSS, JERRY L.

(1948–)

U.S. Astronaut

Jerry Ross was born in Crown Point, Indiana on January 20, 1948. In 1966, he graduated from Crown Point High School and then attended Purdue University, where he received a Bachelor of Science in mechanical engineering in 1970. He was commissioned in the U.S. Air Force, and continuing his education, received a Master of Science degree in mechanical engineering from Purdue in 1972.

He served at Wright-Patterson Air Force Base in Ohio, where he participated in tests of a supersonic ramjet missile and served as project manager for the ASALM strategic air-launched missile.

In 1976, he graduated from the Air Force Test Pilot School as a distinguished graduate and received the school's Outstanding Flight Test Engineer award.

Ross went on to serve with the 6,510th Test Wing where he flight tested the B-1 aircraft, on which he subsequently trained and supervised all Air Force B-1 flight test engineer crew members.

During his distinguished military career, Ross accumulated more than 3,900 hours of flight time. He had risen to the rank of colonel in the U.S. Air Force by the time of his retirement from the service in 2000.

He began his association with NASA in 1979 when he joined the Johnson Space Center (JSC) Payloads Operations Division. NASA selected him for astronaut training in 1980.

Ross's first spaceflight began November 26, 1985, with the STS-61B launch of the shuttle Atlantis. The STS-61B crew deployed three communications satellites, and Ross made two spacewalks with fellow mission specialist Sherwood Spring to conduct tests of assembly procedures for building structures in space. His two STS-61B EVAs totaled more than 12 hours.

Jerry Ross rides the robotic arm of the space shuttle Atlantis during STS-61B on December 1, 1985. [NASA/courtesy of nasaimages.org]

Ross again served as a mission specialist aboard Atlantis during his second spaceflight, STS-27, in December 1988. STS-27 was the third shuttle mission devoted to the classified activities of the U.S. Department of Defense.

After his second flight, Ross served on NASA's Astronaut Selection Board.

During his third space mission, STS-37 in April 1991, he and crew mate Jerome Apt made an unscheduled spacewalk that salvaged the mission of the Compton Gamma Ray Observatory (GRO), which was deployed by the STS-37 crew as part of NASA's "Great Observatories" program. Although the deployment appeared flawless, the observatory's antenna would not extend. To fix the problem, Ross and Apt made impromptu repairs during a spacewalk of 5 hours and 32 minutes. Their effort saved the day, and the Compton GRO operated successfully for almost nine years, through June 2000.

Ross and Apt also made a scheduled EVA during STS-37, during which they tested equipment and procedures for assembling structures in space.

As part of his fourth space mission, Ross and his fellow STS-55 crew mates endured a pad abort on March 22, 1993. Just three seconds before the shuttle Columbia was scheduled to lift off, an incomplete ignition was detected in the number three Space Shuttle Main Engine (SSME), and the launch was halted. The fault was later traced to a liquid oxygen leak.

With the pad abort drama behind them, the STS-55 crew launched on April 26, 1993. As payload commander for the flight, which was the second Spacelab mission devoted to scientific research devised by scientists in Germany, Ross helped carry out a program of 88 experiments in a wide range of disciplines, which the crew completed by working around the clock in alternating shifts.

In November 1995, Ross flew in space a fifth time, as a mission specialist for STS-74, the second docking flight of an American space shuttle and the Russian space station Mir. The STS-74 crew brought equipment and supplies to Mir and delivered a docking module designed to streamline future shuttle-Mir docking missions.

During his sixth spaceflight, STS-88, Ross put his spacewalking expertise to good use. Launched aboard the shuttle Endeavour in December 1998, STS-88 was the first mission devoted to the assembly of the International Space Station (ISS). The STS-88 crew delivered the ISS Unity Node and connected it to the previously-launched Zarya Control Module.

Built in Russia with U.S.-based financial backing, the Zarya module had been launched earlier from the Baikonur Cosmodrome in Kazakhstan. It provided the station's electrical power, communications, and propulsion systems during the initial stage of ISS operations. The Unity module provided the necessary connection points for the later attachment of additional ISS modules and components.

Ross and fellow mission specialist James Newman made three spacewalks during STS-88 while connecting the Unity module. On December 7, they attached the new module's cables to the existing Zarya Control Module during an EVA of more than 7 hours. Then, two days later, they spent another 7 hours working in space while installing the new module's antennas. On December 12, they finished their orbital construction work by attaching tools and hardware to the Unity module. During their three STS-88 spacewalks, Ross and Newman spent more than 21 hours in EVA.

On April 8, 2002, Jerry Ross became the first human being to be launched into space seven times when he lifted off aboard the shuttle Atlantis for the STS-110 ISS assembly mission. The STS-110 crew delivered and installed the space station's S-Zero (S0) Truss, an effort that required four spacewalks, all of which were performed for the first time from the ISS Quest Airlock rather than from the space shuttle.

Ross and Lee Morin made two of the four spacewalks, alternating with Steven Smith and Rex Walheim over the course of six days.

Jerry Ross spent more than 1,393 hours in space during his remarkable career as an astronaut. He is the first person to be launched into space seven times, and the first American to make nine spacewalks. His EVA total of 58 hours and 18 minutes set a new record for the highest career total spacewalking time by an American astronaut.

Since his seventh spaceflight, he has served as chief of NASA's Vehicle Integration Test Office at the Johnson Space Center (JSC), and as Chief Astronaut of the NASA Engineering and Safety Center (NESC).

RUKAVISHNIKOV, NIKOLAI N.

(1932–2002)

Soviet Cosmonaut

A veteran of three spaceflights, Nikolai Rukavishnikov was born in Tomsk, Russia on September 18, 1932. In 1957, he graduated from the Moscow Physics and Engineering Institute and then began his career as a physicist and engineer at the Korolev Design Bureau.

Rukavishnikov was chosen for training as a cosmonaut in 1967. His assignments included work on the Soviet manned lunar landing program, the Zond Program.

His first spaceflight began April 22, 1971, when he launched as research engineer of Soyuz 10, with Vladimir Shatalov serving as commander and Aleksei Yeliseyev serving as flight engineer. They traveled to the Salyut 1 space station, which had been launched unmanned on April 19, where they were scheduled to become the station's first resident crew.

Rukavishnikov and his crew mates were able to rendezvous and dock with Salyut 1, but could not transfer from their Soyuz to the space station. The source of their problem has, in the years since the flight, been alternately identified as a malfunction in the environmental control system on one craft or the other, or as a mechanical defect.

After exhausting all reasonable solutions that they and their support teams on the ground could suggest, the Soyuz 10 cosmonauts were forced to abandon their scheduled long-duration mission. They returned to Earth on April 24, 1971, in the first nighttime landing of the Soviet space program. The ill-fated mission ended with a dramatic flourish when a vapor leak inside the Soyuz 10 Descent Module nearly suffocated Rukavishnikov, Shatalov, and Yeliseyev. They were fortunate to survive the harrowing conclusion of their mission without permanent damage to their health.

Rukavishnikov fared far better on his second spaceflight, Soyuz 16, in December 1974. A rigorous test of the systems, equipment, and procedures that would later be used during the Apollo-Soyuz Test Project (ASTP) mission in July 1975, the Soyuz 16 mission was an important step in the development of the historic ASTP flight, which featured the first docking of Russian and American spacecraft.

Following their successful Soyuz 16 test, Rukavishnikov and Filipchenko served as the backup crew for the Soviet portion of the ASTP flight itself. Rukavishnikov also served as a member of the backup crews of Soyuz 21 and Soyuz 28.

He made his third flight in space as commander of Soyuz 33, which launched on April 10, 1979, with guest cosmonaut Georgi Ivanov, the first citizen of Bulgaria to fly in space. The Soyuz 33 mission plan called for them to make a brief visit to the Salyut 6 space station, but a serious malfunction in their spacecraft left them fortunate to survive the abbreviated flight. They were in the vicinity of the Salyut 6 station when Soyuz 33's main engine failed, an unprecedented crisis that forced them to rely on the spacecraft's backup engine during their return to Earth. Rukavishnikov and Ivanov spent a nerve-wracking day in orbit while waiting for a revised flight plan, and then returned to Earth by using their backup engine. The improvised return resulted in a harrowing ballistic re-entry, with the cosmonauts enduring forces as great as 10Gs and then landing 180 miles away from their originally planned landing site.

After his third space mission, Rukavishnikov continued his academic career, and in 1980 he received the degree of candidate of technical sciences.

During his three space missions, Nikolai Rukavishnikov spent more than nine days in space. He left the cosmonaut corps in 1987, but continued to serve the Soviet space program as an employee of NPO Energia. He died of a heart attack on October 19, 2002.

RUNCO, MARIO, JR.

(1952–)

U.S. Astronaut

Mario Runco was born on January 26, 1952 in the Bronx, New York. In 1970, he graduated from Cardinal Hayes High School in the Bronx, and then attended the City College of New York, where he received a Bachelor of Science degree in Earth and planetary science in 1974. In 1976, he earned a Master of Science degree in atmospheric physics from Rutgers University.

Runco began his career as a research hydrologist for the U.S. Geological Survey, and then, in 1977, he became a New Jersey State Police officer. In June 1978 he joined the U.S. Navy. He completed Navy Officer Candidate School in Newport, Rhode Island and served as a research meteorologist at the Naval Research Lab in Monterey, California.

His naval assignments included service aboard the Amphibious Assault Ship USS *Nassau* and as a laboratory instructor at the Naval Postgraduate School, and in December 1985 he was named Commanding Officer of Oceanographic Unit 4 aboard the USNS Chauvenet. At the time of his selection as an astronaut candidate in 1987, he was serving at Pearl Harbor, Hawaii as Fleet Environmental Services Officer.

Runco first flew in space aboard the shuttle Atlantis during STS-44 in 1991. The STS-44 flight featured some classified payloads flown for the U.S. Department of Defense and some unclassified payloads. The crew completed most of their planned activities before the flight had to be cut short because of the failure of one of the shuttle's three orbiter inertial measurement units, which endangered the shuttle's navigational capabilities. Fortunately, Atlantis landed without incident on December 1, 1991.

During his second spaceflight, STS-54 in January 1993, Runco and fellow mission specialist Gregory Harbaugh made a spacewalk of nearly 4.5 hours in the open payload bay of the space shuttle Endeavour while testing EVA techniques that would later be used in the construction of the International Space Station (ISS).

The STS-54 crew also deployed the fifth Tracking and Data Relay Satellite (TDRS-F) and conducted an elementary school exercise concerning the "Physics of Toys," which was taped and later made available as an engaging educational video.

On May 19, 1996, Runco launched on his third space mission, STS-77, aboard Endeavour. During the flight, he operated the shuttle's Remote Manipulator System (RMS) robotic arm to deploy two satellites, the SPARTAN free-flying satellite, which was used in a test of an inflatable antenna structure, and the Passive Aerodynamically-Stabilized Magnetically-Damped Satellite (PAMS), which was used in rendezvous tests.

Runco also filmed additional segments for the Physics of Toys project that became the basis of a sequel to the video that had resulted from the STS-54 mission. The new sequences also led to his making several appearances on the children's television show "Sesame Street" after the conclusion of STS-77.

In three spaceflights, Mario Runco spent more than 22 days in space, including more than 4 hours in EVA.

After STS-77, he continued his NASA career as an Earth and planetary scientist.

RUSSIAN FEDERAL SPACE AGENCY

(1957–)
Russian National Space Program

The origins of the Russian space program, and to some degree, the entire modern age of space exploration, can be traced back to the ideas of a Russian teacher, Konstantin Tsiolkovskii, who was born in 1857. Ill as a child with scarlet fever, Tsiolkovskii struggled with his early education and pursued much of his initial learning on his own with help only from his parents. He eventually prepared himself well enough to begin a formal education, and in time he became a teacher, dedicating himself to helping others to learn.

In the early 1880s, Tsiolkovskii began serious studies of several technologies that would ultimately prove essential for spaceflight. Combining the unique perspective he had gathered from his years of nontraditional education and the extensive technological expertise he had gathered during his formal

The Soyuz TM-31 launch vehicle is moved to the launch pad at the Baikonur Cosmodrome in Kazakhstan, October 29, 2000. [NASA/courtesy of nasaimages.org]

studies, he published a number of innovative academic papers that described many aspects of rocket technology that could be used for spaceflight.

Among the topics covered in his work, it was the first proposed use of liquid oxygen and liquid hydrogen that was perhaps the most significant. When Tsiolkovskii first described the use of liquid rocket fuel, in his 1903 paper "Investigating Space with Reaction Devices" (his first published academic work), the idea was a major innovation; up to that time, rockets had been fueled by solid propellants. In most cases, the propellant of choice was usually gunpowder, as rocketry was generally considered an adjunct of military activity. Tsiolkovskii's suggestion that liquid fuel could be used as a propellant went hand in hand with his vision of sending rockets into space, rather than utilizing them solely as weapons.

Tsiolkovskii was also the first theorist to describe a multi-stage launch vehicle, another major contribution to the body of scientific work that led to the birth of the space age, and he envisioned the rudimentary details of satellites, space stations in Earth orbit, and life support systems for future space travelers.

Although they might have seemed fantastic to those who first heard them, Tsiolkovskii's remarkable theories about rocketry and spaceflight strongly influenced the generation of young Russian engineers and scientists who began their careers in the late 1920s and 1930s. The pioneering teacher and space theorist lived long enough to see his ideas embraced by his followers; at the time of his death in 1935, Tsiolkovskii was already being hailed for his wide-ranging contributions and would ultimately be accorded far-reaching recognition as the "father of astronautics."

Russian interest in the expanded use of rocket technology grew quickly in the late 1920s and early 1930s, thanks in large part to the Soviet government's desire to develop more sophisticated weaponry for the Red Army. Two major figures in the later Soviet space program, Sergei Korolev and Valentin Glushko, began their careers during that time, each working on the development of new rocket technology in military-sponsored programs.

In 1933, Soviet political leaders founded the Scientific Research Institute of Jet Propulsion, and the nation launched its first liquid fuel rocket. The pace and magnitude of the Soviet achievements in rocketry science seemed remarkable for the time and promising for the future, but a change in the country's political climate scuttled much of the progress by the end of the decade, as a result of government purges under Soviet leader Josef Stalin.

As a means of consolidating his power, Stalin ordered large-scale executions, contrived legal proceedings with coerced and faked evidence, the mass imprisonment of leading citizens, and the creation of a pervasive climate of fear and intimidation. Although neither had done anything to warrant such treatment, both Korolev and Glushko were imprisoned in a gulag, and in 1938 the Institute of Jet Propulsion was closed.

Faced with a nearly overwhelming challenge by invading German forces in the early years of World War II, the Soviet government turned to those engineers and scientists who had survived the purges for help in designing weapons systems for military use.

Soviet interest in the innovative use of rocket technology continued with the end of World War II. While U.S. forces captured the large majority of the scientists who had developed Germany's wartime V-2 vengeance weapon liquid fuel rockets, the Soviet Union secured the V-2 production and laboratory sites at Peenemunde and Nordhausen, along with many of the technicians who had worked at the facilities throughout the development of the V-2 program. The Peenemunde site was largely abandoned by war's end, and the Americans were first to reach Nordhausen; as a result, the Soviets began the postwar period at a decided disadvantage that was gradually overcome by the application of large amounts of research and development expenditures and the extensive use of Soviet military resources.

At the same time, a change in the internal political climate within the Soviet Union resulted in the release and rehabilitation (the dismissal of formal charges and restoration of an individual's reputation) of those scientists who had survived the purges of the late 1930s. These included Korolev and Glushko, who both went on to assume important positions in the post-war Soviet development of rocket technology.

Following the manufacture of several dozen V-2s at Nordhausen, the Soviet rocket program was relocated to Kapustin Yar in May 1946. The first Soviet V-2 rocket successfully lifted off from the Kapustin Yar launch facility the following year, on October 30, 1947.

Korolev played an important role in the development of Soviet rocket technology from the very beginning of the post-war program. He oversaw the initial V-2 production in Germany and led the German and Soviet teams that launched the first Soviet V-2s. Most importantly, and in a pattern that would remain a constant for the rest of his life, he proved himself capable of responding to the demands of Soviet political leaders by designing and rapidly developing amazing, yet practical, space-related technologies, vehicles, and missions.

The earliest example of Korolev's exceptional abilities was evident in the development of the SS-1, SS-1A, and SS-2 rockets, the latter of which had a range of 348 miles (556 kilometers) when it was first operationally tested in 1950. Originally intended as missiles for use by the Red Army and also as a means of developing the technology necessary to create weapons capable of reaching Western Europe, these early rockets led to the SS-3, which was the first Soviet missile capable of carrying a nuclear payload. The Soviet Union tested its first nuclear weapon in August 1949.

The emphasis on developing missiles primarily for military use continued as relations between the Soviet Union and the United States deteriorated during the early- and mid-1950s. With the death of Stalin in 1953 and the installation of Nikita Khrushchev as the nation's new political leader, however, interest in the use of rockets as space launch vehicles slowly began to take hold. The change was likely exacerbated by the realization that the rough parity of Soviet and American nuclear capability prevented either nation from actually attempting to use such weapons to gain military advantage, and by the promise that a successful space venture could give either nation a significant advantage in the struggle to prove its technical superiority over the other.

In the competition to convince nonaligned countries—that is, those countries that had not taken a side in support of the Soviet Union or the United States in the two superpowers' global struggle for political and economic influence—the demonstration of spaceflight capabilities was seen by leaders in both nations as compelling evidence of technical and military superiority and, by extension, evidence of the superior nature of the social and political climate that could produce such capabilities.

With such high stakes in mind, the Soviets set out to build a state-of-the-art launch facility to host their spaceflight activities. In the mid-1950s they began construction of what would in time become known as the Baikonur Cosmodrome, near Tyuratam. The new site would be the vital center of the Russian space program. In its initial years, the facility also served as the birthplace of the SS-6 launcher, which would also become a long-term constant in the Soviet and Russian space effort as arguably the most successful launch vehicle in the first half-century of the space age.

The rocket that would become the workhorse of the Russian space program, the SS-6 Sapwood (as it was commonly known in the West, where it was also sometimes referred to as the "A booster;" while the Soviets referred to the launcher as the R-7) was an intercontinental ballistic missile (ICBM). It featured one large central stage with four engines and four small stages that each contained an additional four stages. The smaller stages were attached to the bottom of the central stage, along with 12 steering rockets. As a single entity, the SS-6 launch vehicle was capable of delivering a warhead as heavy as 9,000 pounds (4,082 kilograms).

A first attempt to launch the SS-6 was made on May 15, 1957; it ended in failure less than a minute after lift-off. Several other failed attempts followed, and then, on August 3, 1957, the Soviets achieved the first completely successful SS-6 launch. Several more tests confirmed the vehicle's reliability, and with Khrushchev's approval and support, Korolev assigned the engineers of his design bureau the task of adapting the SS-6 for use as a launch vehicle for spaceflight.

For his part, Korolev personally oversaw the design and manufacture of Sputnik 1, a tiny aluminum globe that would be the Soviet entry in the International Geophysical Year (IGY) competition to be the first nation to launch a satellite into Earth orbit. At 10:28 P.M. Moscow time on October 4, 1957, Sputnik 1 was successfully propelled into orbit by a modified SS-6, inaugurating the space age.

An impressive scientific and engineering achievement, the launch of Sputnik 1 demonstrated the Soviets' technical capabilities, but also confirmed U.S. intelligence estimates that rated the Soviet Union's spaceflight capabilities as roughly on par with those of the United States.

Popular reaction to Sputnik 1 was far from analytical or dispassionate, however. Fanned by Khrushchev's carefully constructed and frequently pronounced propaganda, the success of Sputnik 1 and its immediate successors (Sputnik 2, launched with a dog, Laika, on November 3, 1957, and the instrument-laden Sputnik 3, which successfully reached orbit on May 15, 1958) gave rise to the popular notion in both the United States and the Soviet Union that the Soviets were far advanced in their development of space-related technology and, by extension, in the development of missiles for potential military use.

In reality, the nascent U.S. space program had for the most part kept pace with the early Soviet program. The first successful U.S. satellite, Explorer 1, launched on January 31, 1958, not quite four months after Sputnik 1 was launched.

Khrushchev kept up an intense propaganda campaign about the perceived Soviet space advantage throughout the 1950s, while at the same time constantly increasing pressure on Korolev and the engineers of the Soviet space program to design and execute ever-more audacious adventures in spaceflight.

Initially, the combination of political pressure and technological acumen garnered impressive results. Korolev's Design Bureau launched probes to explore the Moon, Mars, and Venus, while simultaneously moving inexorably closer to a manned launch attempt.

Launched on January 2, 1959, Luna 1 was the first spacecraft to travel beyond Earth orbit. Its successor, Luna 2, became the first craft to impact the lunar surface when it crash-landed on the Moon on September 14, 1959. Luna 3, launched on October 4, 1959, transmitted the first images of the far side of the Moon.

Soviet attempts to send probes to Mars and Venus were almost entirely unsuccessful throughout the early- and mid-1960s. After years of intensive effort, however, they were able to successfully land a probe on Venus in December 1970, with Venera 7. The Mars 3 spacecraft was the first Soviet-built vehicle to land on Mars, in December 1971, but it failed less than a minute after landing.

At the same time that they were developing their early, unmanned satellite programs, the Soviets were also working on a manned launch capability.

On April 12, 1961, Soviet cosmonaut Yuri Gagarin became the first human being ever to fly in space.

He launched in a man-sized sphere known as a Vostok spacecraft, atop a modified SS-6 Sapwood (an additional stage was added to the top of the original SS-6 for manned launches; the modified launcher was commonly referred to as the "A-1" in the West). The highly automated Vostok was designed to minimize the piloting duties of the crew member, as at the time, very little was known about the impact that exposure to the space environment might have on a human being. Fortunately, Gagarin completed his Vostok 1 mission in good health and good spirits, completing a single orbit and returning to Earth safely.

Five other Vostok flights were made in the two years following Gagarin's historic mission. Fueled by the constant exhortations by Khrushchev to design and execute landmark spaceflights in advance of the United States, Korolev achieved a remarkable series of space successes that included the first day-long flight (Vostok 2), the first group flight of two spacecraft (Vostok 3 and Vostok 4), the longest one-person spaceflight (Vostok 5), and the first flight in space by a female crew member (Valentina Tereshkova in Vostok 6).

Thus the early years of the Soviet program of human spaceflight were characterized by intense political pressure, explicit competition with the United States, and an emphasis on the realization of symbolically important milestones even at the expense of the development of long-term goals.

During the Vostok program, the peculiar mix of political context and technical advancement appeared to work well, as Khrushchev pushed Korolev and his engineers from one successful mission to the next. In the aftermath of U.S. President John F. Kennedy's May 1961 proposal that the U.S. should try to land an American on the Moon by the end of the 1960s, however, the successful Soviet space achievements seemed to fade in significance.

With the end of the Vostok program, the Vostok vehicle was transformed into the Voskhod spacecraft. In anticipation of the first manned flight of the U.S. Gemini program, which would feature a two-man crew, Khrushchev instructed Korolev to launch a flight with three cosmonauts. The result was the Voskhod 1 flight, which launched on October 12, 1964. While the flight was in progress, and after Khrushchev contacted the crew in orbit to congratulate them on their historic flight, Soviet political leaders decided to make a change in the nation's highest office. They replaced Khrushchev with Leonid Brezhnev, who carried out his first official duty as the country's new leader when he welcomed the cosmonauts back to Earth.

The momentum of the Khrushchev years continued with the second Voskhod flight, however, and Voskhod 2 resulted in the last significant milestone of the early Soviet space program. Launched on March 18, 1965, the Voskhod 2 mission profile included an attempt at the world's first extravehicular activity (EVA), or spacewalk, as the exercise quickly became known.

Alexei Leonov successfully achieved the first EVA in history during the Voskhod 2 flight, and safely returned to Earth in good health despite experiencing a great deal of difficulty in getting back into the spacecraft at the end of his spacewalk.

With the change in political leadership and the end of the Voskhod program, the future course of the Soviet space effort was dramatically altered. The previous emphasis on symbolic missions was replaced with a new dedication to developing the technology necessary for beating the United States to be the first nation to land one of its citizens on the Moon. Free of the distracting pressures of the program's early years, Korolev looked forward to concentrating the efforts of his Design Bureau on the Soviet Moon program.

But Korolev was not fated to lead his nation to the Moon. He died on January 14, 1966, as a result of complications from minor surgery. His death was an enormous loss for the Soviet space program, a loss whose magnitude was only fully understood years after his death. In all the time he spent as the crucial figure in the Soviets' emergence as a leading space power, Korolev was publicly identified only as the "Chief Designer of Carrier Rockets and Spacecraft" because the Soviet government feared that revealing his name might compromise his safety. As a result, the oppressive emphasis on secrecy that characterized the totalitarian Soviet state's approach to its space program obscured public knowledge of Korolev's remarkable legacy for many years.

Similar deceptions were used by the Soviets to deflect attention from a variety of failures and mishaps that occurred during the early years of their space program. A particularly poignant example involved the tragic death of Valentin Bondarenko, who was killed during training just weeks before Yuri Gagarin's historic flight in 1961. Not wanting to detract from the accomplishment of Gagarin's launch, Soviet officials decided to avoid any mention of Bondarenko's death. The young cosmonaut trainee was quietly buried in his hometown, and no account of his passing was acknowledged for more than two decades, until the mid-1980s policy of glasnost (openness) took hold.

In a similar vein, unmanned spacecraft that failed in Earth orbit were given the prefix of "Cosmos" in an effort to disguise the true nature of their intended missions.

A tragedy that the Soviets did not try to hide resulted from the first flight of the next-generation Soyuz spacecraft, Soyuz 1, in April 1967. Although the Soyuz would go on to be a reliable spacecraft through several iterations over many decades, Soyuz 1 proved tragically beyond the control of its occupant, cosmonaut Vladimir Komarov, and mission controllers on the ground. The spacecraft crash-landed, killing Komarov and imperiling the Soviet push toward the Moon.

It was the failure of the massive N-1 launch vehicle, however, that definitively cost the Soviets their chance to attempt a manned lunar mission. Beginning in February 1969, the Soviets tried to test-launch the N-1 four times, with each attempt ending in spectacular, fiery failure. The last N-1 launch attempt was made on November 23, 1972.

Unmanned lunar missions progressed much more successfully for the Soviets. They failed to soft-land Luna 15 on the lunar surface at the same time as the manned American Apollo 11 landing (they had apparently intended Luna 15 to land and collect a sample while Neil Armstrong and Buzz Aldrin were on the surface), but subsequently succeeded in landing several vehicles on the Moon that gathered lunar surface samples and returned them to Earth (including Luna 16 in September 1970 and Luna 20 in February 1972). The Soviet Luna 17 (November 1970) and Luna 21 (January 1973) missions carried small rovers that explored the lunar surface via remote control by mission managers on Earth.

With the successful manned lunar landings of the U.S. Apollo program, the Soviet space program again changed direction, moving away from any further attempt to mount a manned Moon flight and focusing instead on the establishment of space stations in Earth orbit.

The Soviets launched the world's first space station, Salyut 1, on April 19, 1971, and the Soyuz 11 crew, Georgi Dobrovolski, Vladislav Volkov, and Viktor Patsayev, followed on June 6, 1971. They became the first human beings ever to live and work in space for an extended period, and through a series of entertaining television broadcasts, they endeared themselves to a large following of their fellow citizens on Earth. Their amazing odyssey came to a stunning finale, however, when a malfunction in their Soyuz spacecraft caused them to be suffocated during re-entry.

As had been the case with the Soyuz 1 accident four years earlier, the June 1971 death of the Soyuz 11 crew was an enormous shock for the Russian people, and a serious setback for the Soviet space program.

Under the guidance of Vasily Mishin, who had succeeded Korolev as the Soviet Chief Designer, the Soviets continued the Salyut program and at the same time launched several military space stations that were designed to gather surveillance photographs and data about the United States and other nations considered to be enemies of the Soviet state. To conceal the true nature of the military stations, which were referred to as Almaz space stations within the Soviet military and space establishment, the spy vehicles were publicly referred to as Salyut stations, and the activities of their crew members were disguised in public pronouncements.

In July 1975, in the midst of the Salyut era, the Soviet Union and the United States mounted a remarkable joint mission known as the Apollo-Soyuz Test Project. The flight was the result of an agreement signed by U.S. President Richard Nixon and Soviet leader Leonid Brezhnev, which was itself a symbolic highlight of President Nixon's landmark visit to the Soviet Union in 1972.

Years of unprecedented cooperation between the two space superpowers preceded the singular mission, which began on July 15, 1975 with the launch of Soyuz 19, manned by cosmonauts Alexei Leonov and Valeri Kubasov, at the Baikonur Cosmodrome. Eight hours later, the final U.S. Apollo spacecraft launched from the Kennedy Space Center (KSC) in Florida carrying NASA astronauts Thomas Stafford, Donald "Deke" Slayton, and Vance Brand. The two spacecraft docked on July 17, and over the next several days, the crew members shared meals, exchanged gifts, and participated in live television broadcasts. They also took turns docking and undocking the two spacecraft and testing the innovative docking mechanism that had been jointly designed for the flight by Russian and American engineers.

After 47 hours of docked operations, Leonov and Kubasov returned to Earth on July 21, 1975; Stafford, Slayton, and Brand returned on July 24.

The forward-looking ASTP mission was an exceptional achievement for its time and a major political event that helped to defray tensions between the rival superpowers during the period following the conclusion of American involvement in the Vietnam War and prior to the Soviet invasion of Afghanistan.

While the earlier Salyut space stations were equipped with a single docking port, the two final Salyuts—Salyut 6 and Salyut 7—each had two docking ports. The expanded docking capability made long-duration missions possible by enabling the delivery of extra fuel and supplies via unmanned cargo carriers known as Progress spacecraft. While the cosmonauts' Soyuz spacecraft occupied one port, the second could accommodate a Progress craft, which would be flown under the control of mission managers on Earth and automatically docked with the station for a brief period before being undocked and de-orbited over the Pacific Ocean.

Even as crews spent longer periods in orbit, however, their Soyuz spacecraft was limited to a maximum of 90 days in space during which their on-board systems could be relied upon to operate safely. As a result, additional Soyuz flights were required to switch a fresh spacecraft with a vehicle that had been docked at the station for several months.

Soviet officials took advantage of the brief switching flights to provide opportunities for citizens from Soviet-bloc nations to visit Salyut 6 and Salyut 7 as part of the Intercosmos program. The program had previously involved scientists and engineers from nations aligned with or friendly to the Soviet Union in unmanned space projects. The manned portion of the program began in March 1978, when Vladimir Remek of Czechoslovakia visited Salyut 6 aboard Soyuz 28.

The Intercosmos program was gradually expanded over the years to include visitors from nonaligned nations, including France and India, and continued with visits to the Mir space station.

In the long history of the Russian space program to date, the Mir space station stands out as the paramount achievement and ultimate expression of years of innovative engineering. Building on the systems and techniques developed for the Salyut program, the Soviets launched Mir on February 20, 1986.

The Soviets, and later, the Russian Republic, successfully operated Mir in Earth orbit for more than 15 years. Twenty-eight long-duration crews visited the station, and a total of 78 extravehicular activities (each involving two crew members) were conducted from the space station.

The Mir core module weighed 20.4 tons when it was launched in 1986, and with the addition of six expansion modules over the course of 10 years, the station grew to a total weight of more than 100 tons when it was fully expanded. With a docked Soyuz and Progress cargo spacecraft, the finished station measured 107 feet long by 90 feet wide.

Mir ultimately outlived the totalitarian Soviet regime that launched it into orbit and emerged as a potent symbol of international cooperation in space.

In addition to the scientific and technical expertise it represented and supported, Mir also played a key role in Russian history as a pivotal focus of the country's relationship with other nations and was a treasured national resource during the transition following the fall of the Soviet Union. The government of Mikhail Gorbachev used the station's launch to demonstrate its policy of glasnost, as it allowed the launch to be televised live and distributed detailed information about the station's structure and purpose.

Several major challenges emerged for the Russian space program in the aftermath of the dissolution of the Soviet Union. Most pressing was the uncertainty about the future of the primary launch facility, the Baikonur Cosmodrome, which was located in the Republic of Kazakhstan. To the credit of leaders in both countries, a long-term lease arrangement was quickly worked out, ensuring the continued safe operation of one of the world's great space centers.

Economic concerns also posed a serious threat to planned Russian space ventures and ultimately forced space officials to abandon the Russian space shuttle program and mothball the completed Buran shuttle before it could fly with a crew. Financial concerns also scuttled plans for Mir 2, a hoped-for successor to the landmark Mir station.

Mir itself became the centerpiece of the Russian Republic's emergence as a driving force in the modern era of international cooperation in space. Joining with its former space rival the United States in a program of cooperative human spaceflight in the 1990s, the Russian space program hosted a succession of U.S. astronauts aboard Mir for a period of more than two years. The two nations also used the shuttle-Mir program, in which U.S. space shuttles docked with Mir to carry out resupply and crew transfer missions, as a springboard to their joint development of the International Space Station (ISS).

The ISS draws upon the long experience and expertise of Russia's engineers and scientists. Its central core, the Zarya Functional Cargo Block module, was built in Russia and launched from Baikonur on November 20, 1998. The ISS Zvezda Service Module is also of Russian design; Zvezda provides the station's primary living area.

Perhaps the most telling symbol of the resilience of the Russian space program, the Soyuz spacecraft, now in its fourth iteration, as the Soyuz TMA-series, is regularly used to ferry visitors to and from the ISS, even as the venerable Soyuz nears its 40th year of service as Russia's manned space vehicle.

RYUMIN, VALERI V.

(1939–)

Soviet Cosmonaut

Valeri Ryumin was born in Komsomolsk-na-Amure, Russia on August 16, 1939. He graduated from the Kaliningrad Mechanical Engineering Technical College in 1958, and served as a tank commander in the Russian Army for three years. In 1966, he graduated from the Moscow Forestry Engineering Institute, having specialized in spacecraft control systems as a student in the institute's department of electronics and computing technology.

After graduation, he became an employee of the Energia Rocket/Space Corporation (RSC Energia), where he helped develop the Salyut and Mir space stations.

Ryumin was selected for cosmonaut training in 1973. His first spaceflight was the ill-fated Soyuz 25 mission to the Salyut 6 space station in October 1977. When they arrived at Salyut 6, Ryumin and Vladimir Kovalyonok were chagrined to find that their Soyuz 25 spacecraft could not achieve a hard dock with Salyut 6. The problem persisted despite the best efforts of the crew and their Earth-bound support team, and the mission had to be cut short. The cosmonauts landed on October 11, 1977, after a frustrating 2 days and 45 minutes in space.

Ryumin next served as a member of the backup crew for Soyuz 29.

On February 25, 1979 he launched with commander Vladimir Lyakhov aboard Soyuz 32 for a long-duration visit to Salyut 6, whose docking problems had since been solved. Their flight went well, and the cosmonauts settled into a routine of living and working at Salyut 6 for several weeks. Less than a month into their stay, however, one of the station's three fuel tanks malfunctioned. Although the mishap was not life-threatening, it meant that any change of the station's orbit could be achieved only by a boost from a Soyuz or Progress spacecraft, since the main engine on Salyut 6 could no longer be used.

Further difficulties arose with the planned visit of Soyuz 33, which was scheduled to bring the first of several cosmonauts from Soviet-bloc nations to Salyut 6 as part of the Intercosmos program. Soyuz 33, commanded by Nikolai Rukavishnikov and also carrying Georgi Ivanov of Bulgaria, lost the use of its main engine just before its planned rendezvous with Salyut 6, on April 10, 1979. After a tense two days in space, Rukavishnikov and Ivanov were able to return to Earth on their backup engine, after enduring a 10G ballistic re-entry and landing some 180 kilometers away from their intending landing site. The flight had achieved the goal of making Ivanov the first Bulgarian citizen to fly in space, but aboard Salyut 6, Ryumin and Lyakhov would have no visitors. Future Intercosmos missions were put off while the Soyuz 33 malfunction was studied.

Far more worrisome than the loneliness of the crew on the space station, the Soyuz 33 failure also meant that the Soyuz 32 craft, which had taken Ryumin and Lyakhov to Salyut 6, could not be switched with Soyuz 33 as scheduled. Because the original Soyuz spacecraft was capable of spending no more than three months in space before its systems would deteriorate to an unsafe level of operation, the switch to a fresh Soyuz was a necessary milestone for long-duration flights.

Rather than cut short the planned six-month stay for Ryumin and Lyakhov on Salyut 6, mission controllers improvised a solution: Soyuz 34 was launched unmanned on June 6, 1979 and docked automatically with the space station while Soyuz 32, also unmanned, returned to Earth on June 13.

Two unmanned cargo supply spacecraft, Progress 6 and Progress 7, also visited the station. The second Progress ship deployed the KRT-10 radio telescope in one of the two Salyut 6 docking ports. Although it functioned well and allowed the crew to achieve their planned objectives, the telescope's dish antenna became stock in the docking port, creating yet another challenge for the crew, one that had to be solved just four days before their scheduled return to Earth.

Improvisation again carried the day as Ryumin emerged from Salyut 6 with a pair of wire cutters and, in an unplanned EVA of about 90 minutes, cut the antenna loose and cleared the docking aperture.

With their work aboard the station complete, Ryumin and Lyakhov returned to Earth on Soyuz 34 on August 19, 1979, after more than 175 days in space.

A series of unexpected events brought Ryumin back to Salyut 6 the following year. The prime crew for Soyuz 35 was thrown into crisis when flight engineer Valentin Lebedev was injured in an accident. Neither the prime commander, Leonid Popov, nor the backup flight engineer, Boris Andreyev, had flown in space before, and Soviet flight rules decreed that at least one long-duration crew member must have prior experience in space. Despite the hardships of his previous stay, Ryumin volunteered for the flight, and he and Popov launched aboard Soyuz 35 on April 9, 1980.

During their six-month stay on Soyuz 6, Ryumin and Popov were visited by four Soyuz crews, which included three Intercosmos flights and the crew of the first manned test of the new Soyuz T series of spacecraft, Soyuz T-2.

Ryumin's second space mission ended on October 11, 1980 when he and Popov landed in Soyuz 37 after more than 184 days in space.

During his two remarkable long-duration missions to Salyut 6, Ryumin logged over 359 days in space within two years' time.

After his second long-duration mission, Ryumin served as flight director for Salyut 7 and then served in the same capacity for flights to the Mir space station. In 1992, he was named director of the Russian portion of the U.S.-Russian space shuttle/Mir program.

Ryumin next flew in space in June 1998 aboard the space shuttle Discovery during STS-91—the ninth and final docking of a space shuttle and Mir.

With his fourth flight, Ryumin bore witness to the vast progress of international cooperation in space. His own career is a testament to the enormous changes that took place in Russia and the United States from the dawn of the Salyut era to the final years of Mir.

Valeri Ryumin spent more than 371 days in space during his four space missions.

INDEX

Stewart, Robert, 1033–34, 1043

Still, Susan, 1119–20, 1121–22

Strategic Defense Initiative (SDI), 749, 1195–96

Strekalov, Gennady M., 1196–99; Mir space station, 633–34, 634; Soyuz Mission 42: Soyuz T-3, 929–30; Soyuz Mission 49: Soyuz T-8, 938–39; Soyuz Mission 51: Soyuz T-10-1, 942, 943; Soyuz Mission 53: Soyuz T-11, 945–46; Soyuz Mission 67: Soyuz TM-10, 967; Soyuz Mission 78: Soyuz TM-21, 978–82

Student-Tracked Atmospheric Research Satellite for Heuristic International Networking Experiment (STARSHINE 2): Godwin, Linda M., 351, 1154

Sturckow, Rick, 1132–34, 1151–52, 1173–74, 1199–1200

Suisei, 1200–1201

Sullivan, Kathryn D., 1037, 1056–57, 1067–68, 1201–3

Sun, 426, 1244. See also Ulysses

Surveyor 1, 1205–6

Surveyor 2, 1206

Surveyor 3, 1206–7; Apollo 12, 49, 51, 52; Bean, Alan L., 109; Conrad, Charles "Pete," Jr., 199–200

Surveyor 4, 1207

Surveyor 5, 1207–9

Surveyor 6, 1209–10

Surveyor 7, 1210–11

Surveyor lunar exploration program, 1203–11

Swanson, Steven, 1173–74

Swigert, John L., Jr., 53–61, 1211–14

Switzerland, 686–89

SYNCOM IV-3, 1038–39, 1042

Taikonauts. See Chinese taikonauts

Tamayo-Mendez, Arnaldo, 419, 928–29

Tani, Daniel M., 1016, 1152–55, 1175, 1215–16

Tanner, Joseph R., 1216–18; Space Shuttle Mission 66: STS-66, 1092; Space Shuttle Mission 82: STS-82, 1117–18; Space Shuttle Mission 101: STS-97, 1143–44; Space Shuttle Mission 116: STS-115, 1171–72

TAS-1 (Technology Applications and Science-01), 1123

Taurus-Littrow, 80, 81–83

TD-1A satellite, 244

TDRS (Tracking and Data Relay Satellite), 1030, 1063–64, 1075–76, 1098–1100

Teacher in Space project, 1047, 1174

Technology Applications and Science-01 (TAS-1), 1123

Tempel 1 (comet), 218

TEMPUS, 1122

Tereshkova, Valentina V., 1218–21, 1299–1302

Tethered Satellite System (TSS-1), 432–33, 554, 1072

Tethered Satellite System (TSS-1R), 17, 377, 1106. See also Space Shuttle Mission 75: STS-75

Thagard, Norman E., 1221–23; Mir space station, 634; Soyuz Mission 78: Soyuz TM-21, 978–82; Space Shuttle Mission 7: STS-7, 1031; Space Shuttle Mission 17: STS-51B, 1039–40; Space Shuttle Mission 29: STS-30, 1052; Space Shuttle Mission 45: STS-42, 1066–67

Thiele, Gerhard, 254, 1138–40

Thirsk, Robert, 152, 153, 1110–12

Thomas, Andrew S. W., 1223–26; Mir space station, 635; Soyuz Mission 83: Soyuz TM-26, 992; Space Shuttle Mission 77: STS-77, 1109–10; Space Shuttle Mission 89: STS-89, 1127–28; Space Shuttle Mission 91: STS-91, 1129–30; Space Shuttle Mission 103: STS-102, 1146–48; Space Shuttle Mission 114: STS-114, 1166–69

Thomas, Donald, 1088–89, 1119–20, 1121–22

Thornton, Kathryn C., 1226–28; Hubble Space Telescope, 7; Space Shuttle Mission 32: STS-33, 1054; Space Shuttle Mission 47: STS-49, 1068–70; Space Shuttle Mission 59: STS-61, 1083–84; Space Shuttle Mission 72: STS-73, 1102–3

Thornton, William, 1031–32, 1039–40

Thuot, Pierre J., 1228–29; Intelsat VI satellite retrieval, 7; Space Shuttle Mission 34: STS-36, 1055–56; Space Shuttle Mission 47: STS-49, 1068–70; Space Shuttle Mission 61: STS-62, 1086

Titan II: Gemini 1, 290–91; Gemini 2, 291–92; Gemini 4, 298–99; Gemini 6-A, 308, 309; Gemini program, 287,